AF553512

Biophysical Chemistry

Biophysical Chemistry

M. Satake Y. Hayashi
M.S. Sethi S.A. Iqbal

DISCOVERY PUBLISHING HOUSE
New Delhi-110 002 (INDIA)

First Published – 1997

Reprinted – 2016

ISBN: 978-81-7141-329-4

Bio Physical Chemistry

Published by:

DISCOVERY PUBLISHING HOUSE PVT. LTD.

4383/4B, Ansari Road, Darya Ganj

New Delhi-110 002 (India)

Phone: +91-11-23279245, 43596064-65

Fax: +91-11-23253475

E-mail: discoverypublishinghouse@gmail.com

sales@discoverypublishinggroup.com

web: www.discoverypublishinggroup.com

Printed at:

Infinity Imaging Systems

Delhi

Preface

The teaching of Chemistry at the introductory stage becomes each day a more challenging task as the subject matter becomes more diverse and more complex. These challenges have evoked a series of responses—the present set of introductory chemistry monographs is one such. The teaching of chemistry recognises a number of problems that confront those who select text books. In order to overcome these problems, this volume **"Bio Physical Chemistry"**—one of about fifty in the Chemistry Monograph Series—is introduced. Each volume is independent of the others deals with one of Chemistry topics and constitutes and complete entity. Each volume is more comprehensive than can be possible in a single volume text. It is intended to provide a range of topics to cover most undergraduate and chemistry main courses of study. These volumes can be used to enrich the more conventional courses of study.

Suggestions for improvement are welcome and shall be gratefully acknowledged.

Authors

Contents

1
The Gas Laws

IDEAL GASES

Some of the most important substrates and products of metabolism are gases: for example, oxygen, carbon dioxide, nitrogen and hydrogen. It is therefore important to understand some of their characteristic properties and, since most metabolic reactions take place in an aqueous medium, we should also examine the behaviour of gases in solution.

The Gaseous State

The gaseous state is the simplest of the three fundamental states of matter (gas, liquid and solid). A gas differs from matter in the liquid or solid state in that it possesses no intrinsic volume, which means that, it fully occupies any enclosed space into which it is introduced. This and other properties peculiar to the gaseous state can be interpreted in terms of what is known as the Kinetic Theory of Gases. When talking of a 'gas' we are usually discussing an 'ideal gas' whose behaviour is perfectly predicted by the various Gas Laws. All 'real gases' whether they be elemental, e.g. helium or chlorine, or compound, e.g. carbon dioxide or ammonia, differ to some extent from the imaginary ideal gas, but it is much more convenient to define the properties of an ideal gas and note particular deviations from this ideal, than to attempt an individual examination of the behaviour of every known gas as though these had no properties in common.

Kinetic Theory of Gases

According to the Kinetic Theory, the ideal gas is composed of extremely small particles (its molecules) that are in continuous, random and independent motion. During their random motion the molecules of gas collide incessantly with the walls of the container, and it is this continuous bombardment of the confining walls that is recognized as the pressure of the gas. The component 'particles' of the ideal gas are completely elastic and rebound with an energy equal to that which they possessed when entering the collision. This appears to be very reasonable, for were it not so, the pressure of a gas kept at constant volume and temperature in any container would progressively decrease over the course of time. Furthermore, the molecules of an ideal gas must occupy nil volume (which confirms one's suspicion that the ideal gas is a useful fiction).

By virtue of the random, independent motion of its molecules, when a gas of a certain density is introduced into a larger space than that which it formerly occupied at the same temperature, the molecules redistribute themselves in such a way that each has maximum freedom of movement. The gas then fully occupies the new volume, with a corresponding decrease in its density. This tendency for gaseous molecules to move from a zone of high density to another of lower density, and so achieve a mean equilibrium density, is expressed in the force of diffusion. It follows that constraint must be placed upon a gas to increase its density—the force of compression.

The effect on a gas of changes in its temperature may also be interpreted in terms of the Kinetic Theory. Input of heat increases the kinetic energy of the molecules, enhances their tendency to move even further apart from one another, and thus provokes expansion of the gas at constant pressure. Decreasing the temperature decreases the mobility of the molecules and the tendency at constant pressure is for the gas to contract. In a sense, therefore, increasing the pressure and lowering the temperature tend towards the same end, namely decrease in the volume of the gas.

It follows that the condition of an ideal gas is affected by three interdependent variables: (i) volume, (ii) pressure and (iii) temperature. Examination of the effect of changes of pressure and/or temperature on the volume of a given mass of ideal gas has resulted in the establishment of certain fixed relations between these factors that are known as Ideal Gas Laws. For the most part these laws bear the names of their proponents.

Boyle's Law

If the temperature is not allowed to change, then by Boyle's Law, 'the volume of a given mass of gas is inversely proportional to the pressure exerted upon it'.

This means that an isothermal increase of pressure will proportionately decrease the volume of a quantity of gas, and vice versa.

$$V \propto \frac{I}{P} \quad \text{where} \begin{cases} V = \text{Volume} \\ P = \text{Pressure} \end{cases}$$

$$\text{or} \quad PV = \text{Constant}$$

EXAMPLE 1.1 *A balloon, which was perfectly elastic up to its bursting volume of 1.68 dm³, was filled at sea level with 1 dm³ of a light ideal gas. To what atmospheric pressure may it rise before bursting? (Assume no change in temperature; atmospheric pressure at sea level was 101 kPa.)*

Solution

P_1 = Pressure at sea level = 1.01×10^5 Nm^{-2} $\quad P_2$ = ?

$V_1 = 1.0$ dm^3 $\quad V_2 = 1.68$ dm^3

By Boyle's Law, $\quad PV$ = Constant, for a given mass of gas at a constant temperature

$$\therefore \quad P_1V_1 = P_2V_2$$

so that $\quad 1 \times 1.01 \times 10^5 = 1.68 \times P_2$

or $\quad \dfrac{1.01 \times 10^5}{1.68} = P_2$

whence $\quad P_2 = 6.01 \times 10^4$ N m^{-2}

The balloon will burst when the external pressure is 60.1 kPa

---

Charle's Law

If the pressure is maintained constant, then by Charle's Law, 'a given mass of gas will increase in volume by approximately 1/273rd of its volume at 0°C for every 1°C rise in its temperature' (and vice versa).

$$V_t = V_0 + t\left(\frac{V_0}{273}\right) \quad \text{where} \begin{cases} V_t = \text{Volume at } t\,°\text{C} \\ V_0 = \text{Volume at } 0\,°\text{C} \end{cases}$$

or $$V_t = V_0\left(1+\frac{t}{273}\right)$$

Kelvin proposed that a new scale of temperature be adopted whose zero would be the temperature at which an *ideal* gas would occupy nil volume. If the plot of V versus t°C for an ideal gas is extrapolated to the point where $V = 0$, it is found that this point is – 273.15°C, so the zero point (0 K) on the Kelvin scale equals – 273.15°C and the exact Charle's increment is 1/273.15 of the gas volume at 0°C per °C change in temperature. The degree Kelvin is equal in magnitude to the degree Celsius (i.e. 1 K = 1°C) so that,

$$t\text{°C} = (t + 273.15\ )\ \text{K}$$

and in all but the most exact calculations we can assume that t°C equals (t + 273) K.

Using the Kelvin scale of temperature we can express the Charle's relationship in another way. Let us consider the volumes V_1 and V_2 occupied by the same mass of gas at temperatures t_1°C and t_2°C.

Then, $$V_1 = V_0\left(1+\frac{t_1}{273}\right) \text{ and } V_2 = V_0\left(1+\frac{t_2}{273}\right)$$

whence $$\frac{V_1}{V_2} = \frac{273+t_1}{273+t_2}$$

Expressing temperatures in K, when T_1 and T_2 are temperatures of the gas in K,

$$\frac{V_1}{V_2} = \frac{T_1}{T_2} \text{ and } \frac{V_1}{T_1} = \frac{V_2}{T_2}$$

In general terms, for a given mass of gas maintained at constant pressure V/T = Constant.

EXAMPLE 1.2 *Gas evolved during the fermentative growth of a bacterial culture had a volume of 580 cm^3 when measured at a laboratory temperature of 17° C. What was the volume of this gas at the growth temperature of 37° C? (Assume that the gas volumes were measured at a constant pressure.)*

Solution

$t_1 = 17°C \quad \therefore T_1 = (17 + 273) = 290 \text{ K}$

$t_2 = 37°C \quad \therefore T_2 = (37 + 273) = 310 \text{ K}$

$V_1 = 580 \text{ cm}^3$

$V_2 = ?$

By Charle's Law, $\frac{V}{T} = \text{Constant}$

$$\therefore \quad \frac{V_1}{T_1} = \frac{V_2}{T_2} \text{ and } \frac{580}{290} = \frac{V_2}{310}$$

$$\text{or,} \quad V_2 = \frac{580 \times 310}{290}$$

$$= 620 \text{ cm}^3$$

$\therefore$ the fermentation gas would have occupied a volume of 620 cm^3 at 37° C.

Equation of State

The Laws of Boyle and Charle's can be combined to yield an expression which predicts the volume change that results from changes in the temperature and pressure of a given mass of ideal gas namely

$$PV/T = \text{Constant}$$

This equation is called an *equation of state*; it describes a quantity of ideal gas solely in terms of its pressure, volume and temperature.

EXAMPLE 1.3 *The gas pressure in a reaction vessel of fixed volume was to be reduced to 1 kPa. The available vacuum pump could lower the pressure only to 1.5 kPa at the laboratory temperature of 17° C. Could the required vacuum be obtained by then cooling the vessel in an ice-salt mixture at – 25° C?*

Solution

From the Laws of Boyle and Charle, $PV/T = \text{Constant}$

$$\text{or} \quad \frac{P_1 V_1}{T_1} = \frac{P_2 V_2}{T_2}$$

[Initially]	*[After cooling]*
$P_1 = 1.5$ kPa	$P_2 = ?$
V_1 = volume of the reaction vessel	$V_2 = V_1$ = volume of the reaction vessel
$T_1 = (17 + 273) = 290$ K	$T_2 = (-25 + 273) = 248$ K

Substituting these values in the above equation,

$$\frac{15 \times V_1}{290} = \frac{P_2 \times V_1}{248}$$

whence

$$\frac{15 \times 248}{290} = P_2 = 1.28 \text{ kPa}$$

Thus the cooling of the vessel to -25°C would not achieve the requisite partial vacuum.

Standard Temperature and Pressure (STP)

The volume occupied by a given quantity of a gas is dependent on the prevailing temperature and pressure. It therefore follows that if the volume of gas is to define its mass, its temperature and pressure must be defined. For this purpose, gas volumes may be recorded at a standard temperature and pressure (STP) defined as 273.15 K and standard atmospheric pressure (101 325 Pa). Of course, the volumes need not in practice be measured at 273.15 K and 101 325 Pa, for so long as the temperature and pressure of the gas are recorded when its volume is measured, the volume that it *would* have occupied at STP can always be calculated by applying the equation PV/T= Constant.

Avogadro's Law

The mole (symbol, mol) is a convenient unit in which to express the mass of any quantity of a compound (it equals the molecular weight of the compound in grammes). One mole of any compound contains 6.023×10^{23} molecules (the Avogadro number) and it has been discovered that one mole of any gas at STP will occupy a volume of 22.414 dm^3 or 22.414 L. This means that 22.414 dm^3 of any gas at STP contains 6.023×10^{23} molecules, and, since PV/T is constant for any given mass of an ideal gas, it follows that '*equal volumes of all ideal gases at the same temperature and pressure contain the same number of molecules*'. This statement is known as *Avogadro's Law*.

The Ideal Gas Equation

For a given mass of any ideal gas PV/T is constant and when the 'given mass of gas' equals 1 mole the value of this constant is unique. Furthermore, as 1 mole of any one gas contains precisely the same number of molecules as are present in 1 mole of any other gas, this unique value of PV/T is a universal molar Gas Constant. If this molar Gas Constant is represented by the symbol R, then for any quantity of gas containing n mole of gas PV/T equals nR, or

$$PV = nRT$$

This expression is known as the *ideal gas equation.*

Since real gases do not always behave even approximately in accordance with the predictions of the ideal gas equation, modifications of this equation (particularly that proposed by van der Waals) are sometimes employed. Yet most gases depart so little from ideality at low pressures that the ideal gas equation can then be applied to them in its unmodified form.

The Value of the Molar Gas Constant (R)

R equals PV/T for 1 mol of ideal gas which occupies a volume of 22.414 dm^3 at STP.

$$\therefore \quad \frac{PV}{T} = R \quad \text{when} \begin{cases} P = 101.325 \text{ kPa} = 101.325 \text{ N m}^{-2} \\ V = 22.414 \text{ dm}^3 = 22.414 \times 10^{-3} \text{ m}^3 \\ T = 273.15 \text{ K} \end{cases}$$

$$\therefore \quad R = \frac{101.325 \times (22.414 \times 10^{-3})}{273.15} \text{ N m K}^{-1} \text{ mol}^{-1}$$

$$= 8.3143 \text{ J K}^{-1} \text{ mol}^{-1} \quad (\text{since } 1 \text{ N m} = 1 \text{ J})$$

The units in which R is measured are units of energy per degree of temperature per mole, so that in older texts employing non-SI units you will find the value of R expressed in these units, e.g.

$$R = 1.987 \text{ cal. degree}^{-1} \text{ mole}^{-1}$$

or

$$R = 0.082 \text{ litre-atmospheres degree}^{-1} \text{ mole}^{-1}.$$

EXAMPLE 1.4 *A 5-dm^3 bomb calorimeter was to be filled with sufficient oxygen under pressure to support the complete combustion of 36 g of glucose. When filled with oxygen at room temperature from the only available cylinder, the final pressure was only 7.1×10^5 Pa. Would this be sufficient to allow of complete combustion of the sugar? (Room temperature was 290 K, i.e. 17°C; glucose, $C_6H_{12}O_6$.)*

Solution

$P = 7.1 \times 10^5 \text{ N m}^{-2}$ $\qquad PV = nRT$

$$V = 5 \text{ dm}^3 = 5 \times 10^{-3} \text{m}^3 \therefore n = \frac{PV}{RT} = \frac{(7.1 \times 10^5) \times (5 \times 10^{-3})}{8.314 \times 290} \text{ mol}$$

$$R = 8.314 \text{ J K}^{-1} \text{ mol}^{-1} \qquad = \frac{3550}{2411} \text{ mol}$$

$T = 290$ K

$n = ?$ mol $\qquad = 1.47$ mol

The complete combustion of glucose yields CO_2 and water;

$$C_6H_{12}O_6 + 6O_2 = 6O_2 + 6H_2O$$

Six moles of O_2 would be required for the complete combustion of 1 mole of glucose. Thus for the 0.2 mol of glucose provided, 1.2 mol of oxygen would be required, and the 5 dm^3 of O_2 at 7.1×10^5 Pa, being equal to 1.47 mol of O_2, would prove more than sufficient.

Partial Pressures

It is characteristic of mixtures of gases that the component gases in many respects behave completely independently of one another. Thus it is possible to attribute the total pressure exerted by the gas mixture to the sum of the pressure contribution of each gas—this being the pressure that the gas would exert if it alone occupied the volume of the mixture at that temperature.

This statement, that 'the pressure of the mixture is the sum of the partial pressures of the component gases', is known as *Dalton's Law of Partial Pressures.*

Thus $\qquad P_t = P_a + P_b + P_c + \ldots$

where P_t = total pressure of gas mixture and P_a, P_b, P_c . . . are partial pressure of components *A*, *B*, *C* etc.

From this and the Ideal Gas Law, it follows that the partial pressure of any gas in a mixture stands in the same ratio to the total pressure of the mixture as does its quantity in moles to the total number of moles of all gases present;

i.e. $$\frac{P_a}{P_t} = \frac{n_a}{n_t} \quad \text{where} \begin{cases} P_a = \text{partial pressure of gas (A)} \\ P_t = \text{total pressure of gas mixture} \\ n_a = \text{number of moles of gas (A)} \\ n_t = \text{total number of moles of all gases in the mixture} \end{cases}$$

EXAMPLE 1.5 *In clinical studies involving gas mixtures, their composition is frequently expressed as the volume per cent (vol. %) contribution of each component gas, which is that percentage of the total volume that is occupied by the stated gas (all volumes reduced to STP and referring to dry gas). Thus, alveolar air from the human lung contains nitrogen 80.5, oxygen 14.0 and carbon dioxide 5.5 vol. %. If the pressure in the lung is 1.01×10^5 Pa, and the vapour pressure of water is 6.25×10^3 Pa, calculate the partial pressures exerted by these major constituents.*

Solution

Actual pressure exerted by the dry gases = Total pressure – contribution of water vapour

$$= (1.01 \times 10^5) - (6.25 \times 10^3) \text{ Pa}$$
$$= 94.75 \text{ kPa}$$

By Avogadro's Hypothesis, the number of moles of gas is proportional to the volume it occupies at STP and since, from Dalton's Law plus the Ideal Gas Law, $\dfrac{P_a}{P_t} = \dfrac{n_a}{n_t}$,

then, $$\frac{P_a}{P_t} = \frac{V_a}{V_t} \quad \text{where} \begin{cases} V_a = \text{volume at STP of gas (A)} \\ V_t = \text{total volume at STP of mixture} \end{cases}$$

Thus,

Partial pressure of N_2

$$P_{N_2} = \frac{V_{N_2} \times P_t}{V_t} = \frac{80.5 \times (94.75 \times 10^3)}{100} = 76.27 \times 10^3 \text{ Pa}$$

Partial pressure of O_2

$$P_{O_2} = \frac{V_{O_2} \times P_t}{V_t} = \frac{14.0 \times (94.75 \times 10^3)}{100} = 13.27 \times 10^3 \text{ Pa}$$

Partial pressure of CO_2

$$P_{CO_2} = \frac{V_{CO_2} \times P_t}{V_t} = \frac{5.5 \times (94.75 \times 10^3)}{100} = 52.11 \times 10^3 \text{ Pa}$$

Gas Density

The density of a gas (as of any other state of matter) is its mass divided by its volume,[†] i.e.

$$\therefore \quad d = \frac{m}{V} \quad \text{where} \begin{cases} d = \text{density in kg m}^{-3} \\ m = \text{mass in kg} \\ V = \text{volume in m}^3 \end{cases}$$

Since the volume occupied by a given mass of gas is affected by changes in pressure and temperature, its density will also be affected (though inversely). In the Ideal Gas Equation, $PV = nRT$, the term n represents the number of moles of gas considered,

$$\text{Thus,} \quad n = \frac{m}{M} \quad \text{where} \begin{cases} m = \text{mass in kg} \\ M = \text{molecular weight in kg mol}^{-1} \end{cases}$$

or $\quad m = nM$

Substituting for m in the density equation (above):

$$d = \frac{nM}{V}$$

† In SI, density is usually reported in kg m^{-3} or g cm^{-3}

For a given mass of gas at a fixed temperature and pressure, both n and V are constant. Therefore, under these conditions the density of a gas is directly proportional to its molecular weight.

The variable density of a gas must not be confused with its *vapour density* which is a constant and is defined as being equal to the fraction $\frac{\text{mass of a given volume of gas}}{\text{mass of an equal volume of hydrogen}}$, both masses being measured under identical conditions of pressure and temperature.

Thus,
$$\text{vapour density of a gas} = \frac{\text{mass of } n \text{ mol of gas}}{\text{mass of } n \text{ mol of } H_2}$$
$$= \frac{\text{mass of 1 mol of gas}}{\text{mass of 1 mol of } H_2}$$

But molecular weight of $H_2 = 2$,

$\therefore$ vapour density = $\frac{1}{2}$ molecular weight of the gas

---

EXAMPLE 1.6 *250 cm³ of a gas evolved by a photo-synthesising culture of green algae, weighed 0.335 g at 303 K.*

Determine the molecular mass of the gas (molar gas constant, R = 8.314 J K⁻¹ mol⁻¹).

Solution

$\therefore$ mass of 250 cm^3 of unknown gas sampled at 303 K and 105.9 kPa

$$= (48.19 - 47.855) \text{ g} = 0.335 \text{ g}$$

By the Ideal Gas Equation,

$$PV = nRT \text{ whence, } PV = \frac{mRT}{M} \text{ and } M = \frac{mRT}{PV}$$

$$\text{Here,} \begin{cases} P = 105.9 \text{ kPa} = 1.059 \times 10^5 \text{ N m}^{-2} \\ V = 250 \text{ cm}^3 = 2.5 \times 10^{-4} \text{ m}^3 \\ m = 0.335 \text{ g} \\ T = 303 \text{ K} \\ R = 8.314 \text{ J K}^{-1} \text{ mol}^{-1} = 8.314 \text{ N m K}^{-1} \text{ mol}^{-1} \\ M = \text{molecular weight in g mol}^{-1} \end{cases}$$

Substituting these values in the equation given above,

$$M = \frac{0.335 \times 8.314 \times 303}{(1.059 \times 10^5) \times (2.5 \times 10^{-4})} \text{ g mol}^{-1} = 31.87 \text{ g mol}^{-1}$$

Diffusion of Gases

It has been mentioned that the expansion of a gas to fill a newly available space is the consequence of diffusion, which impels molecules of gas to quit a region of high density for one of lower density until homogeneity has been achieved. *Graham's Law of Diffusion* states that 'the rate of diffusion of a gas is inversely proportional to the square root of its density',

$$r \propto \frac{1}{\sqrt{d}} \quad \text{where} \begin{cases} r = \text{rate of diffusion} \\ d = \text{density} \end{cases}$$

But the density of a gas at a fixed temperature and pressure is directly proportional to its molecular weight. Hence the rate of diffusion of a gas is inversely proportional to the square root of its molecular weight. For two gases whose rates of diffusion are measured at the same pressure and temperature,

$$r_1 \propto \frac{1}{\sqrt{M_1}} \text{ and } r_2 \propto \frac{1}{\sqrt{M_2}}$$

$$\therefore \qquad \frac{r_1}{r_2} = \sqrt{\frac{M_2}{M_1}} \quad \text{where} \begin{cases} r_1 = \text{rate of diffusion of gas 1} \\ M_1 = \text{M. Wt. of gas 1} \\ r_2 = \text{rate of diffusion of gas 2} \\ M_2 = \text{M. Wt. of gas 2} \end{cases}$$

The process of effusion (issuance of a gas through a small hole in its container) obeys the same laws as simple diffusion. By either means the molecular weight of a gas may be experimentally determined if a ' reference' gas of known molecular weight is available. In practice it is usual to measure the time (t) taken by a given volume of gas to effuse through a small hole in its container. This will be inversely proportional to its rate of diffusion; i.e.

$$t \propto \frac{1}{r} \quad \text{and} \quad \frac{r_1}{r_2} = \sqrt{\frac{M_2}{M_1}}, \quad \therefore \frac{t_1}{t_2} = \sqrt{\frac{M_1}{M_2}}$$

EXAMPLE 1.7 *A certain volume of the gas evolved by a photosynthesizing algal culture took 231 s to stream through a small hole. Under precisely the same conditions, an equal volume of argon took 258 s. Calculate (a) the molecular weight, and (b) the vapour density of the unknown gas (M. Wt. of argon = 40).*

Solution

From Graham's Law $\frac{t_1}{t_2} = \sqrt{\frac{M_1}{M_2}}$

For the unknown gas:	For argon:
$t_1 = 231$ s	$t_2 = 258$ s
$M_1 = ?$	$M_2 = 40$

$$\frac{231}{258} = \sqrt{\frac{M_1}{40}}$$

whence, $\left(\frac{231}{258}\right)^2 \times 40 = M_1$

$\therefore \quad 0.802 \times 40 = M_1$

$\therefore \quad 32.08 = M_1$

Thus, (a) molecular weight of the unknown gas = 32

(b) vapour density of the unknown gas $= \frac{1}{2}$ molecular weight

= 16

Vapour Pressure

Liquids and solids in a closed space are in equilibrium with their vapour. The partial pressure of this vapour (known as the *vapour pressure* or, more correctly, the *saturation vapour pressure*) is dependent on the temperature, but is always significant in the case of liquids. When considering the composition of gases maintained above a liquid phase, it must be remembered that the liquid will contribute its vapour pressure as a component of the total measurable gas pressure.

SOLUBILITY OF GASES IN LIQUIDS

Any gas will be to some extent soluble in every liquid. The rate at which it dissolves will depend on several factors, e.g. temperature, pressure and the surface area of the gas-liquid interface, but equilibrium will be established between a given volume of liquid and an excess of the gas only when the liquid is fully saturated with gas. The amount of gas that will then be in solution will again depend on the prevailing temperature and pressure, but will also depend upon the degree of solubility of the gas in that liquid.

Henry's Law

The law states that 'the solubility (c) of a gas is directly proportional to the pressure of the gas' (at a given temperature).

Components of a gas mixture behave independently of each other, and the mass of each gas which at constant temperature dissolves in a given volume of liquid will be directly proportional to its partial pressure in the mixture.

We may express Henry's Law in the form of the following equation,

$$C_A = kP_a \quad \text{where} \begin{cases} C_A = \text{solubility of a gas A} \\ k = \text{a constant specific to the solvent and gas at a given temperature} \\ P_a = \text{partial pressure of the gas} \end{cases}$$

Henry's Law is yet another example of an Ideal Gas relationship applicable to gases at low pressures and therefore to dilute solutions of many real gases. Gross deviations from the Law at low partial pressures of a gas are generally indicative of chemical interaction between gas and liquid, or of association or dissociation of the gas molecules in solution.

EXAMPLE 1.8 *Calculate the quantity of nitrogen (in g) dissolved by 100 cm^3 of blood plasma when this is aerated at 311 K (i.e. 38°C) and 102.7 kPa. 1.2 cm^3 of N_2 dissolves in 100 cm^3 of plasma at 311 K and* *standard atmospheric pressure (i.e. 101.325 kPa); air consists of 78% N_2 by volume.*

Solution

100 cm^3 of plasma at 311 K and 101.325 kPa dissolve 1.2 cm^3 of N_2 at STP

Partial pressure of N_2 in air at 102.7 kPa = 78/100 × 102.7 kPa

From Henry's Law, $C_A = kP_a$ **or** $\dfrac{C_1}{C_2} = \dfrac{P_1}{P_2}$

Solution of pure N_2 at 101.325 kPa   Solution of N_2 from air at 102.7 kPa

$$C_1 = 1.2 \text{ cm}^3 \text{ N}_2/100 \text{ cm}^3 \text{ plasma} \qquad C_2 = ?$$

$$P_1 = 101.325 \text{ kPa} \qquad P_2 = \frac{78 \times 102.7}{100} \text{ kPa}$$

$$\therefore \qquad \frac{1.2}{C_2} = \frac{101.325 \times 100}{1.027 \times 78}$$

$$\text{whence} \qquad C_2 = \frac{1.2 \times 1.027 \times 78}{101.325 \times 100}$$

$$= 0.949 \text{ cm}^3 \text{ of N}_2/100 \text{ cm}^3 \text{ plasma}$$

But at STP 24.414 dm^3 of N_2 has a mass of 28 g

∴ 0.949 cm^3 of N_2 will have a mass of $\dfrac{0.949 \times 28}{24.414 \times 10^3}$ **g**

$$= 1.186 \times 10^{-3} \text{ g } (= 1.186 \text{ mg})$$

Thus, 100 cm^3 of blood plasma will dissolve 1.186 mg N_2 when aerated under the given conditions.

Temperature and Solubility of a Gas

Heat is generally liberated when gases dissolve in water. Since most gases have a positive heat of solution, it follows that the higher is the temperature the lower will be the solubility of a gas in an aqueous medium, provided that no chemical interaction occurs. The decrease in

solubility that will result from a given rise in temperature can be predicted from the quantity of heat that is liberated when 1 mole of the gas is added to such a large volume of the solution that it causes no appreciable change in its concentration. This heat is called the *differential heat of solution* of the gas, and its value can be determined experimentally.

REAL GASES

For a gas to conform precisely at all pressures and temperatures to the predictions made by Boyle's and Charle's Laws, its molecules would have to be devoid of volume and exert no mutual forces of attraction or repulsion. It is not surprising therefore that the *ideal gas* is a notional concept, of value because it approximately describes the actual properties of *real gases* at low pressures and moderate temperatures, and facilitates interpretation of their aberrant behaviour as the pressure is increased or the temperature lowered.

We have seen that according to the ideal gas law,

$$PV = nRT \quad \text{or} \quad \frac{PV}{nRT} = 1$$

Figure 1.1 shows the manner in which three real gases are experimentally observed to deviate from this prediction. Whereas for 1 mol of ideal gas at 273.15 K, PV/nRT should be unity irrespective of the prevailing pressure, we see that both 'positive' and 'negative' deviations from this prediction are evidenced by different real (i.e. actual) gases. The nature and extent of these deviations should in turn be interpretable in terms of those molecular properties of these real gases which are the root cause of their non-ideal behaviour. Consider neon, whose value of PV/nRT becomes increasingly greater than 1.0 as the pressure is increased (Fig. 1.1). The low boiling point of this gas (27 K) indicates that the force of attraction between its molecules is extremely weak, so that the 'positive' deviation in PV/nRT is likely to be almost entirely attributable to the non-ideality of its molecules in possessing a far from negligible volume. Intermolecular attraction is evidently more important in the case of oxygen, whose boiling point (90 K) is substantially higher, while carbon dioxide condenses to the solid state at 195 K and the greater attraction between its molecules is mirrored in the even more pronounced 'negative' deviation of PV/RT (see Fig. 1.1).

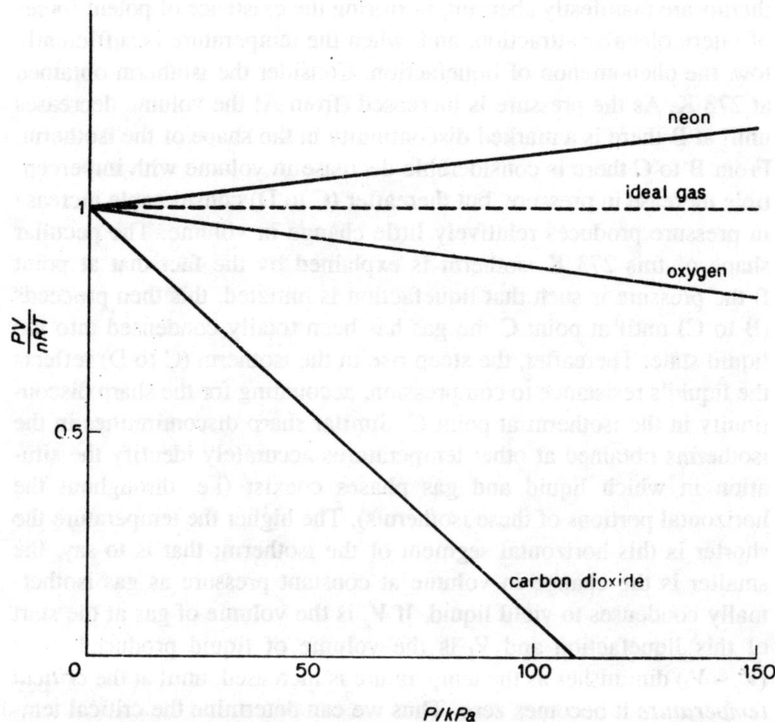

Fig. 1.1 'Positive' and 'negative' deviations from the ideal gas law (at 273 K).

Boyle's Law predicts that at a given temperature a given increase in pressure will proportionately decrease the volume of a quantity of ideal gas, thus if one plots *P versus V* for an ideal gas maintained isothermally (i.e. at constant temperature) one predictably obtains a hyperbolic graph predictably, since whenever *y* changes in inverse proportion to *x* the plot of *x* versus *y* is a hyperbola of this type. Series of corresponding values of *P* and *V* obtained for an ideal gas at different temperatures will plot as a family of these hyperbolas. Each of these is called an *isotherm* (or *isothermal*) since it represents the *P* vs. *V* behaviour of the gas at a given constant temperature.

In contrast, Fig. 1.2 shows the actual isotherms obtained when 1 mole of carbon dioxide was held at six different temperatures between 273 K and 323 K. The isotherm obtained at 323 K most nearly resembles the ideal hyperbolic plot. At lower temperatures the isotherms are manifestly aberrant, mirroring the existence of potent forces of intermolecular attraction, and, when the temperature is sufficiently low, the phenomenon of liquefaction. Consider the isotherm obtained at 273 K. As the pressure is increased (from A) the volume decreases until at B there is a marked discontinuity in the shape of the isotherm. From B to C there is considerable decrease in volume with imperceptible increase in pressure, but thereafter (C to D) considerable increase in pressure produces relatively little change in volume. The peculiar shape of this 273 K isotherm is explained by the fact that at point B the pressure is such that liquefaction is initiated, this then proceeds (B to C) until at point C the gas has been totally condensed into the liquid state. Thereafter, the steep rise in the isotherm (C to D) reflects the liquid's resistance to compression, accounting for the sharp discontinuity in the isotherm at point C. Similar sharp discontinuities in the isotherms obtained at other temperatures accurately identify the situation in which liquid and gas phases coexist (i.e. throughout the horizontal portions of these isotherms). The higher the temperature the shorter is this horizontal segment of the isotherm; that is to say, the smaller is the change in volume at constant pressure as gas isothermally condenses to yield liquid. If V_g is the volume of gas at the start of this liquefaction and V_l is the volume of liquid produced, then $(V_g - V_l)$ diminishes as the temperature is increased, until at the *critical temperature* it becomes zero. Thus we can determine the critical temperature (T_c) by plotting values of $(V_g - V_l)$ against the values of T at which these were obtained, and then extrapolating to $(V_g - V_l) = 0$. In the case of carbon dioxide the critical temperature emerges as 304.2 K (i.e. 31.1°C), and as shown in Fig. 1.2 the isotherm of carbon dioxide obtained at 304 K is the first (as the temperature is increased) not to display a recognizable horizontal segment. The isotherm obtained at the critical temperature therefore has zero slope only at a single point (the *critical point*); the pressure at the critical point is the *critical pressure* (P_c) and the molar volume is the *critical volume* (V_c). Below the critical temperature, application of pressure to a real gas decreases its volume and increases its density until intermolecular cohesive forces cause condensation of a portion of the gas to produce liquid. The two distinct phases are simultaneously present (and at a given temperature and pressure are in equilibrium with each other) over a visible boundary which we know as the fluid meniscus. If the temperature were increased to the critical temperature this meniscus would disappear, for the density of the gas would then precisely equal

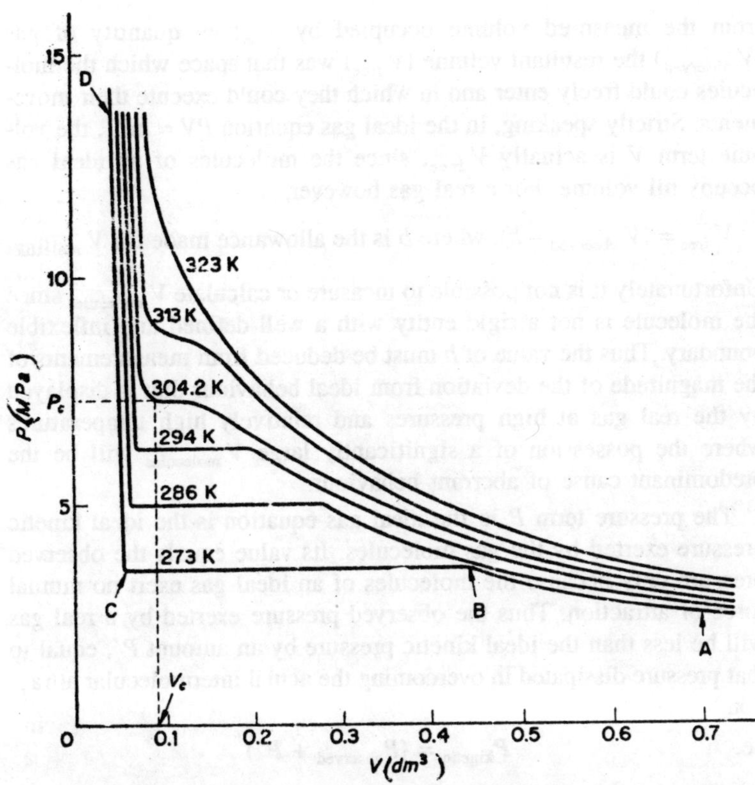

Fig. 1.2 Isotherms of CO_2 (1 mole).

that of the liquid. Thus, above the critical temperature (as at the critical point) it is invidious to talk of liquid or gas at all, one can only talk of a homogeneous hyperfluid state.

Van der Waals Equation

If the real gas behaves non-ideally because its molecules (a) occupy a significant volume, and (b) exert upon each other a significant force of attraction, it might be possible to modify the ideal gas equation by introducing terms which compensate for these factors, so as to produce a somewhat more complex equation which better predicts the behaviour of real gases. In 1873, van der Waals proposed such a modified gas equation which he derived as follows. He argued that if the volume occupied by the gas molecules themselves ($V_{\text{molecules}}$) was subtracted from the measured volume occupied by a given quantity of gas ($V_{observed}$) the resultant volume (V_{free}) was that space which the molecules could freely enter and in which they could execute their movements. Strictly speaking, in the ideal gas equation $PV = nRT$, the volume term V is actually V_{free}, since the molecules of an ideal gas occupy nil volume. For a real gas however,

$$V_{free} = (V_{observed} - b), \text{ where } b \text{ is the allowance made for } V_{molecules}$$

Unfortunately it is not possible to measure or calculate $V_{molecules}$ since the molecule is not a rigid entity with a well-defined and inflexible boundary. Thus the value of b must be deduced from measurements of the magnitude of the deviation from ideal behaviour that is displayed by the real gas at high pressures and relatively high temperatures where the possession of a significantly large $V_{molecules}$ will be the predominant cause of aberrant behaviour.

The pressure term P in the ideal gas equation is the ideal kinetic pressure exerted by the gas molecules. Its value equals the observed pressure only because the molecules of an ideal gas exert no mutual force of attraction. Thus the observed pressure exerted by a real gas will be less than the ideal kinetic pressure by an amount P', equal to that pressure dissipated in overcoming the actual intermolecular attraction,

i.e. $$P_{kinetic} = (P_{observed} + P')$$

Since the force of intermolecular attraction should be inversely proportional to the square of the volume, P' should similarly decrease in magnitude as the volume increases,

i.e. $$P' = \frac{a}{V^2}$$

where a is an empirical constant characteristic for each real gas. The value of a is obtained by measuring the extent by which the gas deviates from ideal behaviour at relatively low temperatures where the effect of intermolecular attraction predominates.

The effective kinetic pressure of a real gas is therefore given by the equation,

$$P_{kinetic} = \left(P_{observed} + \frac{a}{V^2}\right)$$

Van der Waals concluded that it was the product of the kinetic pressure and the free volume which was constant for a given quantity of gas at a set temperature, i.e. for 1 mol of real gas,

$$P_{\text{kinetic}} \times V_{\text{free}} = RT$$

Thus when the observed pressure is P and the observed volume is V,

(i) *for an ideal gas* (1 mol)

$$\left.\begin{array}{r} P_{\text{kinetic}} = P \\ V_{\text{free}} = V \end{array}\right\} \qquad \therefore PV = RT$$

and for a quantity of n mol, $PV = nRT$

(ii) *for a real gas* (1 mol)

$$\left.\begin{array}{r} P_{\text{kinetic}} = \left(P + \dfrac{a}{V^2}\right) \\ V_{\text{free}} = (V - b) \end{array}\right\} \qquad \therefore \left(P + \frac{a}{V^2}\right)(V - b) = RT$$

while for n mol of a real gas this Van der Waals equation becomes

$$\left(P + \frac{an^2}{V^2}\right)(V - nb) = nRT$$

Values of the Van der Waals constants a and b have been determined for most gases. Generally those gases with the larger molecules have the larger values of b. Similarly, since the larger molecule will have the larger surface, it is to be expected that it will exert a more powerful attractive force on its neighbour, so that a larger value of a might also be anticipated. But complications can arise from the fact that the shape of the molecule will also be a determinant of its surface area and other forces such as hydrogen bonding will unduly increase the magnitude of a.

Van der Waals' equation only approximately describes the behaviour of a real gas, but, even so, it represents a considerable advance over the attempt to impose the ideal relationship on real gases at high pressures or low temperatures. Fortunately, since the biologist is generally interested in the behaviour of gases at low pressures and at near room temperature, he can often assume that any difference between real and ideal behaviour may be disregarded.

PROBLEMS

(Assume where necessary that STP is 273 K and 101.3 kPa, and that the gas constant R = 8.314 J K^{-1} mol^{-1}.)

1. A perfectly elastic, hydrogen balloon of 2 m diameter was released at sea level. What would be its diameter when it had risen to 3050 m above sea level? (Assume no change in temperature. Atmospheric pressures are: 101.3 kPa at sea level, and 68.1 kPa at 3050 m above sea level. Volume of a sphere is $4/3\pi r^3$.)

2. Convert the following volumes of gas into volumes at STP:
   (i) 450 cm^3 at 303 K and 102.7 kPa;
   (ii) 25 cm^3 at 310 K and 12.156×10^5 Pa;
   (iii) 25 cm^3 at 256 K and 72 kPa.

3. What volume would be occupied by 1 dm^3 (at STP) of an ideal gas at:
   (i) 303 K and 102 kPa?
   (ii) 288 K and 2.026×10^6 Pa?
   (iii) 258 K and 2.026×10^5 Pa?

4. A small cylinder contained 500 cm^3 of gas at 1.52 MN m^{-2} and 291 K. If this gas were dissolved in 10 dm^3 of water, what would be the molarity of the solution?

5. An insoluble gas, produced during fermentation by a bacterial culture, was collected above water at 30°C and 750 mm Hg pressure. If it occupied 430 cm^3 under these conditions, calculate the volume of the dry gas at STP.
   (Vapour pressure of water at 303 K = 4.266 kPa; 1 mm Hg = 133.32 Pa)

Solution to Problems

1. Diameter = 2.284 m
2. (i) 411 cm^3 (ii) 264 cm^3 (iii) 18.9 cm^3
3. (i) 1101 cm^3 (ii) 52.7 cm^3 (iii) 472 cm^3
4. 31.4 m mol dm^{-3}
5. 366 cm^3 at STP

2

Thermodynamics

INTRODUCTION

Thermodynamics is concerned with the *bulk behaviour of substances*. Certain empirical laws are used to derive equations which can then be used to calculate the final results of processes. It does not deal with the rates of such processes; this aspect is encompassed in the subject of kinetics. To the biochemist, the importance of thermodynamics lies in its ability to predict the position of equilibrium in a system. For example, we might wish to study the equilibrium position of, and energy changes involved in, the reaction:

$$ATP + H_2O \rightarrow ADP + \text{phosphate.}$$

As we shall see later, determination of the energy changes would involve measuring the concentrations of the various substances taking part in the reaction when the system has come to equilibrium. In this system, however, the equilibrium concentration of ATP is too small to be accurately measured.

However, we can make measurements of the equilibrium concentrations of reactants in the reactions:

(1) Glutamate + NH_4^+ + ATP $\rightarrow$ Glutamine + ADP + phosphate

and

(2) Glutamine + H_2O $\rightarrow$ Glutamate + NH_4^+.

Using thermodynamics we could then predict the position of equilibrium and calculate the energy changes involved in the reaction of interest, i.e.

$$ATP + H_2O \rightarrow ADP + \text{phosphate}$$

(which is the sum of reactions (1) and (2) above).

The position of equilibrium in a biochemical system is often of crucial importance. Many processes are regulated by the binding of one molecule to another. For instance, the catalytic activity of many enzymes is affected by the binding of small molecules:

$$\underset{\left(\substack{\text{active}\\ \text{form}}\right)}{\text{Enzyme}} \quad + \quad \underset{\left(\substack{\text{regulator}\\ \text{molecule}}\right)}{\text{R}} \quad \rightleftharpoons \quad \underset{\left(\substack{\text{inactive}\\ \text{form}}\right)}{(\text{Enzyme . R})}.$$

An example of such a system might be, for instance, pyruvate dehydrogenase which can be inhibited by the binding of acetyl CoA (which is in fact a product of the reaction).

Knowledge of the thermodynamic quantities involved in this reaction allows us to predict the position of equilibrium under any specified set of conditions (concentration of reactants, temperature, etc.) and hence to calculate the amount of enzyme in the active form under these conditions.

Some Basic Definitions

A *system* consists of matter which is capable of undergoing a change. A complete definition includes the matter contained and the pressure, volume and temperature.

The *surroundings* are anything in contact with the system which can influence its state.

A *process* is any change which is taking place in the system.

Systems can interact with their surroundings via a flow of heat or work, or matter. Systems such as the cell which interact via a flow of matter are called *open* systems but initially we will consider *closed* systems where only heat and work changes occur. If two systems, or a system and its surroundings, are at the same temperature they are said to be in thermal equilibrium.

THE FIRST LAW OF THERMODYNAMICS

Statement of the First Law of Thermodynamics

The First Law is the law of conservation of energy and defines a quantity known as *internal energy*. If a system is isolated (that is it does not interact with its surroundings) the First Law states that its energy will remain constant irrespective of any processes taking place.

Suppose however the system interacts with its surroundings. The First Law states that 'a change in internal energy ΔE occurs such that

$$\boxed{\Delta E = \Delta q - \Delta w}$$

Δq is the heat *absorbed* by the system during the interaction and Δw is the work done *on* the system *by the* its surroundings.' Any surplus heat energy is gained as internal energy so no energy is destroyed or created. We should note that the change in internal energy is independent of the path taken to go from the initial to the final state:[†]

$$\Delta E = E_f - E_i.$$

If this were not so, a cyclic process could create or destroy energy and perpetual motion would be possible.

The symbols Δ in the above equations refer to large measurable changes in the quantities E, q, w. We could, of course, also express the First Law in terms of very small changes in these quantities:

$$\delta E - \delta q + \delta w.$$

Applications of the First Law of Thermodynamics

In biochemical systems several forms of work are carried out. Some examples are: the mechanical work performed by muscles, the electrical work required to charge nerve membranes, and the chemical work in synthesis of large molecules, or to produce the light emitted by glow worms. The First Law of Thermodynamics can be applied to each of these processes, provided the appropriate Δw (work) term is used. However initially we will consider only the work done against a pressure when chemical reactions proceed.

Consider a small change (δV) in volume when such a reaction proceeds. The work done (δw) against the pressure is given by

$$\delta w = -P\delta V.$$

(Note that if δV is small enough, the pressure P can be considered as effectively constant.)

Now from the First Law

[†] Properties of a system which depend only on the initial and final states of the system are termed *state functions*. Examples that we shall meet include internal energy, enthalpy, entropy, and Gibbs free energy.

$$\delta E = \delta q - P\delta V.$$

For large measurable changes

$$\Delta E = \Delta q - \int P\,dV.$$

The pressure must be known as a function of volume before the work term (Δw) can be calculated by integration.

Reactions are usually studied at constant volume or pressure to simplify the integral above. *At constant volume dV will be zero, so that $\Delta E = \Delta q_v$, while at constant pressure*

$$\int_{V_i}^{V_f} P\,dV = P(V_f - V_i) = P\Delta V$$

and so

$$\Delta E = \Delta q_p - P\Delta V.$$

Clearly measuring Δq_v, the heat absorbed at constant volume, e.g. in a bomb calorimeter, gives ΔE for the reaction directly. From the above equation however, Δq_p, the heat absorbed at constant pressure, equals $\Delta E + P\Delta V$. For convenience we will define an energy function H (known as *enthalpy*) such that

$$\boxed{H = E + PV}$$

Then at constant pressure,

$$(H_f - H_i) = (E_f - E_i) + P(V_f - V_i)$$

or

$$\Delta H = \Delta E + P\Delta V$$

for any reaction.

Consequently ΔH (the change in enthalpy of the reaction) equals Δq_p. Since most chemical reactions are studied at constant pressure, enthalpy changes (ΔH) rather than internal energy changes (ΔE) are usually quoted. For ideal gases a simple relationship between ΔH and ΔE exists. Since $PV = nRT$, then for a reaction at constant temperature and pressure, which involves a change Δn in the number of moles of gas, we have

$$P\Delta V = \Delta n\,(RT)$$

$$\boxed{\Delta H = \Delta E + \Delta n\,(RT)}$$

For reactions in solution, the volume changes are normally negligible and so $\Delta H = \Delta E$.

EXAMPLE 2.1 *The oxidation of solid lactic acid was studied in a bomb calorimeter at 291 K. The heat released was 1367 kJ mol^{-1}; find ΔH for the process.*

Solution

Now Δq_v is the heat *absorbed* at constant volume and so

$$\Delta E = \Delta q_v = -\,1367 \text{ kJ mol}^{-1}$$

$$CH_3 \cdot CHOH \cdot CO_2H(s) + 3O_2(g) \rightarrow 3CO_2(g) + 3H_2O(l).$$

Δn equals the change in the number of moles of gaseous components and here Δn is zero. So ΔH is also -1367 kJ mol^{-1} for this reaction.

EXAMPLE 2.2 *In a bomb calorimeter, the combustion of fumaric acid released 1330 kJ mol^{-4}; calculate ΔH for this process (T = 291 K). R = 8.31 J K^{-1} mol^{-1}.*

Solution

$$\underset{HO_2C}{\overset{H}{}}\!\!\!\!\!\!\!\!\!\!\!\!>C{=}C<\!\!\!\!\overset{CO_2H}{\underset{H}{}}\ (s) + 3O_2(g) \rightarrow 4CO_2(g) + 2H_2O(l).$$

Here Δn is + 1, so

$$\Delta H = -\,1330 + \frac{(1)\,(8.31)\,(291)}{1000} \text{ kJ mol}^{-1}$$

$$= -\,1327.6 \text{ kJ mol}^{-1}$$

These examples show that the complete oxidation (combustion) of an organic compound leads to the release of a large amount of energy.† For instance in the cell, much of the energy used in metabolism is obtained via the oxidation of pyruvate to carbon dioxide in the tricarboxylic acid cycle. In this case the energy released is utilized in

† This amount of energy is large compared with that of most biochemical reactions.

a series of steps, resulting in the conversion of 15 moles of ADP to ATP per mole of pyruvate oxidized.

Thermochemistry

Thermochemistry is the practical application of the First Law to chemical reactions. For convenience several different types of heat of reaction are defined.

The *enthalpy of reaction* is defined as the heat *absorbed* during a reaction at constant pressure. Consequently ΔH is always negative for exothermic reactions (i.e. reactions in which heat is evolved).

The *enthalpy of combustion* is the heat *absorbed* at constant pressure when one mole of substance is converted to its combustion products, e.g. CO_2, H_2O etc.

An important quantity is the *enthalpy of formation* (ΔH_f) of a compound. This is the heat absorbed at constant pressure when a compound is formed from its elements in their most stable form. The value of ΔH_f at 298 K at 1 atm. pressure is known as the *standard enthalpy of formation* (denoted by ΔH°_f). This quantity is often difficult to measure directly but this difficulty can be overcome by making use of ***Hess's Law of Constant Heat Summation***. This is a restatement of the First Law and says that ΔH for any chemical reaction is independent of the reaction path (i.e. enthalpy is a *state function* and depends only on the initial and final states). For example, if one requires ΔH°_f for glucose, the following reactions can be considered:

(1) $C(\text{graphite}) + O_2(g) \rightarrow CO_2(g) \quad \Delta H^\circ_1 = -393.1 \text{ kJ mol}^{-1}$

(2) $H_2(g) + \frac{1}{2}O_2(g) \rightarrow H_2O\ (l) \quad \Delta H^\circ_2 = -285.5 \text{ kJ mol}^{-1}$

(3) $C_6H_{12}O_6(s) + 6O_2(g) \rightarrow 6CO_2(g) + 6H_2O(l)$

$$\Delta H^\circ_3 = -2821.5 \text{ kJ mol}^{-1}.$$

We are interested in the reaction

$$6C(\text{graphite}) + 6H_2(g) + 3O_2(g) \rightarrow C_6H_{12}O_6(s)\ (\Delta H^\circ_f).$$

From Hess's Law, $\Delta H^\circ_f = 6\Delta H^\circ_1 + 6\Delta H^\circ_2 - \Delta H^\circ_3 = -1250.1 \text{ kJ mol}^{-1}$. So glucose is an *exothermic*† compound (i.e. ΔH°_f is negative). An *endothermic* compound (such as ethene) has a positive value of ΔH°_f (in this case 54.3 kJ mol^{-1}).

---

† ΔH°_f is the enthalpy of the products *minus* that of the reactants. Thus a negative value of ΔH°_f indicates that the system has lost enthalpy (to the surroundings) and vice versa. This *sign convention* applies throughout this book.

THE SECOND LAW OF THERMODYNAMICS

Introduction

The First Law of thermodynamics deals simply with the energy balance in a given process or reaction. According to this law a process (reaction) can occur in either direction, provided that the internal energy of the system remains constant. However, it is found in practice that many processes will only occur in one direction. For instance, heat will flow from a system at one temperature to another system at a lower temperature, but the reverse flow does not occur. It is observed that a beaker of water will not suddenly divide its internal energy so that half the water freezes while the other half heats up (or boils!) (although from the First Law there is nothing to suggest that this cannot happen). Clearly we need some criterion to tell us about the likelihood of a process occurring in a certain direction. This essentially is the subject of the second law of thermodynamics.

However before we discuss the second law we must introduce a new concept—that of the entropy of a system.

The Meaning of Entropy

The simplest way to introduce entropy is to think of it as measuring the degree of disorder (or randomness) of a system. Thus we find that the entropy of water (at 273.15 K) is considerably greater than that of ice (at 273.15 K), reflecting the more ordered arrangement of the molecules in the ice structure. Similarly the entropy of water vapour at 373.15 K is greater than that of the liquid at this temperature, again reflecting the greater freedom of motion (randomness) of the molecules in the gaseous phase. If we consider the melting of ice, at 273.15 K, for example, we achieve the increase in entropy (or disorder) by the input of heat energy (known as the *latent heat* of melting or fusion). It follows that entropy and heat changes must be related in some way.

Statement of the Second Law

The Second Law relates the change in the *entropy* of a system (dS) during some process (or reaction) to the heat absorbed by the system (dq) at a temperature, T(K). The law can be expressed as

$$\boxed{dS \geq dq/T}$$

The equality in the above definition (i.e. when $dS = dq/T$) refers to a process which is *reversible* whereas the inequality refers to an *irreversible* process.†

Reversible and Irreversible Processes

A process is said to be *reversible* (or carried out *reversibly*) when a system can be taken from state A to state B and then back to state A, with the surroundings also being restored to their original state. *There are thus no permanent changes in either the system or its surroundings.* Reversible processes do not occur in nature but they represent the limits of certain types of processes. Consider, for example, a gas enclosed by a piston with the pressure on the two sides of the piston equal. If this equilibrium is perturbed by lowering the opposing pressure, the gas will expand; conversely the gas will contract if the opposing pressure is increased. When the changes in the opposing pressure are made infinitesimally small, the process of expansion or contraction approaches reversibility. A reversible process thus corresponds to a series of equilibrium states in which one state is converted to another by an infinitesimal change in pressure or other variable such as temperature, concentration, etc. From this it follows that reversible processes occur infinitely slowly.

Irreversible processes are those which proceed spontaneously, i.e. at a finite rate. Examples of such processes include the sudden expansion of a gas into a vacuum, the flow of heat from a hot bath to a cold bath, and the flow from matter from a region of high concentration to one of low concentration.

Most chemical reactions are carried out irreversibly, i.e. the reactants are mixed and the reactions allowed to proceed. In these situations we note from the second law that $dS > dq/T$. This inequality provides a measure of the spontaneity of, or the tendency of, the reaction to proceed in a given direction. *So long as* $dS > dq/T$ *the reaction will proceed.*

The use of thermodynamics would be very limited if it only al-

† Of course heat is transferred from the *surroundings* to the system and hence the entropy of the surroundings must decrease. Actually in a reversible process the entropy is conserved (i.e. the entropy lost by the surroundings equals the entropy gained by the system), while in an irreversible process the entropy gained by the system is *greater* than that lost by the surroundings (i.e. there is overall a *net* increase in entropy).

lowed us to consider inequalities. Of course most processes *are* carried out irreversibly, but the assumption of reversibility allows us to derive equations which are found to hold very well in practice.

The Concept of Free Energy

The statement of the Second Law tells us that a reaction or a chemical process will proceed so long as $dq < T\,dS$. The vast majority of reactions are carried out at constant temperature and pressure and under such conditions

$$dq = dH$$

i.e. the heat absorbed is equal to the change in enthalpy for the process (see Chapter 1). Substituting for dq, we thus have $dH < T\,dS$ or $dH - T\,dS < 0$. So long as this condition is valid the reaction will proceed. The importance of the quantity $(dH - T\,dS)$ as a measure of the extent to which a reaction will proceed is such that it is given a special symbol (dG), i.e.

$$dG = dH - T\,dS.$$

Thus if a process occurs then $dG < 0$. The process will continue until an equilibrium is attained. At equilibrium, the second law states that

$$dq = T\,dS$$

so that under our conditions of constant temperature and pressure

$$dH = T\,dS.$$

We then have that at ***equilibrium***

$$\boxed{dG = dH - T\,dS = 0}$$

We are now in a position to make a statement about the possibility of a process occurring.

'Processes will occur at constant temperature and pressure so long as the changes in G are negative, i.e. $dG < 0$. The equilibrium condition is that $dG = 0$.' The symbol G is termed the ***Gibbs free energy***.

Fig. 2.1 is a pictorial illustration of the above general statement.

In the same way as we refer to the enthalpy and entropy of a substance, we can refer to its Gibbs free energy. This is defined simply as:

$$G = H - TS$$

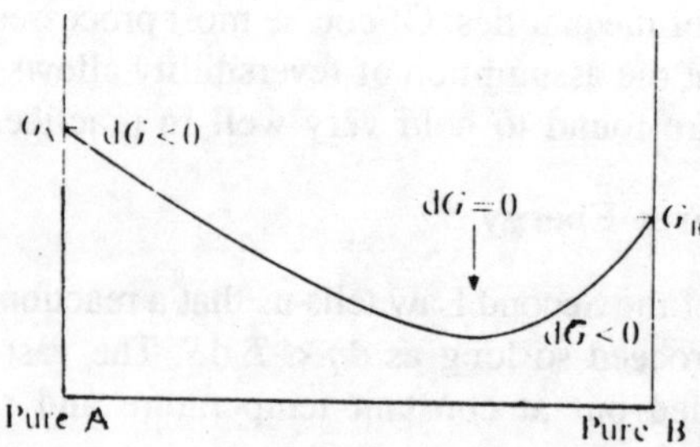

Fig. 2.1 The change in Gibbs free energy as the reaction A $\rightleftharpoons$ B proceeds. At equilibrium G is a minimum (i.e. $dG = 0$).

For a *reaction* at constant temperature and pressure the change in Gibbs free energy (ΔG) is given by

$$\boxed{\Delta G = \Delta H - T\Delta S}$$

where ΔH and ΔS are the corresponding changes in enthalpy and entropy.

From our previous discussions it is clear that a reaction will proceed if $\Delta G < 0$ and will be at equilibrium if $\Delta G = 0$. ΔG is a measure of the work (or energy) that can be obtained from a reaction. At equilibrium no work can be obtained (since there is now no tendency for the reaction to proceed).

An everyday example is afforded by a battery. Here a finite voltage is obtained only because the chemical reaction in the battery is not at equilibrium. As the reaction proceeds towards equilibrium, the voltage drops, until, at equilibrium, it becomes zero (i.e. the battery is said to be 'run down').

Like enthalpy and entropy, G is a *state function*—that is, changes in its value depend only on the initial and final states of the system. This means we can add and subtract values of ΔG for different reactions, in the same way that we did for changes in enthalpy.

For most processes, ΔG is dominated by either the ΔH or the $T\Delta S$ term, and the process is said to be enthalpy or entropy controlled. If ΔH for a reaction is large and negative, ΔG will be negative unless ΔS is also large and negative. Consequently highly exothermic reactions (ΔH negative) are almost invariably favourable on free energy grounds (i.e. ΔG negative) and ΔH dominated. However, for some processes ΔH is very small and the entropy term dominates. Examples of such processes are metal ion complex formation with polydentate ligands such as $EDTA^{4-}$, the formation of certain compounds such as CS_2 and some cases of protein denaturation. These processes are all favourable around room temperature because the $T \Delta S$ term is large enough to make ΔG negative even though ΔH is positive.

If the metal ion complexing is considered, the process could be for example:

$$Mg(H_2O)_6^{2+} + EDTA^{4-} \rightarrow [Mg(EDTA)(H_2O)]^{2-} + 5H_2O$$

The value of ΔH is about + 12.5 kJ mol^{-1} but the entropy term is highly favourable (i.e. positive) as each $EDTA^{4-}$ ligand releases five water molecules from an ordered solvation shell. This makes the ΔG for the process negative, i.e. complex formation is favoured. Similarly if CS_2 formation is considered.

$$C(s) + 2S(s) \rightarrow CS_2(l)$$

entropy considerations always favour liquid formation from two ordered solid phases. The $T \Delta S$ term is sufficiently large to overcome the positive value of ΔH, so CS_2 can be formed at room temperature.

At 313 K and pH 3, chymotrypsin can be denatured. ΔH is highly unfavourable, 62.7 kJ mol^{-1}, but ΔS is 1839 J K^{-1} mol^{-1} at 313 K and so ΔG is – 513 kJ mol^{-1}. As a result denaturation is highly favourable.†

The denaturation of a protein involves the destruction of a large number of non-covalent bonds (e.g. H-bonds) which are responsible for maintaining the highly ordered *tertiary* structure of the active form. Denaturation has thus increased the disorder (or randomness) of the system, and it is for this reason that the *denatured* form is often referred to as the *random coil* form.

Standard States

The value of G of a substance in its standard state is denoted by G°. The *standard state* of a substance (which is then a common reference point) *is defined as that substance under a pressure of 1 atmosphere.* This definition does not include temperature, so we can talk about the standard free energies of a compound at 298 K or 310 K or any other temperature and denote them by G°_{298} or G°_{310}, etc.

† The active form is actually stable because of the high 'activation energy' required to denature it.

For a reaction we can then write ΔG° for the change in free energy when reactants in their standard state are converted to products in their standard state. The superscript ($^\circ$), included in the enthalpy and entropy terms, has a similar meaning.

$$\Delta G^\circ = \Delta H^\circ - T\Delta S^\circ$$

EXAMPLE 2.3 *Given that ΔG° for the hydrolysis of ATP to ADP and phosphate is -30.5 kJ mol^{-1} and ΔG° for the hydrolysis of ADP to AMP and phosphate is -31.1 kJ mol^{-1} what is the ΔG° for the process below*

$$ATP + AMP \rightarrow 2ADP$$

under these conditions? This is known as a 'coupled' reaction. (The hydrolysis of ATP is linked to the phosphorylation of AMP.)

Solution

We have

(i) $\mathrm{ATP} \rightarrow \mathrm{ADP} + P_i$[†] $\qquad \Delta G^\circ = -30.5\ \mathrm{kJ\ mol^{-1}}$

(ii) $\mathrm{ADP} \rightarrow \mathrm{AMP} + P_i$ $\qquad \Delta G^\circ = -31.1\ \mathrm{kJ\ mol^{-1}}$

Since G is a *state function*, we can add or subtract the ΔG's. Hence by subtraction of (2) from (1)

$$\mathrm{ATP} + \mathrm{AMP} \rightarrow 2\mathrm{ADP} \qquad \Delta G^\circ = 0.6\ \mathrm{KJ\ mol^{-1}}$$

Thus the equilibrium in this overall reaction lies slightly to the left.

EXAMPLE 2.4 *At 310 K, the ΔG° for ATP hydrolysis is -30.5 kJ mol^{-1}. Calorimetric measurements give $\Delta H^\circ = -20.1$ kJ mol^{-1}. What is the value of ΔS° for this process?*

Solution

Now $\qquad \Delta G^\circ = \Delta H^\circ - T\Delta S^\circ$

$$\therefore \qquad \Delta S^\circ = \frac{\Delta H^\circ - \Delta G^\circ}{T}$$

---

† P_i is often used as an abbreviation for phosphate.

$$= \frac{-20.1 + 30.5}{310}(1000)\ \text{J K}^{-1}\ \text{mol}^{-1}$$

$$\Delta S^\circ = 33.5\ \text{J K}^{-1}\ \text{mol}^{-1}.$$

Bioenergetics

Bioenergetics is the study of energy transformation in living systems. As scientists are beginning to understand many biological and biochemical phenomena at the molecular level, they are learning to apply thermodynamics to the study of living systems.

The 'Biochemical' Standard State

In many processes of biochemical interest, there is a net uptake or release of protons as the reaction proceeds, e.g. the hydrolysis of ATP at a pH around 8 can be formulated as:

$$H_2O + ATP^{4-} \rightarrow ADP^{3-} + HPO_4^{2-} + H^+$$

Strictly speaking, ΔG° is the change in free energy when the reactants in their standard states are converted to products in their standard states. For the solutes, i.e. ATP^{4-}, ADP^{3-}, HPO_4^{2-}, and H^+ the standard state is defined as a one molar solution of these ions at 1 atmosphere pressure). For H^+ a 1 mol dm^{-3} solution would correspond to a pH = 0. This standard state is not of much interest to the biochemist who is normally concerned with reactions in solution where the pH is around 7. For this reason, it is convenient to define a new 'biochemical' standard state, where all the components are present in their standard states *except* H^+ which is present at 10^{-7} mol dm^{-3} (pH = 7).

Consequently, the change in the standard Gibbs free energy according to these two conventions will be different for reactions involving uptake or liberation of hydrogen ions. We shall therefore replace ΔG° with $\Delta G^{\circ\prime}$ in discussing biochemical processes. Consider the reaction

$$A + B \rightarrow C + xH^+$$

The Gibbs free energy change (ΔG) for the process is given by

$$\Delta G = \Delta G^\circ + RT \ln \frac{\frac{[C]}{1\,M}\left(\frac{[H^+]}{1\,M}\right)^x}{\frac{[A]}{1\,M}\,\frac{[B]}{1\,M}}$$

where 1 M represents the physical chemists' standard state of solute in solution. Since the biochemists' standard state for H^+ ions is 10^{-7} M, the Gibbs free energy change for the same process is now given by

$$\Delta G = \Delta G^{\circ\prime} + RT \ln \frac{\dfrac{[C]}{1\,M}\left(\dfrac{[H^+]}{10^{-7}\,M}\right)^x}{\dfrac{[A]}{1\,M}\dfrac{[B]}{1\,M}}$$

Note that regardless of the convention for the standard state, ΔG must remain unchanged. From the last two equations we obtain

$$\Delta G^{\circ} = \Delta G^{\circ\prime} + xRT \ln \frac{1}{10^{-7}}$$

If $x = 1$ and $T = 298$ K, then

$$\Delta G^{\circ} = \Delta G^{\circ\prime} + 40.0 \text{ kJ}$$

This means that for reactions producing H^+ ions, ΔG° is greater than $\Delta G^{\circ\prime}$ by 40.0 kJ per mole of H^+ ions released. Hence the reaction is more spontaneous at pH 7 than at pH 0. On the other hand, if H^+ ion appears as the reactant,

$$C + x\,H^+ \rightarrow A + B$$

we can show that

$$\Delta G^{\circ} = \Delta G^{\circ\prime} - 40.0 \text{ kJ}$$

and this reaction will be more spontaneous at pH 0 than at pH 7. For reactions not involving H^+ ions, ΔG° is equal to $\Delta G^{\circ\prime}$.

ATP — The Currency of Energy

Adenosine-5′-triphosphate (Fig. 2.2) is the primary energy source for numerous biological reactions, ranging from protein synthesis and ion transport to muscle contraction and electrical activities in nerve cells. The energy required to carry out these processes is derived from the hydrolysis reaction at pH 7:

$$ATP^{4-} + H_2O \rightarrow ADP^{3-} + P_i + H^+$$

where P_i denotes the inorganic phosphate HPO_4^{2-}. This reaction is accompanied by a decrease in the standard Gibbs free energy of as

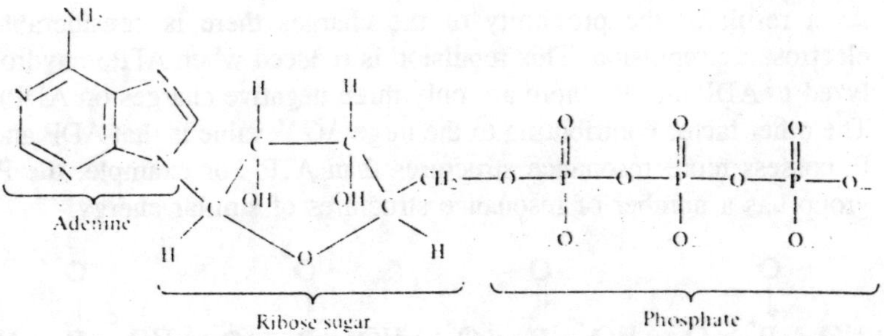

Adenine

Ribose sugar

Phosphate

Fig. 2.2 Structure of adenosine-5′-triphosphate (ATP). Upon hydrolysis, ATP loses the end phosphate group to form adenosine-5′-diphosphate (ADP). ADP may undergo further hydrolysis to form adenosine -5′-monophosphate (AMP).

much as 25 to 40 kJ per mole of ATP hydrolyzed. The exact value depends on the pH temperature, and the counter metal ions present. One of the most accurately studied systems is the Mg–ATP complex, which, at pH = 7 and $T = 310$ K, has the value $\Delta G^\circ = -30.5$ kJ. The hydrolysis reaction may proceed further as follows:

$$ADP^{3-} + H_2O \rightarrow AMP^{2-} + P_i + H^+ \qquad \Delta G^{\circ\prime} \simeq -30 \text{ kJ}$$

$$AMP^{2-} + H_2O \rightarrow \text{adenosine} + P_i \qquad \Delta G^{\circ\prime} \simeq -14 \text{ kJ}$$

However, for most cases we are only concerned with the hydrolysis of ATP to ADP.

The large decrease in the Gibbs free energy as a result of the hydrolysis of ATP and ADP prompted some biochemists to use the term *high-energy phosphate bond* to describe these compounds. The usage of this term was unfortunate because it implied that the P—O bond in these molecules is somehow different from the normal covalent bond, which is untrue. Then why such a large negative $\Delta G^{\circ\prime}$? To answer this question, we must examine the structures of ATP and its hydrolysis products, ADP and P_i, since $\Delta G^{\circ\prime}$ depends on the *difference* in the standard Gibbs free energies of the products and reactants. At least two factors must be taken into consideration: electrostatic repulsion and resonance stabilization. At pH 7 the triphosphate unit of ATP carries four negative charges:

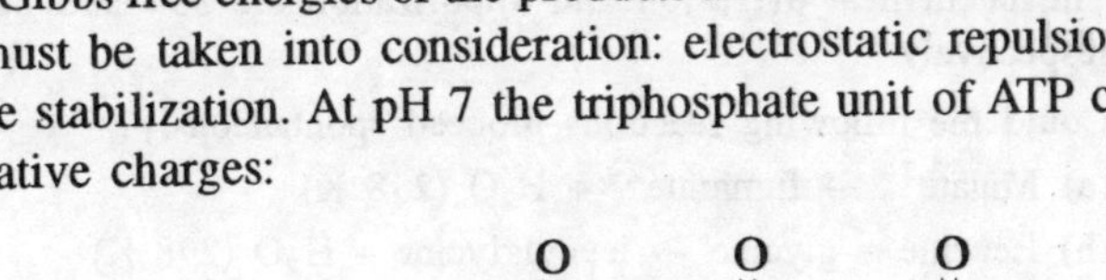

$$\text{adenine}-\text{ribose}-O-\underset{\underset{O^-}{|}}{\overset{\overset{O}{\|}}{P}}-O-\underset{\underset{O^-}{|}}{\overset{\overset{O}{\|}}{P}}-O-\underset{\underset{O^-}{|}}{\overset{\overset{O}{\|}}{P}}-O^-$$

As a result of the proximity of the charges there is considerable electrostatic repulsion. This repulsion is reduced when ATP is hydrolyzed to ADP and P_i (there are only three negative charges on ADP). The other factor contributing to the large $\Delta G^{\circ\prime}$ value is that ADP and P_i possess more resonance structures than ATP. For example, the P_i group has a number of resonance structures of similar energy:

$$\text{HO}-\overset{\overset{\displaystyle \text{O}^-}{|}}{\underset{\underset{\displaystyle \text{O}^-}{|}}{\text{P}^+}}-\text{O}^- \leftrightarrow \text{HO}-\overset{\overset{\displaystyle \text{O}}{\|}}{\underset{\underset{\displaystyle \text{O}^-}{|}}{\text{P}}}-\text{O}^- \leftrightarrow \text{HO}-\overset{\overset{\displaystyle \text{O}^-}{|}}{\underset{\underset{\displaystyle \text{O}^-}{|}}{\text{P}}}=\text{O} \leftrightarrow \text{HO}-\overset{\overset{\displaystyle \text{O}^-}{|}}{\underset{\underset{\displaystyle \text{O}}{\|}}{\text{P}}}-\text{O}^-$$

On the other hand, the terminal portion of ATP has fewer significant resonance structures per phosphate group. The following resonance structure for ATP is improbable because one of the oxygen atoms has three bonds and the fact that there is a positive charge on the oxygen atom adjacent to a positively charged phosphorus atom:

$$\text{adenine}-\text{ribose}-\text{O}-\overset{\overset{\displaystyle \text{O}}{\|}}{\underset{\underset{\displaystyle \text{O}^-}{|}}{\text{P}}}-\text{O}-\overset{\overset{\displaystyle \text{O}^-}{\|}}{\underset{\underset{\displaystyle \text{O}^-}{|}}{\text{P}^+}}-\text{O}^+=\overset{\overset{\displaystyle \text{O}^-}{\|}}{\underset{\underset{\displaystyle \text{O}^-}{|}}{\text{P}^+}}-\text{O}^-$$

Finally, to a less extent, the release of the steric crowding among oxygen atoms on adjacent phosphate groups in ATP upon hydrolysis can also contribute to the large decrease in the standard Gibbs free energy.

PROBLEMS

1. Calculate the changes in entropy for the following processes:

   (a) Ice → water (273.15 K).

   (b) Water → water vapour (373.15 K).

   The latent heat of fusion and vaporization are 334 and 2424 kJ kg^{-1} respectively.

2. Could the following reactions proceed spontaneously?

   (a) $Malate^{2-} \rightarrow fumarate^{2-} + H_2O$ (298 K)

   (b) Leucine + glycine → leucylglycine + H_2O (298 K)

   (c) $Oxaloacetate^{2-} + H_2O \rightarrow pyruvate^- + HCO_3^-$ (298 K)

(d) Ice → water (263 K).

Values for the standard Gibbs free energies of formation at 298 K in kJ mol^{-1} are as follows:

$malate^{2-}$	−844	leucine	−341
$fumarate^{2-}$	−604	glycine	−373
$oxaloacetate^{2-}$	−796	leucylglycine	−464
$pyruvate^{-}$	−474	water	−237
HCO_3^-	−586		

3. The heat of neutralization of chloracetic acid by OH^- is −62.3 kJ mol^{-1}. The ΔG° for ionization (at 298 K) is 17.1 kJ mol^{-1}. Calculate the ΔS° for ionization of chloracetic acid. (For $H^+ + OH^- \rightarrow H_2O$; $\Delta H^\circ = -56.8$ kJ mol^{-1}.)

4. The transfer of methane from an inert (non-polar) solvent to water is often used as a model for consideration of 'hydrophobic' forces. Given the following values (at 298 K):

CH_4 (inert solvent) → CH_4(g) $\Delta G^\circ = -14.6$ kJ mol^{-1}
$\Delta H^\circ = +2.1$ kJ mol^{-1}

and

CH_4 (g) → CH_4 (aqueous solution) $\Delta G^\circ = +26.3$ kJ mol^{-1}
$\Delta H^\circ = -13.4$ kJ mol^{-1}

calculate the ΔG° and ΔH° for transfer of CH_4 from an inert solvent to water. Comment on the differences between the two quantities.

5. The binding of inhibitor to trypsin was studied. $\Delta G^{\circ\prime}$ for the formation of the complex was found to be − 20.9 kJ mol^{-1} (at 298 K). Calorimetric measurements showed that the enthalpy change was zero. What is the $\Delta S^{\circ\prime}$ for the process below?

Trypsin + Inhibitor → (Trypsin. Inhibitor complex).

Comment on these data.

6. The enzyme glutamine synthetase catalyses the reaction

$$\text{Glutamate} + NH_4^+ + \text{ATP} \xrightarrow{Mg^{2+}} \text{glutamine} + \text{ADP} + P_i$$

For this reaction $\Delta G^{\circ\prime} = -15.3$ kJ mol^{-1} at 310 K. $\Delta G^{\circ\prime}$ for the hydrolysis of ATP (to yield ADP and P_i) is − 30.5 kJ mol^{-1}. Calculate the value of $\Delta G^{\circ\prime}$ for the hydrolysis of glutamine to yield glutamate and NH_4^+

7. The value of $\Delta G^{\circ\prime}$ for the hydrolysis of phosphocreatine

$$\text{Phosphocreatine} + H_2O \rightarrow \text{creatine} + P_i$$

is − 37.6 kJ mol^{-1} at 310 K. Could this reaction be used to favour the synthesis of ATP from ADP and P_i? (ΔG° for the hydrolysis of ATP is -30.5 kJ mol^{-1} at 310 K.)

8. The enthalpy change in unfolding ribonuclease at pH 6 is +209 kJ mol^{-1}. If the entropy change on unfolding is +554 J K^{-1} mol^{-1}, calculate the free energy of unfolding at 298 K. Comment on the magnitude of the numbers given above.

9. The heats of combustion of fumaric and maleic acids are -1335.9 and -1359.1 kJ mol^{-1} respectively at 298 K. Calculate the heat of formation of each of these isomers from its elements and also the heat of isomerization, i.e. maleic → fumaric at 298 K. (Heats of formation CO_2 and H_2O are -393.1 and -285.5 kJ mol^{-1} respectively at 298 K.)

10. The values for ΔE on combustion of glucose and stearic acid are -2880 and -11.360 kJ mol^{-1} respectively (at 310 K). Evaluate the ΔH for each of these processes. (Glucose is $C_6H_{12}O_6$ and stearic acid is $C_{18}H_{36}O_2$: both are solids.) $R = 8.31$ J K^{-1} mol^{-1}.

    In the light of your results comment on the relative suitability of glycogen (a polymer made up of glucose units) and fatty acids as energy reserves in the body.

11. The heats of combustion of glucose and pyruvic acid at 298 K and 1 atm pressure are -2821.5 and -1170.4 kJ mol^{-1} respectively. Calculate the enthalpy change for the conversion of glucose to pyruvic acid in the reaction:

$$\text{Glucose} + O_2 \rightarrow 2\ \text{pyruvic acid} + 2H_2O.$$

12. On arousal from hibernation, a hamster can raise its body temperature by as much as 30 K. Assuming that the heat required for this arises from the combustion of fatty acids calculate the weight of fatty acid which would need to be oxidized in order to warm up a 100 g hamster. (Use the data provided in question 10 for the combustion of stearic acid and assume that the specific heat of the hamster tissue is 3.30 J K^{-1} g^{-1}.)

13. (a) Using Hess's Law calculate the lattice energy of $MgCl_2$, i.e. ΔH for the process

$$MgCl_2\ (s) \rightarrow Mg^{2+}\ (g) + 2Cl^-\ (g),$$

given the following data:

$Mg(s) \rightarrow Mg(g)$	$\Delta H = 167.2$ kJ mol^{-1}
$Mg(g) \rightarrow Mg^{2+}(g)$	$\Delta H = 2182.0$ kJ mol^{-1}
$Cl_2(g) \rightarrow 2Cl.(g)$	$\Delta H = 241.6$ kJ mol^{-1}
$Cl.(g) \rightarrow Cl^-(g)$	$\Delta H = -364.9$ kJ mol^{-1}

and that the heat of formation of $MgCl_2$ is -639.5 kJ mol^{-1}

(b) The heat of solution of anhydrous solid $MgCl_2$ is -150.5 kJ mol^{-1}.

Using the result from part (a) above, calculate the heat of hydration of the gaseous ions.

(c) If a value of -383.7 kJ mol^{-1} is assumed for the heat of hydration of a gaseous Cl^- ion, what is the heat of hydration for gaseous Mg^{2+}? The corresponding value for gaseous Ca^{2+} is -1560 kJ mol^{-1} . Comment.

14. One of the uses of thermochemical measurements is in the determination of 'bond energies'. The C–H bond energy in methane for example would be one quarter of the enthalpy changes of the reaction

$$CH_4(g) \rightarrow C(g) + 4H(g).$$

Determine this bond energy given the following data:

$$C(s) + 2H_2(g) \rightarrow CH_4(g) \qquad \Delta H^\circ_1 = -74.8 \text{ kJ mol}^{-1}$$
$$H_2(g) \rightarrow 2H.(g) \qquad \Delta H^\circ_2 = +434.7 \text{ kJ mol}^{-1}$$
$$C(s) \rightarrow C(g) \qquad \Delta H^\circ_3 = +719.0 \text{ kJ mol}^{-1}$$

assuming that the C–H bond energy is the same in methane and ethene and that the ΔH°_f for ethene is $+54.3$ kj mol^{-1}, calculate the C $=$C bond energy in ethene.

15. The ΔU for hydrogenation of ethene (to give ethane) is -133.8 kJ mol^{-1} (at 298K). Calculate the value of ΔH for this process. The value of ΔH for the hydrogenation of benzene is -208.2 kJ mol^{-1} (at 298 K). Compare these results and work out by how much benzene is more stable than a hydrocarbon containing three isolated double bonds. (This quantity is known as the resonance energy of benzene.)

Solution to Problems

1. ΔH°_f (fumaric) $= -807.6$ kJ mol^{-1}
ΔH°_f (maleic) $= -784.4$ kJ mol^{-1}
ΔH°_f (maleic $\rightarrow$ fumaric) $= -23.2$ kJ mol^{-1}
Thus fumaric acid in which the carboxyl groups are *trans* to each other is the more stable isomer.

2. $\Delta H = \Delta U + \Delta n\,(RT)$
For glucose, Δn =O, so $\Delta H = -2880$ kJ mol^{-1}
For stearic acid, $\Delta n = -8$, so $\Delta H = -11381$ kJ mol^{-1}

Since the molecular weights of glucose and stearic acid are 180 and 284 respectively, this means that the ΔH values are -16 and -40.1 kJ g^{-1} respectively. Also fatty acids and fats (esterified fatty acids) require much less associated water for storage, and these two factors make fat a much more efficient energy reserve than carbohydrate. If a bird had to store all its energy in the form of glycogen, it has been calculated that the extra weight would render it incapable of leaving the ground!

3. $\Delta H = -480.7$ kJ mol^{-1}
4. The heat required to warm up the hamster = (30)(3.3)(100) J = 9.9 kJ. Now from question 2 combustion of 1 g of fatty acid will provide 40.1 kJ. So approximately 0.25 g of fatty acid will be required (or 0.25 per cent of the animal's body weight). Measurements of the amount of brown adipose tissue (the fat tissue largely responsible for heat production) suggest that in hibernators it constitutes about 2 per cent of the body weight. This would seem to provide adequate energy reserves for the arousal.
5. (a) Form the data

$$Mg(s) \rightarrow Mg^{2+}(g) \qquad \Delta H^\circ = +2349.2 \text{ kJ mol}^{-1}$$

and

$$Cl_2(g) \rightarrow 2Cl^-(g) \qquad \Delta H^\circ = -488.2 \text{ kJ mol}^{-1}$$

so that

$$Mg(s) + Cl_2(g) \rightarrow Mg^{2+}(g) + Cl_2^-(g) \qquad \Delta H^\circ = +1861 \text{ kJ mol}^{-1}$$

and using

$$Mg(s) + Cl_2(g) \rightarrow MgCl_2(s) \qquad \Delta H^\circ = -639.5 \text{ kJ mol}^{-1}$$

we have

$$Mg^{2+}(g) + 2Cl^-(g) \rightarrow MgCl_2(s) \qquad \Delta H^\circ = -2500.5 \text{ kJ mol}^{-1},$$

i.e. the lattice energy of $MgCl_2$ is 2500.5 kJ mol^{-1}

(b) Since $MgCl_2(s) \rightarrow MgCl_2(aq)$, $\Delta H = -150.5$ kJ mol^{-1}, the heat of hydration of $Mg^{2+} + 2Cl^-$ ions (gaseous) is – 2651 kJ mol^{-1}.

(c) Assuming that for

$$Cl^-(g) \rightarrow Cl^-(aq) \qquad \Delta H = -383.7 \text{ kJ mol}^{-1}$$

we obtain the heat of hydration of Mg^{2+} (g) as –1883.6 kJ mol^{-1}. This value is numerically greater than that for Ca^{2+} (g)(–1560 kJ mol^{-1}) principally because of the smaller ionic radius, r, of Mg^{2+}. The theory suggests that the $-\Delta H_{\text{hydration}}$ is proportional to r^{-2}. This preferential hydration of Mg^{2+} has important implications in biochemistry. Mg^{2+} usually occurs as a cofactor in enzyme reactions especially those involved in transfer of phosphoryl groups. In these cases the Mg^{2+} ion remains hydrated. However, Ca^{2+} has a very important role in acting as a 'structural cement' in many cases, e.g. in bones, shells, etc. In these cases the Ca^{2+} ion is not hydrated.

6. ΔH for the process $CH_4(g) \rightarrow C(g) + 4H.(g)$ is given by $\Delta H^\circ_3 + 2\Delta H^\circ_2 - \Delta H^\circ_1$ i.e. + 1663.2 kJ mol^{-1} , i.e. the bond energy for C—H is 415.8 kJ mol^{-1}. From the data, the ΔH for the process

$$C_2H_4(g) \rightarrow 2C(g) + 4H.(g)$$

can be calculated as 2253.1 kJ mol^{-1}

This process involves breaking four C—H bonds and the C=C bond. Subtracting the contribution of the C—H bonds, this leaves the C=C bond energy as 589.9 kJ mol^{-1}

7. For the hydrogenation of ethene, $\Delta n = -1$, so $\Delta H = -136.3$ kJ mol^{-1} Comparing this result with that given for benzene, we calculate that benzene is 200.7 kJ mol^{-1} more stable than would be expected for three isolated double bonds.

8. (*a*) $\Delta S = 1.22$ J K^{-1} g^{-1}
   $= 22.0$ J $K^{-1} mol^{-1}$
   (*b*) $\Delta S = 6.5$ J K^{-1} g^{-1}
   $= 117.0$ J K^{-1} mol^{-1}

9. The ΔG° values for these reactions are evaluated from the ΔG_f° values.
   (a) $\Delta G^\circ = +3$ kJ mol^{-1} , i.e. would not proceed spontaneously.
   (b) $\Delta G^\circ = +13$ kJ mol^{-1} i.e. would not proceed spontaneously.
   (c) $\Delta G^\circ = -27$ kJ mol^{-1} i.e. could proceed spontaneously.
   (d) Clearly this reaction would not proceed spontaneously below 273.15 K. For the ice → water transition, ΔS is positive (water has a greater degree of randomness than ice). Now at 273.15 K, $\Delta H = T\,\Delta S$ since $\Delta G = 0$ (equilibrium). Thus, assuming ΔH and ΔS are both independent of temperature, at temperatures below 273.15 K, ΔG is positive and the transition will not occur spontaneously. When T is above 273.15 K, ΔG is negative and ice will melt spontaneously.

10. For

$$ClCH_2CO_2H + OH^- \rightarrow ClCH_2CO_2^- + H_2O;\quad \Delta H^\circ = -62.3 \text{ kJ mol}^{-1}$$

so that for the ionization of $ClCH_2CO_2H$

$$ClCH_2CO_2H \rightarrow ClCH_2CO_2^- + H^{\circ +};\quad \Delta H^\circ = -5.5 \text{ kJ mol}^{-1}$$

Since ΔG° for the ionization is 17.1 kJ mol^{-1} we have ΔS° ionization = $(\Delta H^\circ - \Delta G^\circ)/T = -75.8$ J K^{-1} mol^{-1}, i.e. the ionization of chloracetic acid is very unfavourable on entropy grounds. The ordering of the solvent molecules by the ions produced (i.e. negative ΔS) is much more significant than the increase in entropy which would be expected from the increase in the number of species on dissociation.

11. From the data, and using the fact that G and H are both state functions, we deduce that for

$$CH_4(\text{inert solvent}) \rightarrow CH_4(\text{aqu})$$

$$\Delta G^\circ = 11.7 \text{ kJ mol}^{-1}$$

$$\Delta H^\circ = -11.3 \text{ kJ mol}^{-1}$$

From these values we can calculate ΔS° for the transfer as -77.2 J K^{-1} mol^{-1}. This transfer process is thought of as a model for 'hydrophobic interactions', i.e. to explain why non-polar molecules (or amino acid side chains) prefer non-polar environments to aqueous solution. These values suggest that the 'hydrophobic' effect is primarily due to the entropy term, and this has been explained on the basis that hydrocarbon (or non-polar) molecules would lead to an 'ordering' of water molecules if placed in the aqueous phase (i.e. negative ΔS). In proteins, the amino acids with non-polar side chains are usually 'buried' in the interior of the molecule, away from the solvent water.

12. For the process, $\Delta S°$ is 70.2 J K^{-1} mol^{-1}. This reaction is therefore dominated by the entropy term, which is positive even though a complex is formed from two species. By reference to the previous example, we might suggest that 'hydrophobic' interactions between the inhibitor and the enzyme are important.
13. $\Delta G°' = -15.2$ kJ mol^{-1}.
14. For the 'coupled reaction'

$$\text{Creatine phosphate} + \text{ADP} \rightarrow \text{Creatine} + \text{ATP}$$

$$\Delta G°' = -7.1 \text{ kJ mol}^{-1}$$

i.e. the creatine phosphate hydrolysis *could* be used to favour the synthesis of ATP from ADP.
15. $\Delta G° = +43.9$ kJ mol^{-1}. The values of ΔH and ΔS are very large because they refer to one mole of protein which is a large quantity of protein. Thus for unfolding many hydrogen bonds, etc. must be broken (high $\Delta H°$), but this would result in a great deal of flexibility of the molecule (high $\Delta S°$).

3

Chemical Equilibrium

INTRODUCTION

Biochemical systems are rarely at equilibrium as matter is constantly entering and leaving cells, resulting in a *steady state* situation rather than a true equilibrium. (The term steady state is often used to describe a situation in which the concentration of a substance remains almost constant, i.e. the rate of its formation equals the rate of its breakdown.) In these circumstances equilibrium thermodynamics is not strictly applicable and non-equilibrium thermodynamics has been developed to describe such processes.† However, using equilibrium thermodynamics, a good idea of which cell reactions are favourable *in vitro* can be obtained, thus providing the biochemist with the means to understand some of the energy balances involved in cellular processes.

† The difference between the two types of thermodynamics can be described as follows. Equilibrium thermodynamics considers processes which are reversible (e.g. reactions at equilibrium). In these circumstances the entropy of a system increases by an amount $dS = dq/T$, where dq is the heat transferred. This entropy is lost by the *surroundings* so that the *net* change in entropy (i.e. in the system and its surroundings) equals zero. Now for irreversible processes such as systems not at equilibrium, $dS > dq/T$ and there is a ***net increase of entropy*** accompanying the process. Non-equilibrium thermodynamics shows that the rate of increase of entropy in the system and its surroundings is a ***minimum*** when the system is in a *steady state*. Thus non-equilibrium thermodynamics states that for systems away from equilibrium, the *steady state* is the *most ordered* state. To a very good approximation we can consider living cells as being in a steady state with the concentration of intermediary metabolites remaining approximately constant. It is important to note that when a reaction is at equilibrium no useful work can be obtained from it, and it is only when non-equilibrium conditions prevail (such as in a steady state) that useful work can be obtained.

Equilibrium and Non-Equilibrium Reactions

If we examine a biochemical pathway in detail we often find that the concentrations of metabolites are such that some of the individual steps are approximately at equilibrium, whereas others are maintained far away from equilibrium. It seems that the finding that a given step is not at equilibrium, constitutes good evidence that this step may be a control point in the overall pathway. Examples in the glycolytic pathway include:

$$\text{G-6-P} \rightarrow \text{F-6-P (catalysed by phosphohexoseisomerase)}$$

which is essentially at equilibrium and

$$\text{F-6-P + ATP} \rightarrow \text{FBP + ADP (catalysed by phosphofructokinase)}$$

which is far from equilibrium.

Phosphofructokinase is considered to be one of the key regulatory enzymes in glycolysis.

The relationship between ΔG° and equilibrium constant

ΔG provides a measure of the tendency of a reaction to proceed towards equilibrium (where $\Delta G = 0$). In the following text an equation linking the standard free energy change for a reaction (ΔG°) and the equilibrium constant of the reaction is derived. It is then possible to *predict* the position of equilibrium in reactions, since the values of the standard free energies of reactants and products are often available in tables or can be measured.

The starting point is the relationship

$$G = H - TS.$$

which applies to each component in a system.

Now since

$$H = E + PV,$$

$$G = E + PV - TS;$$

differentiating (i.e. considering a small change)

$$dG = dE + P\,dV + V\,dP - T\,dS - S\,dT.$$

Now from the First Law, for any process

$$dE = dq - P\,dV,$$

where dq is the heat absorbed by the system and $P\,dV$ is the work done by the surroundings on the system. From the Second Law for a *reversible* process

$$dS = dq/T$$

i.e

$$dq = T\,dS.$$

Then we have

$$dE = T\,dS - P\,dV.$$

Substituting in the expression for dG, we obtain

$$dG = V\,dP - S\,dT.$$

At constant temperature (i.e. $dT = 0$)

$$dG = V\,dP.$$

We can integrate this equation if we know V as a function of P. For instance let us apply this equation to a perfect gas, for which $PV = nRT$

$$dG = \frac{nRT}{P}\,dP.$$

Integrating this between the standard state (where $P = P^\circ = 1$ atm and $G = G^\circ$) and any arbitrary value of P, then we obtain

$$\int_{G^\circ}^{G} dG = \int_{P^\circ}^{P} \frac{nRT}{P}\,dP$$

$$\boxed{G - G^\circ = nRT \ln (P/P^\circ)}\,. \tag{3.1}$$

If we now consider the reaction

$$aA + bB \rightleftharpoons cC + dD$$

in which A, B, C, and D are all gases, we can apply the above equation to each component of the reaction. Collecting terms and rearranging this gives

$$\Delta G - \Delta G^\circ = RT \ln \frac{(P_C/P_C^\circ)^c\,(P_D/P_D^\circ)^d}{(P_A/P_A^\circ)^a\,(P_B/P_B^\circ)^b} \tag{3.2}$$

where

$$\Delta G^\circ = G^\circ\,(\text{products}) - G^\circ\,(\text{reactants}).$$

Suppose the reaction has come to *equilibrium*, then $\Delta G = 0$, and the values of P_A, etc., would be their *equilibrium values*. Hence

$$-\Delta G^\circ = RT \ln \frac{(P_C/P_C^\circ)^c\,(P_D/P_D^\circ)^d}{(P_A/P_A^\circ)^a\,(P_B/P_B^\circ)^b}.$$

ΔG° is the change in free energy when 1 mole of the reactants is converted to 1 mole of products, each *in their standard states*. At a given temperature, this is a constant and so the R.H.S. of the equation must also be a constant. We write the equation as

$$-\Delta G^\circ = RT \ln K_P \qquad (3.3)$$

where K_P is known as the *equilibrium constant*, and is defined as

$$K_P = \frac{(P_C/P_C^\circ)^c\,(P_D/P_D^\circ)^d}{(P_A/P_A^\circ)^a\,(P_B/P_B^\circ)^b}.$$

We have retained the terms in P° for each component in this equation (even though $P^\circ = 1$ atm). This emphasizes the point that the term for each component is the *ratio* of its pressure at equilibrium to that in the standard state. K_P is therefore dimensionless.*

Applications to Solutions

The equation linking ΔG° and K_P has been derived for reactions involving ideal gases. We can use a similar relationship for reactions in solution if we make the following modifications:

(*a*) *Definitions of standard state*

For *gases* the standard state is defined as the pure gas at 1 atmosphere pressure, while for *liquids* it is the pure liquid at 1 atmosphere

---

* Throughout this book we shall treat equilibrium constants as being dimensionless, but shall include the standard state(s) referred to as a bracketed quantity, e.g. where other workers write $K = 5$ mol dm^{-3}, this would be written as $K = 5$ (mol dm^{-3}).

** In this book we have used natural logarithms (recall that $\ln x \equiv \log_e x = 2.303 \log_{10} x$.)

pressure. For *solutes* (whether ionic or not) the standard state is usually defined as a 1 mol dm^{-3} solution, under a pressure of 1 atmosphere. For solids, the standard state is defined as the pure solid at 1 atmosphere pressure. Thus if a solid is present, it is always in its standard state, and hence terms involving the solid do not appear in the equilibrium constant expression.

(b) Definition of equilibrium constant

For solution, the equilibrium constant is now defined in terms of concentrations rather than pressures. We shall denote the equilibrium constant by *K*.

These points are illustrated in the examples below.

EXAMPLE 3.1

Consider the reaction

$$\underset{(s)}{CaCO_3} \rightleftharpoons \underset{(s)}{CaO} + \underset{(g)}{CO_2}$$

In this reaction $CaCO_3$ and CaO are both in their standard states, since they are solids (and will not appear in the expression for the equilibrium constant). The standard state of CO_2 is 1 atmosphere. Hence the equilibrium constant becomes

$$K = \frac{\left(P_{CO_2}\right)}{\left(P^{\circ}_{CO_2}\right)}.$$

At 298 K the standard free energies of formation of $CaCO_3$, CaO and CO_2 are –1130.3°, –593.6, and –394.2 kJ mol^{-1}, respectively. Calculate P_{CO_2}.

Solution

ΔG° for the reaction is + 142.5 kJ mol^{-1}.

Using

$$\Delta G^{\circ} = -RT \ln K \tag{3.3}$$

we have

$$\Delta G^{\circ} = -RT \ln (P_{CO_2}/P^{\circ}_{CO_2}),$$

i.e.
$$\ln (P_{CO_2}/P^{\circ}_{CO_2}) = -\frac{\Delta G^{\circ}}{RT}$$

$$= -\frac{142500}{(8.31)(298)}$$

$$= -57.5$$

$$\therefore \quad (P_{CO_2}/P^{\circ}_{CO_2}) = 1.07 \times 10^{-25}$$

$$\therefore \quad P_{CO_2} = 1.07 \times 10^{-25} \text{ atm}, \quad \text{since } P^{\circ}_{CO_2} = 1 \text{ atm.}$$

EXAMPLE 3.2 *In the reaction*

$$\textit{citrate}^{3-} \rightarrow \textit{cis-aconitate}^{3-} + H_2O$$

the standard free energies of formation at 298 K are −1167.1, −921.7, − 237.0 kJ mol^{-1} for citrate, cis-aconitate and H_2O. What is the equilibrium constant for the reaction, and what concentration of citrate will be in equilibrium with 0.4 mmol dm^{-3} cis-aconitate?

Solution

From the ΔG°_{f} value, the ΔG° for the reaction is + 8.4 kJ mol^{-1}. Using

$$-\Delta G^{\circ} = RT \ln K, \tag{3.3}$$

we obtain

$$K = 0.034 \text{ (mol dm}^{-3}\text{)}$$

The true equilibrium constant for the reaction (K_{eq}) can be written

$$K_{eq} = \frac{([\textit{cis}\text{-aconitate}]/[\textit{cis}\text{-aconitate}]^{\circ})([H_2O]/[H_2O]^{\circ})}{([\text{citrate}]/[\text{citrate}]^{\circ})},$$

where [ ] refers to concentrations and [ ]$^{\circ}$ to the standard state concentration. In dilute solutions, however, the concentration of water is so large (55.5 mol dm^{-3}) that it can be considered to be effectively constant and the equilibrium constant for the reaction in aqueous solution (K) is obtained by dividing K_{eq} by the term ($[H_2O]/[H_2O]^{\circ}$)

$$K = \frac{([\textit{cis}\text{-aconitate}]/[\textit{cis}\text{-aconitate}]^{\circ})}{([\text{citrate}]/[\text{citrate}]^{\circ})}.$$

Since the standard states of the solutes citrate and cis-aconitate are 1 mol dm^{-3} solutions, this can be simplified to give:

$$K = \frac{[cis\text{-aconitate}]}{[\text{citrate}]}.$$

Now K = 0.034 (mol dm^{-3}) and [*cis*-aconitate] = 0.4 mmol dm^{-3}.

$\therefore$ [citrate] = 11.8 mmol dm^{-3}.

Note. In general we do not write out the full expressions for the equilibrium constant (i.e. including the standard states), but use equations such as the simpler one for K above. This is illustrated again in the next example.

EXAMPLE 3.3 *A solution is 0.05 mol dm^{-3} in glyceraldehyde-3-P (G-3-P). After addition of triosephosphate isomerase the concentration of G-3-P was 0.002 mol dm^{-3} (T = 298 K).*

Calculate the $\Delta G°$ for the reaction catalysed by this enzyme, i.e. glyceraldehyde-3-P $\rightarrow$ dihydroxyacetone-P.

Solution

At equilibrium the [dihydroxyacetone-P] is 0.048 mol dm^{-3}. The equilibrium constant K for the reaction is given by

$$K = \frac{[\text{dihydroxyacetone-P}]}{[\text{glyceraldehyde-3-P}]}$$

$$= \frac{[0.048]}{[0.002]}$$

$$K = 24 \text{ (mol dm}^{-3}\text{)}.$$

Substituting in:

$$\Delta G° = -RT \ln K, \tag{3.3}$$

we have

$$\Delta G° = -8.31 \times 298 \times \ln 24$$

$$\Delta G° = -7.87 \text{ kJ mol}^{-1}.$$

EXAMPLE 3.4 *Under certain conditions, the intracellular levels of ADP and P_i were found to be 3 mmol dm^{-3} and 1 mmol dm^{-3} respectively. Assuming $\Delta G°$ = −30.5 kJ mol^{-1} for ATP hydrolysis, calculate the concentration of ATP which would be present at equilibrium (T = 310 K).*

The actual level of ATP was found to be 10 mmol dm^{-3} under these conditions. What is the value of ΔG for the reaction

$$ATP \rightarrow ADP + P_i$$

under these conditions?

Solution

Using the equation

$$\Delta G^\circ = -RT \ln K \tag{3.3}$$

we have

$$K = 1.3 \times 10^5 \text{ (mol dm}^{-3}\text{)}$$

for the hydrolysis of ATP. The equilibrium constant for the reaction can be written as:

$$K = \frac{[\text{ADP}][\text{P}_\text{i}]}{[\text{ATP}]}.$$

Substituting [ADP] = 3×10^{-3} mol dm^{-3} and [P_i] = 10^{-3} mol dm^{-3}

$$[\text{ATP}] = \frac{3 \times 10^{-6}}{1.3 \times 10^5} \text{ mol dm}^{-3}$$

$$= 2.3 \times 10^{-11} \text{ mol dm}^{-3}.$$

Since the actual level of ATP is 10 mmol dm^{-3} it is clear that the reaction is not at equilibrium. We can calculate the actual value of ΔG under intracellular conditions using the equation

$$\Delta G - \Delta G^\circ = RT \ln \frac{[\text{ADP}][\text{P}_\text{i}]}{[\text{ATP}]}$$

which is the form of eqn (3.2) applicable to solutes,

$$= RT \ln \frac{3 \times 10^{-3} \times 1 \times 10^{-3}}{10 \times 10^{-3}}$$

$$= -20.9 \text{ kJ mol}^{-1}$$

$$\therefore \quad \Delta G = -30.5 - 20.9 \text{ kJ mol}^{-1}$$

$$= -51.4 \text{ kJ mol}^{-1}.$$

If the level of ATP were 2.3×10^{-11} mol dm^{-3}, the system would be at equilibrium and no work could be obtained from the reaction. Under the actual conditions we see that ΔG is even more negative than ΔG° and hence there is an even greater tendency for the reaction to occur, than if all the components were in their standard states (i.e. 1 mol dm^{-3}). This illustrates the point made previously that the further system is from its equilibrium position, the greater the work (ΔG) that can be obtained. In the cell, the energy (work) made available by ATP hydrolysis is used in various ways (mechanical work, biosynthetic pathways, transport of metabolites against concentration gradients. etc.).

Distinction between ΔG and ΔG°

Before finishing this section we should perhaps just emphasize once again the distinction between ΔG and ΔG° for a reaction. ΔG° is the difference in free energy between products and reactants (each in their standard states). This does *not* refer to the actual reaction at equilibrium (except in the very unusual case where K, the equilibrium constant, equals one). ΔG however, refers to the difference in free energy (products – reactants) at any other concentrations (or pressures) of the components. When $\Delta G = 0$, the reaction is at equilibrium, and the concentrations of the components are those which appear in the equilibrium constant expression, so that, for instance

$$-\Delta G^\circ = RT \ln \frac{[\mathrm{ADP}]_{\mathrm{eq}}\,[\mathrm{P_i}]_{\mathrm{eq}}}{[\mathrm{ATP}]_{\mathrm{eq}}}$$

(since $\Delta G = 0$).

When the reaction is not at equilibrium, the actual value of ΔG is calculated from

$$\Delta G - \Delta G^\circ = RT \ln \frac{[\mathrm{ADP}]\,[\mathrm{P_i}]}{[\mathrm{ATP}]}$$

It may be helpful to have an idea of the magnitude of ΔG° associate with a given value of K. Table 3.1 illustrates this for a temperature of 310 K.

TABLE 3.1

ΔG°_{310} kJ mol^{-1}	+17.80	+11.86	+5.93	0	−5.93	−11.86	−17.80
K	10^{-3}	10^{-2}	10^{-1}	1	10^{1}	10^{2}	10^{3}

Note that K increases by a factor of 10 as ΔG° doubles. Thus even quite large errors in the determination of an equilibrium constant lead to relatively small errors in ΔG°.

The Variation of the Equilibrium Constant with Temperature

At equilibrium we have the relationship

$$-\Delta G^\circ = RT \ln K \tag{3.3}$$

$$\therefore \quad \ln K = -\frac{\Delta G^\circ}{RT}$$

$$\frac{\mathrm{d}(\ln K)}{\mathrm{d}T} = -\frac{1}{R}\frac{\mathrm{d}(\Delta G^\circ/T)}{\mathrm{d}T} \tag{3.4}$$

We simply the R.H.S. of eqn. (3.4) as follows:

Using

$$\Delta G^\circ = \Delta H^\circ - T\Delta S^\circ$$

we have

$$\frac{\Delta G^\circ}{T} = \frac{\Delta H^\circ}{T} - \Delta S^\circ \tag{3.5}$$

so that by differentiation

$$\boxed{\frac{\mathrm{d}(\Delta G^\circ/T)}{\mathrm{d}T} = -\frac{\Delta H^\circ}{T^2}} \tag{3.6}$$

assuming ΔH° and ΔS° are independent of temperature over the range considered[†],

by substitution in eqn (3.4)

$$\boxed{\frac{\mathrm{d}(\ln K)}{\mathrm{d}T} = \frac{\Delta H^\circ}{RT^2}} \tag{3.7}$$

This eqn (3.7) is often referred to as the *Van't Hoff Isochore*. It is more useful in the form obtained by integrating between temperatures T_1 and T_2, viz:

[†] This formula can be derived more generally *without* assuming that ΔS° is constant.

$$\int_{K_1}^{K_2} \mathrm{d}(\ln K) = \int_{T_1}^{T_2} \frac{\Delta H^\circ}{RT^2}\,\mathrm{d}T \tag{3.8}$$

$$\therefore \quad \boxed{\ln\frac{K_2}{K_1} = -\frac{\Delta H^\circ}{R}\left(\frac{1}{T_2}-\frac{1}{T_1}\right)}. \tag{3.9†}$$

Thus a plot of $\ln K$ against $1/T$ gives a straight line of slope $-\Delta H^\circ/R$. This equation can be used in two main ways:

(1) If we know the values of K at various temperatures, ΔH° can be obtained from the slope of the graph of $\ln K$ vs. $1/T$;

(2) If we know K at one temperature and the value of ΔH°, we can calculate K at another temperature.

The assumptions that ΔH° and ΔS° are both independent of temperature are generally satisfactory, if the temperature range considered is fairly small. For more accurate work variations in these quantities with temperature can be taken into account (see Appendix 1) in the integrated expression. Provided the plot of $\ln K$ vs. $1/T$ is linear we can be confident about the validity of these assumptions.

We now consider some examples of the use of this equation.

EXAMPLE 3.5 *The equilibrium constant for a reaction doubles on raising the temperature from 298 K to 308 K. What is the value of ΔH° for the reaction?*

---

† A simpler way of obtaining this equation is to combine eqn (3.3) with eqn (3.5).

$$\ln K = -\frac{\Delta G^\circ}{RT}$$

$$\ln K = -\frac{\Delta H^\circ}{RT} + \frac{\Delta S^\circ}{R}$$

Assuming that ΔH° and ΔS° are constant between two temperatures, T_1 and T_2, we can then write:

$$\ln\frac{K_2}{K_1} = -\frac{\Delta H^\circ}{RT}\left(\frac{1}{T_2}-\frac{1}{T_1}\right)$$

which is eqn (3.9).

Solution

Using eqn (3.9), i.e.

$$\ln \frac{K_2}{K_1} = -\frac{\Delta H^\circ}{R}\left(\frac{1}{T_2} - \frac{1}{T_1}\right)$$

Now $K_2/K_1 = 2$, $T_2 = 308$ K, and $T_1 = 298$ K.

$$\therefore \qquad \ln 2 = -\frac{\Delta H^\circ}{R}\left(\frac{1}{308} - \frac{1}{298}\right)$$

$$\therefore \qquad \Delta H^\circ = 52.9 \text{ kJ mol}^{-1}.$$

This gives us a useful 'rule of thumb'—if an equilibrium constant is doubled for a 10 K temperature rise, the ΔH° is about 53 kJ mol^{-1}. If the equilibrium constant is not so sensitive to temperature, ΔH° is less than this value.

We should also note that if *K increases with increasing temperature* ΔH° is *positive*. This is easily seen by considering the eqn (3.7), i.e.

$$\frac{\mathrm{d}\ln K}{\mathrm{d}T} = \frac{\Delta H^\circ}{RT^2}.$$

EXAMPLE 3.6 *The following data were obtained for the temperature variation of the equilibrium constant of an inhibitor binding to the enzyme carbonic anhydrase. Determine the value of* ΔG°, ΔH°, *and* ΔS° *for this process at 298 K.*

T/K	289	294.2	298	304.9	310.5
$K(\times 10^{-7})$ (mol dm^{-3})	7.25	5.25	4.17	2.66	2.0

Solution

We can express the data in the appropriate form:

T/K	289	294.2	298	304.9	310.5
$1/T \times 10^3$	3.460	3.400	3.356	3.280	3.221
$K \times 10^{-7}$ (mol dm^{-3})	7.25	5.25	4.17	2.66	2.0
ln K	18.10	17.78	17.55	17.10	16.81

From the slope of ln K vs. $1/T$ we can calculate that $\Delta H^\circ = -45.1$ kJ mol^{-1}. Now, at 298 K, $K = 4.17 \times 10^7$ so that using the equation

$$-\Delta G^\circ = RT \ln K, \tag{3.3}$$

we can calculate that

$$\Delta G^\circ_{298} = -43.45 \text{ kJ mol}^{-1}.$$

Now

$$\Delta G^\circ = \Delta H^\circ - T\Delta S^\circ,$$

$$\therefore \qquad \Delta S^\circ = \frac{\Delta H^\circ - \Delta G^\circ}{T},$$

$$\Delta S^\circ = \frac{(-45.1 + 43.45)}{298} 1000,$$

i.e.

$$\Delta S^\circ = -5.53 \text{ J K}^{-1} \text{ mol}^{-1}.$$

EXAMPLE 3.7 *At 298 K the value of the equilibrium constant for the binding of phosphate to aldolase is 540 (mol dm^{-3}). Direct measurements show that the enthalpy change is – 87. 8 kJ mol^{-1}. What is the value of the equilibrium constant at 310 K if ΔH° is assumed independent of temperature?*

Solution

Using eqn (3.9).

$$\ln\left(\frac{K_2}{K_1}\right) = -\frac{\Delta H^\circ}{R}\left(\frac{1}{T_2} - \frac{1}{T_1}\right),$$

$K_1 = 540$, $\Delta H^\circ = -87800$ J mol^{-1}, $T_2 = 310$ K, $T_1 = 298$ K,

$$\therefore \qquad \ln\left(\frac{K_2}{540}\right) = -\frac{(87800)}{8.31}\left(\frac{1}{310} - \frac{1}{298}\right),$$

$$= -1.372,$$

$$\therefore \qquad \frac{K_2}{540} = 0.253.$$

$$K_2 = 137,$$

i.e. the equilibrium constant at 310 K is 137 (mol dm^{-3}).

This is in accord with our expectations, since ΔH° is negative and so K should decrease with increasing temperature. Also since ΔH° is numerically greater than 53 kJ mol^{-1}, the equilibrium constant should change by a factor greater than 2 for this change in temperature, as indeed is the case.

Measurements of the Thermodynamic Functions of Reactions

We are in a position to indicate some of the methods for measuring ΔG°, ΔH°, and ΔS° for a reaction.

Measurement of ΔG°

If the equilibrium constant for a reaction can be measured, ΔG° may be calculated directly from the equation:

$$-\Delta G^\circ = RT \ln K.$$

In some cases the equilibrium lies so far over to one side that direct measurement of K is difficult. ΔG° can then be obtained either by knowing ΔH° and ΔS° (as outlined below) *or* by making measurements on associated or coupled reactions for which the ΔG°s are known. Of course, if the values of the standard free energies of formation of the components of the reaction are available in Tables or can be measured, then ΔG° for the reaction is readily calculated.

Measurement of ΔH°

The value for the enthalpy change of a reaction can be determined by measuring the equilibrium constant at various temperatures and then using the equation:

$$\ln\left(\frac{K_2}{K_1}\right) = -\frac{\Delta H^\circ}{R}\left(\frac{1}{T_2} - \frac{1}{T_1}\right) \quad (3.9)$$

As described above a plot of ln K vs. $1/T$ gives a straight line of slope $-\Delta H^\circ/R$.

Alternatively, ΔH° may be obtained by direct calorimetric measurements at constant pressure. If measurements are made at constant volume (as for instance in a bomb calorimeter) a correction has to be made, as outlined in Chapter 1. This direct determination of ΔH° is becoming more widespread because of the availability of extremely accurate micro-calorimeters.

Again if the standard enthalpies of formation of the components of the reaction are available or can be measured (e.g. using combustion data), the ΔH° for the reaction can be calculated.

Measurement of ΔS°

ΔS° is usually obtained by knowing ΔG° and ΔH° at a certain temperature and then substituting in the equation

$$\Delta G^\circ = \Delta H^\circ - T\Delta S^\circ.$$

However ΔS° can also be calculated if the absolute standard entropies of the components at the temperature in question are known. These are derived using the Third Law of Thermodynamics as outlined in Appendix 1.

A knowledge of ΔG°, ΔH° and ΔS° for a reaction enables us to predict not only the value of the equilibrium constant at a given temperature, but also the variation of this constant with temperature. This is the principal application of thermodynamics with which we are concerned.

PROBLEMS

1. What is the difference between ΔG and ΔG° for a reaction? Calculate ΔG° for the following processes:

   (a) $H_2O\ (1, 373.15\ K) \rightleftharpoons H_2O\ (g, 373.15\ K)$.

   (b) $H_2O\ (1, 310\ K) \rightleftharpoons H_2O\ (g, 310\ K)$.

   The vapour pressure of H_2O at 310 K is 0.062 atm.

2. $\Delta G^{\circ\prime}$ for the reaction

   $$\text{glycerol} + P_i \rightleftharpoons \text{L-glycerol-1-phosphate} + H_2O$$

   is 11.1 kJ mol^{-1} at 298 K. What concentration of L-glycerol-1-phosphate will be present at equilibrium if we start with 1 mol dm^{-3} glycerol and 0.5 mol dm^{-3} phosphate?

3. The conversion of glucose-6-P to fructose-6-P is catalysed by the enzyme phosphoglucose isomerase. For this reaction

   $$\text{G-6-P} \rightleftharpoons \text{F-6-P}; \qquad \Delta G^{\circ\prime} = 2.1\ \text{kJ mol}^{-1}$$

If we start with 0.1 mol dm^{-3} G-6-P what is the final composition of the solution? (T = 298 K.)
If phosphoglucomutase, which catalyses the following reaction, is now added,

$$\text{G-1-P} \rightleftharpoons \text{G-6-P}; \qquad \Delta G^{\circ\prime} = -7.27 \text{ kJ mol}^{-1}$$

what is the new composition of the mixture at equilibrium?

4. The formation of G-6-P from glucose and P_i is unfavourable.

$$\text{glucose} + \text{P1} \rightleftharpoons \text{G-6-P} + H_2O; \qquad \Delta G^{\circ\prime} = +17.1 \text{ kJ mol}^{-1}$$

whereas the hydrolysis of phosphoenolpyruvate (PEP) is highly favourable

$$H_2O + \text{PEP} \rightleftharpoons \text{pyruvate} + P_i; \qquad \Delta G^{\circ\prime} = -55.2 \text{ kJ mol}^{-1}.$$

Show how these reactions can be coupled to the synthesis or hydrolysis of ATP, given that

$$H_2O + \text{ATP} \rightleftharpoons \text{ADP} + P_i; \qquad \Delta G^{\circ\prime} = -30.5 \text{ kJ mol}^{-1}.$$

calculate the equilibrium constants for these coupled reactions. (T = 310 K.)

5. Glycogen phosphorylase catalyses the reaction:

$$(\text{glycogen})_n + P_i \rightleftharpoons (\text{glycogen})_{n-1} + \text{G-1-P}.$$

Assuming that the concentrations of P_i and G-1-P are equals, does the equilibrium lie in favour of glycogen synthesis or degradation? (T = 310 K, $\Delta G^{\circ\prime}$ = + 3.05 kJ mol^{-1}.)
In muscle, the concentrations of P_i and G-1-P are found to be 10 mmol dm^{-3} and 30 μ mol dm^{-3}. On which side does the equilibrium lie in muscle?

6. In a study of the reaction catalysed by phosphofructokinase

$$\text{F-6-P} + \text{ATP} \rightleftharpoons \text{FBP} + \text{ADP}$$

the following data were obtained for metabolite levels in perfused rat heart at 308 K.

F-6-P	60 μ mol dm^{-3}
ATP	5.3 mmol dm^{-3}
FBP	9 μ mol dm^{-3}
ADP	1.1 mmol dm^{-3}

$\Delta G^{\circ\prime}$ for the phosphofructokinase reaction is −17.7 kJ mol^{-1}. Is the reaction at equilibrium in the perfused heart? If not, what is the value of $\Delta G'$?

In the same tissue the concentration of AMP was found to be 95 μmol dm^{-3}. Is the reaction catalysed by adenylate kinase

$$2\ ADP \rightleftharpoons AMP + ATP$$

at equilibrium? ($\Delta G^{\circ\prime} = +2.1$ kJ mol^{-1}.)
If not, what is the value of $\Delta G'$?
Comment briefly on your results.

7. Calculate $\Delta G^{\circ\prime}$ for the hydrolysis of *L*-leucylglycine by peptidase at 298 K. The standard Gibbs free energies of formation in kJ mol^{-1} of the compounds in aqueous solution at 298 K are:

L-leucylglycine	−464	water	−237
glycine	−373	L-leucine	−341

In vivo, new proteins are synthesized by activating amino acids by attachment to the appropriate tRNA. This is accomplished at the expense of ATP hydrolysis. Using the above value of $\Delta G^{\circ\prime}$ for peptide bond hydrolysis, and the following information.

$$(\text{amino acid})_1 + \text{tRNA} + \text{ATP} \rightleftharpoons (\text{amino acyl})_1\text{—tRNA} + \text{AMP} + 2P_i;$$

$$\Delta G^{\circ\prime} = -29.3 \text{ kJ mol}^{-1}.$$

$$H_2O + ATP \rightleftharpoons AMP + 2P_i; \quad \Delta G^{\circ\prime} = -61.0 \text{ kJ mol}^{-1}.$$

Calculate $\Delta G^{\circ\prime}$ for the reaction step involved in protein biosynthesis, i.e.

$$(\text{amino acyl})_1\text{—tRNA} + (\text{amino acid})_2 \rightleftharpoons (\text{amino acid})_1\text{—(amino acid})_2 + \text{tRNA}.$$

What is the equilibrium constant for this reaction?

8. For the oxygenation of haemoglobin (Hb)

$$Hb\ (aq) + O_2\ (g) \rightleftharpoons HbO_2\ (aq)$$

The equilibrium constant is 85.5 at 292 K.
Calculate ΔG° for this process.
At 292 K, in equilibrium with air, the partial pressure of oxygen is 0.2 atm and the solubility of oxygen in water is 0.23 mmol dm^{-3}. From these data calculate ΔG° for the reaction

$$O_2(g) \rightleftharpoons O_2(aq)$$

and hence ΔG° for

$$Hb(aq) + O_2(aq) \rightleftharpoons HbO_2(aq)$$

What assumptions are made in these calculations?

9. The equilibrium constant for the hydrolysis of ATP at 310 K is 1.3×10^5 (mol dm^{-3})

$$H_2O + ATP \rightleftharpoons ADP + P_i; \quad K = 1.3 \times 10^5 \text{ (mol dm}^{-3}\text{)}.$$

If $\Delta H^{\circ\prime} = -20.0$ kJ mol^{-1}, calculate K at 298 K and 273.15 K.

In the mitochondria, ATP synthesis is thought to be achieved by removal of H_2O from the active site of the ATPase enzyme. If the P_i concentration is 10 mmol dm^{-3}, at what concentration of water would the ATP and ADP concentrations be equal? (T = 310 K.)

10. The enzyme phosphorylase b can be activated by the addition of AMP. The dissociation constant (K_d) for the reaction:

$$\text{phosphorylase b.AMP} \rightleftharpoons \text{phosphorylase b} + \text{AMP}$$

varied with temperature as follows:

Temperature/K	285.5	289	300	312.5
$K_d(\times 10^5)$ (mol dm^{-3})	2.75	3.1	4.2	5.9

Calculate ΔH°, ΔS°, and ΔG° for this reaction at 303 K.

State any assumptions you have to make.

11. In the biosynthesis of proteins it is important that the correct tRNA binds to its corresponding amino acid tRNA ligase enzyme. In the case of isoleucine tRNA ligase from *E. coli*, the binding of isoleucine tRNA under certain conditions is characterized by values of $\Delta H^\circ = 0$ kJ mol^{-1} and $\Delta S^\circ = +142$ J K^{-1} mol^{-1}. The binding of valine tRNA under these conditions is characterized by values of $\Delta H^\circ = +33.4$ k J mol^{-1} and $\Delta S^\circ = +225.7$ J K^{-1} mol^{-1}. By what factor is binding of the correct tRNA favoured over incorrect tRNA at 293 K and at 313 K?

Solution to Problems

1. ΔG° refers to standard state conditions.

(a) $\Delta G^\circ = 0$ At 373.15 K the liquid and gas are in equilibrium, so that $\Delta G = 0$. Since both are in their standard states, ΔG° also equals 0.

(b) $\Delta G^\circ = 7.2$ kJ mol^{-1}. This result can be derived by considering the cycle below

H_2O (g), 310 K, 0.062 atm

(i) H_2O (l), 310 K $\nearrow$ H_2O (g), 310 K, 0.062 atm $\swarrow$ (ii) H_2O (g), 310 K, 1 atm

For process (i) $\Delta G = 0$, since liquid and gas are in equilibrium.

For process (ii)

$$\Delta G = RT \ln (P_2/P_1)$$

$$= RT \ln \left(\frac{1}{0.062}\right)$$

$$= 7.2 \text{ kJ mol}^{-1}$$

Thus for the overall process linking liquid and gas in their standard states at 310 K we add these two figures. This gives $\Delta G^\circ = 7.2$ kJ mol^{-1}. Note that the standard state of the gas at 310 K is not a stable state (but this does not matter).

2. For the reaction $K = 0.0113$ (mol dm^{-3}) from the value of $\Delta G^{\circ\prime}$. If x mol dm^{-3} is the concentration of L-glycerol-1-phosphate at equilibrium then

$$\frac{x}{(1-x)(0.5-x)} = 0.0113$$

from which $x = 0.0056$, so that the concentration of L-glycerol-1-phosphate at equilibrium is 5.6 mmol dm^{-3}.

3 (a) For the reaction $K = 0.43$ (mol dm^{-3}), so that at equilibrium

$$[\text{F-6-P}] = 0.03 \text{ mol dm}^{-3}, \quad [\text{G-6-P}] = 0.07 \text{ mol dm}^{-3}.$$

(b) For the second reaction, $K = 18.84$ (mol dm^{-3})

$$\frac{[\text{G-6-P}]}{[\text{G-1-P}]} = 18.84, \quad \frac{[\text{F-6-P}]}{[\text{G-6-P}]} = 0.43$$

and

$$[\text{G-6-P}] + [\text{G-1-P}] + [\text{F-6-P}] = 0.1 \text{ mol dm}^{-3}.$$

The final composition is

$$[\text{F-6-P}] = 0.029 \text{ mol dm}^{-3}, \qquad [\text{G-6-P}] = 0.0674 \text{ mol dm}^{-3}$$

and $[\text{G-1-P}] = 0.0036$ mol dm^{-3}.

4. (a) The synthesis of G-6-P can be coupled to the hydrolysis of ATP. For

$$\text{glucose} + \text{ATP} \rightleftharpoons \text{glucose-6-phosphate} + \text{ADP}$$

$$\Delta G^{\circ\prime} = -13.4 \text{ k J mol}^{-1},$$

i.e. $K = 182$ (mol dm^{-3}).

(b) The synthesis of ATP can be coupled to the hydrolysis of PEP. For

$$\text{PEP} + \text{ADP} \rightleftharpoons \text{pyruvate} + \text{ATP}$$

$$\Delta G^{\circ\prime} = -24.7 \text{ kJ mol}^{-1},$$

i.e. $$K = 1.46 \times 10^4 \text{ (mol dm}^{-3}\text{)}.$$

5. K for the reaction = 0.31 (mol dm^{-3}).
So if [Pi] = [G-1-P], $[(\text{glycogen})_{n-1}]/[(\text{glycogen})_n]$ =0.31, i.e. synthesis of glycogen is favoured.
If [Pi] = 10 mmol dm^{-3} and [G-1-P] = 30 μ mol dm^{-3}, $[(\text{glycogen})_{n-1}]/[(\text{glycogen})_n]$ = 102, i.e. degradation glycogen is favoured.

This result shows that in the muscle cell, the levels of Pi and G-1-P are such that phosphorylase acts to degrade glycogen. There is in fact a separate system (involving UDP-glucose as an intermediate) for the synthesis of glycogen. Confirmation of this role of phosphorylase is provided by sufferers from McArdle's disease (where muscle phosphorylase is absent). These patients have high levels of muscle glycogen (which could not be the case if phosphorylase acted in the direction of glycogen synthesis).

6. The ratio

$$\frac{[\text{FDP}]\,[\text{ADP}]}{[\text{F-6-P}]\,[\text{ATP}]} = 0.031$$

and therefore

$$\begin{aligned}\Delta G' &= \Delta G^{\circ\prime} + RT \ln (0.031)\\ &= -17.7 - 8.89 \text{ kJ mol}^{-1}\\ &= -26.6 \text{ kJ mol}^{-1},\end{aligned}$$

i.e. the reaction is not at equilibrium (where $\Delta G' = 0$)
The ratio

$$\frac{[\text{AMP}]\,[\text{ATP}]}{[\text{ADP}]^2} = 0.416$$

and therefore

$$\begin{aligned}\Delta G' &= \Delta G^{\circ\prime} + RT \ln (0.416)\\ &= 2.1 - 2.24 \text{ kJ mol}^{-1}\\ &= -0.14 \text{ kJ mol}^{-1},\end{aligned}$$

i.e. the reaction is almost at equilibrium.

These results point to the possibility that phosphofructokinase may be a regulatory enzyme in glycolysis. Adenylate kinase is. however, presumably present in sufficient amounts to ensure that equilibrium between the various adenine nucleotides prevails, and is unlikely to represent a control point.

7. ΔG° for the process below is -13 kJ mol^{-1}

$$\text{Leu—Gly} + H_2O \rightleftharpoons \text{Leu} + \text{Gly}$$

i.e. (amino acid)$_1$ – (amino acid)$_2$ + $H_2O \rightleftharpoons$ (amino acid)$_1$ + (amino acid)$_2$ and for the process

$$(\text{amino acid})_1 + \text{tRNA} \rightleftharpoons (\text{amino acyl})_1\text{—tRNA}$$

$\Delta G^\circ = +31.7$ kJ mol^{-1}.
Thus for the chain extension reaction $\Delta G^\circ = -18.7$ kJ mol^{-1}, i.e. $K = 1.9 \times 10^3$ (mol dm^{-3}).

This shows that attaching the tRNA to the amino acid has sufficiently 'activated' it to drive the peptide bond formation reaction over to the right.

8. (a) $\Delta G^\circ = -10.8$ kJ mol^{-1}.
   (b) For $O_2(g) \rightleftharpoons O_2(aq)$

$$K = \frac{2.3 \times 10^{-4}}{0.2} \text{ (mol dm}^{-3}\text{, atm)}$$

$$= 1.15 \times 10^{-3} \text{ (mol dm}^{-3}\text{, atm)}$$

   $\Delta G^\circ = +16.4$ kJ mol^{-1}.
   (c) Thus for $\text{Hb(aq)} + O_2\text{(aq)} \rightleftharpoons \text{HbO}_2\text{(aq)}$
   $\Delta G^\circ = -27.2$ kJ mol^{-1}.

The assumptions here are (a) *reversibility* is assumed in the derivation of the equation $\Delta G^\circ = -RT \ln K$; (b) the solution are assumed *ideal*, i.e. that concentration terms can be used in the equilibrium expression.

9. Using the Van't Hoff isochore we find
   (a) K (298 K) = 1.78×10^5 (mol dm^{-3})
   (b) K (273.15 K) = 3.71×10^5 (mol dm^{-3}).

Note that we assume that ΔH° and ΔS° are constant over this temperature range.

The value of K must be divided by the concentration of water (55 mol dm^{-3}) to obtain the true equilibrium constant for the reaction (K_{eq}). Thus K_{eq} (310 K) $= 1.3 \times 10^5/55$ (mol dm^{-3}) $= 2.36 \times 10^3$ (mol dm^{-3}). Putting [ADP] = [ATP] and [P_i] = 10 mmol dm^{-3}, the concentration of H_2O is calculated as 4.2 μ mol dm^{-3}.

10. The plot of $\ln K_d$ vs. $1/T$ is linear. The slope given $\Delta H^\circ = +20.9$ kJ mol^{-1}. From the graph, when $T = 303$ K, $\ln K_d = -9.99$ (i.e. $K_d = 4.59$

$\times 10^{-5}$ (mol dm^{-3})) and so $\Delta G^{\circ}_{303} = +25.2$ kJ mol^{-1}. Hence $\Delta S^{\circ} = -14.2$ J K^{-1} mol^{-1}. The assumptions are (*a*) *constancy of* ΔH° and ΔS° throughout this temperature range, which seem to be justified by the linearity of the plot; (*b*) *reversibility* of the process, to use the equality in the second law, in the derivation of the equations.

11. The equilibrium constants for binding are:

	293 K	313K
Isoleucine tRNA	2.64×10^7	2.64×10^7
Valine tRNA	6.89×10^5	1.65×10^6

The factors are thus 38 (293 K) and 16 (313 K).

4
Solutions

In general terms, a solution consists of a wholly homogeneous, single phase mixture of two or more substances. The component which determines the phase of the solution, and which is usually present in the highest proportion, is called the *solvent*; the remaining components are known as *solutes* and are considered to be dissolved in the solvent. The physico-chemical behaviour of any solution is determined not only by the nature of its solvent and solutes, but also by their proportions in the solution.

Though this definition of a solution includes mixture of gases, and solutions of substances in a solid, we are more usually concerned with the properties of solutions which have a liquid solvent.

In chapter 3 we noted that we could apply the equation $-\Delta G^\circ = RT \ln K$ to reactions in solutions, even though the equation was initially derived for reactions involving perfect gases. In this Chapter we shall show that there is a relationship between the properties of a solution and the vapour above the solution (Raoult's Law).

Water as a Liquid

The reference condition (or standard state) of a substance is considered to be its state at 298 K and standard atmospheric pressure (i.e. 101 325 Pa). At this temperature and pressure, water is a liquid with a density much greater than any gas, but with no rigid shape. The fluid properties of a liquid suggest that its molecules possess greater liberty than do the molecules of a solid, but undertake considerably less translational movement than do the molecules of a gas which quickly distribute themselves throughout any space into which they are introduced. That the liquid state of water is in many respects intermediate between its solid and gaseous states is simply demonstrated by changing its temperature at standard atmospheric pressure, when it freezes to form ice at approx. 273 K, and boils to form steam at 373 K. There is good reason to believe that in its liquid state the molecules of water are structurally associated, and there is further evidence that a proportion of them are ionized. These, and other unique characteristics, are reflected in the physical properties of water including its density, viscosity, high boiling point, high dielectric constant and electrical conductance; they also affect its behaviour as a solvent. However, for the elementary purposes of this chapter, we need not concern ourselves with many of these special properties of liquid water. Instead we shall be particularly concerned with one property that water shares with all liquids and solids, namely that at a given temperature in a closed vessel it exists in equilibrium with its vapour.

Equilibrium between Phases

An obvious starting point to relate gases and liquids is to consider the vapour above a liquid in an isolated container (Fig. 4.1)

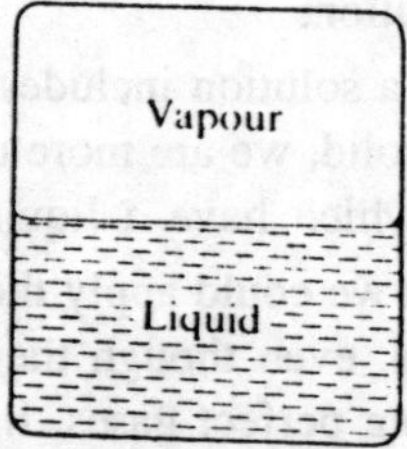

Fig 4.1 Equilibrium between vapour and liquid in a closed container.

Under a given set of conditions (temperature, pressure) an equilibrium will be set up between the two phases. The condition for equilibrium between the phase is that

$$\Delta G = 0$$

i.e. $$G_{\text{liquid}} = G_{\text{vapour}}.$$

This means that the properties of the liquid (such as its free energy) can be related to those of its vapour, provide equilibrium conditions prevail.

Consider a simple example

$$\text{liquid} \rightleftharpoons \text{vapour} \quad (\text{pressure} = p_{\text{vap}}).$$

The equilibrium constant K is given by

$$K = \left(\frac{p_{vap}}{p^{\circ}_{vap}}\right),$$

where p°_{vap} refers to the standard state of the vapour, i.e. 1 atmosphere pressure.

The standard state of the liquid is simply the pure liquid, and we recall from Chapter 3 that components in their standard state (pure liquids, solids, gases at 1 atmosphere, or solutes at 1 mol dm^{-3} concentration) do not appear in the equilibrium constant expression.

VAPOUR PRESSURE

Whenever a liquid (or solid) possesses an interface with a gaseous phase, a proportion of its molecules are lost by vaporization into this gaseous phase. When a quantity of liquid is introduced into a closed vessel, the rate at which its molecules quit the liquid for the gaseous phase will at first exceed the rate at which they recondense from the gaseous into the liquid phase. Consequently, the partial pressure exerted by the vapour molecules progressively increases and promotes their recondensation. Eventually the partial pressure is such that the rate of evaporation exactly equals the rate of condensation. The partial pressure of its vapour at this equilibrium is characteristic of a liquid at a given temperature and external pressure, and is called its *vapour pressure*.†

The Influence of Temperature on the Vapour Pressure of a Liquid

Increasing the temperature of a liquid causes its vapour pressure to rise until it eventually equals the external pressure; then, the liquid boils. It follows that the boiling point of a liquid depends on the prevailing external pressure. Thus, the *normal boiling point* of a liquid, which is the temperature at which it boils under an external pressure of 101 325 Pa, can also be defined as 'that temperature at which its vapour pressure is 101 325 Pa'. It follows that a liquid boils at a lower temperature than its normal boiling point when the external pressure is less than standard atmospheric pressure, a fact that is made use of to distil heat-labile materials under reduced pressure. Conversely, at pressure greater than the standard atmospheric pressure, the boiling point of a liquid is higher than normal. Thus water, whose normal boiling point is 373 K, boils at 283 K at a pressure of 1226 Pa, and at 394 K under a pressure of 203 kPa. This shows that at 283 K the vapour pressure of water is 1226 Pa; at 373 K it is 101 325 Pa, while at 394 K it is 203 kPa.

If we consider the freezing of water at a fixed pressure, we find that it takes place at that temperature (approx. 273 K at standard atmospheric pressure) at which the vapour pressure of water is exactly equal to the vapour pressure of ice. The normal freezing point of a liquid may in fact be defined as 'that temperature at which, under standard atmospheric pressure, its solid and liquid states are in equilibrium'. By increasing the applied pressure, the freezing point of most liquids is raised, but it is a peculiarity of water (and of a few other liquids) that its freezing point is *lowered* by enhanced pressures.

When the temperature of water is increased at standard atmospheric pressure from its normal freezing point (273 K) to its normal boiling point (373 K), its vapour pressure concurrently rises in an exponential manner. In fact, if the logarithm of the vapour pressure of any liquid is plotted against the reciprocal of the temperature in K, a straight line of negative slope is obtained.

The Clausius–Clapeyron equation

$$\ln\left(\frac{p_{\text{vap}}\ (T_2)}{p_{\text{vap}}\ (T_1)}\right) = -\frac{L_{\text{vap}}}{R}\left(\frac{1}{T_2} - \frac{1}{T_1}\right) \tag{4.1}$$

shows how the vapour pressure of a liquid varies with temperature. A plot of ln (vapour pressure) against $1/T$ gives a straight line from which the latent heat of vaporization can be calculated. This equation provides a link between liquid and vapour phase since by making measurements of vapour pressure we can deduce the value of L_{vap}. (This reflects the forces between the molecules in the *liquid state*, since in the vapour there are no such intermolecular forces, if the vapour is assumed to be a perfect gas.)

Effect of Solutes on the Vapour Pressure of Water

When water acts as a solvent, its vapour pressure is lowered by an amount which, in extremely dilute solutions, is directly proportional to the number of solute 'particles' introduced per unit volume of the aqueous solution, but is independent of the specific nature of these particles (i.e. their size, shape, chemical constitution, electrical charge). Coincident with this lowering of the vapour pressure the solution exhibits other properties to a proportional degree. All of these properties, in a sufficiently dilute solution, are determined by the concentration of solute particles regardless of their nature; they are called *colligative properties* of the solution and include,

(a) lowering of the vapour pressure of the solvent;
(b) elevation of the boiling point of the solvent;
(c) depression of the freezing point of the solvent;
(d) exhibition of an osmotic pressure.

SOLUTIONS OF NON-ELECTROLYTES

Raoult's Law—Ideal Solutions

The most important equation linking vapour and liquid is known as *Raoult's law*

$$\boxed{p = p^* \chi_1}, \qquad (4.2)$$

where p is the vapour pressure of the solvent at temperature T, p^* is the vapour pressure of pure solvent at the same temperature and χ_1 is the mole fraction of solvent.

i.e. $$\chi_1 = \frac{\text{number of moles of solvent}}{\text{number of moles of solvent + number of moles of solute}}.$$

This relationship was initially based on experimental measurements of vapour pressure. An *ideal solution* can be defined as one which obeys Raoult's Law.† (In an analogous fashion, an *ideal gas* is defined as one which obeys the equation $PV = nRT$.) On a molecular level, an ideal solution implies that the intermolecular forces (which are responsible for the liquid state) are all uniform. Thus if the components are termed 1 and 2, the forces between 1 and 1, 2 and 2, and 1 and 2, are all identical. In an ideal gas, of course, there are no forces between the molecules.

The uniformity of the intermolecular forces in ideal solutions means that the enthalpy of mixing of the components is zero (i.e. $\Delta H = 0$) and suggests that such behaviour will be observed when the components

† If there is more than one volatile component in the solution, e.g. a solution of benzene in toluene, then Raoult's Law predicts the partial vapour pressure of each component. The total vapour pressure is then the sum of the partial vapour pressures. However we shall be dealing almost exclusively with solutions of non-volatile solutes where this complication does not arise.

are chemically very similar. A further implication is that there is no volume change on mixing the components.

Alternative Definition of an Ideal Solution

Consider the vapour in equilibrium with a solvent.

Now for the vapour we have

$$G_{vap} = G^{\circ}_{vap} + RT \ln p_{vap}, \quad \text{since } p^{\circ}_{vap} = 1 \text{ atm.} \qquad (4.3)$$

To relate G°_{vap} to $G^{\circ}_{solvent}$ we have to consider the following cycle.‡

vapour at pressure p^*

(i) ↗ ↘ (ii)

(iii)

pure solvent → vapour at pressure 1 atm.

The ΔG for step (*i*) = 0 since pure solvent is in equilibrium with vapour at a pressure p^*.

For step (*ii*) (compression of vapour from pressure = p^* to pressure = 1 atm.)

$$\Delta G = -RT \ln p^*/1 = -RT \ln p^*.$$

For step (*iii*) ΔG is the sum of the ΔGs for steps (*i*) and (*ii*)

i.e. $$\Delta G = RT \ln p^*,$$

but since step (*iii*) is conversion of solvent in *its standard state* to vapour in *its standard state,* this is a ΔG°

i.e. $$\Delta G^{\circ} = -RT \ln p^*,$$

so that $$G^{\circ}_{vap} - G^{\circ}_{solvent} = -RT \ln p^*,$$

∴ $$G^{\circ}_{vap} = G^{\circ}_{solvent} - RT \ln p^*. \qquad (4.4)$$

Under any conditions where *equilibrium* between the solvent and its vapour exists, then

$$G_{solvent} = G_{vap},$$

substituting in eqn (4.3) this gives

$$G_{solvent} = G^{\circ}_{vap} = RT \ln p_{vap}.$$

Using eqn (4.4)

$$G_{solvent} = G^{\circ}_{solvent} - RT \ln p^* + RT \ln p_{vap}.$$

Now from *Raoult's Law*

$$P_{\text{vap}} = p^* \chi_1,$$

where χ_1 is the mole fraction of solvent.

$$\therefore \quad G_{\text{solvent}} = G^\circ_{\text{solvent}} - RT \ln p^* + RT \ln p^* + RT \ln \chi_1.$$

$$\therefore \quad \boxed{G_{\text{solvent}} = G^\circ_{\text{solvent}} + RT \ln \chi_1}. \tag{4.5}$$

This fundamental equation can also be used to define an ideal solution,[†] i.e. the free energy of the solvent in an ideal solution is given by the expression above. We have shown how the equation follows from Raoult's Law.

Correlation of Raoult's Law with Henry's Law

Suppose that n_2 mol of a gas are dissolved in N_1 mol of a liquid solvent. Raoult's Law predicts (1) that the vapour pressure of solvent above the solution, at a given temperature, is directly proportional to the mole fraction of the solvent, and (2) that the proportionality constant is equal to the vapour pressure of the pure solvent at the same temperature ; i.e.

$$p = p^* \chi_1 \quad \text{where} \begin{cases} p = \text{vapour pressure of solvent above the solution} \\ p^* = \text{vapour pressure of pure solvent at the same temperature and pressure} \\ \chi_1 = \text{mole fraction of solvent in the solution} \end{cases}$$

Henry's Law states that 'the mass of gas, dissolved in a certain volume of solvent, is directly proportional to the partial pressure of the gas that is in equilibrium with the solution'. Therefore, applying Henry's Law to this same solution, we obtain the following equation,

---

[†] Raoult's law defines an ideal solution in terms of the properties of the *solvent*. There is another law (*Henry's Law*) dealing with the properties of the *solute* in an ideal solution. Henry's Law states the mole fraction of the solute is *proportional* to the vapour pressure of the solute. Clearly this is of most use when the solute is volatile—e.g. oxygen gas dissolved in water. Since we are almost exclusively concerned with non-volatile solutes we shall not consider Henry's Law further.

$$\frac{n_2}{N_1} \propto p_2$$

where p_2 is the partial pressure of the gas in equilibrium with the solution (i.e. the vapour pressure of the gas as solute).

For a sparingly soluble gas that yields a very dilute solution, n_2/N_1 must be approximately equal to $\frac{n_2}{n_2 + N_1}$ (since n_2 is now insignificant in comparison with N_1). The Henry's Law prediction for such a solution is that,

$$\frac{n_2}{n_2 + N_1} \propto p_2$$

where $\frac{n_2}{n_2 + N_1} = \chi_2 =$ mole fraction of solute, and therefore,

$$p_2 = k\chi_2$$

In principle, Henry's Law applies to any solute, so that while Raoult's Law conceives of the ideal solvent as one whose vapour pressure is directly proportional to its mole fraction, Henry's Law similarly sees the ideal solute as one whose vapour pressure is directly proportional to *its* mole fraction. In a sense, therefore, Raoult's Law can be thought of as a 'special' application of Henry's Law in which the proportionality constant has a unique value (p^*) equal to the vapour pressure of the pure component at the same temperature and pressure. In practice, in the solution in which the solute behaves ideally according to Henry's Law, the solvent also behaves ideally according to Raoult's Law (though the reverse is not necessarily true and Raoult's Law ideality on the part of the solvent does not mean that the solute must obey Henry's Law).

Both Raoult's Law and Henry's Law are *limiting* laws in that they define the ideal behaviour of the components of a solution as their behaviour in an infinitely dilute solution; as a real solution approaches the limit of infinite dilution so will its components behave more ideally. The ideal state of the solvent envisaged by Raoult's Law is 'solvent at a mole fraction of unity', which is of course pure solvent. As a solute behaves ideally only in an infinitely dilute solution, its ideal state is more difficult to define. It is therefore more usual to define a hypothetical, reference solution in which, although the solute is present at unit concentration, it still behaves as if it were present in an infinitely dilute solution. This reference solution, which is taken as the standard state of the solute and in which the solute is said to be present at *unit activity,* is defined on the mol dm^{-3} scale of concentrations as 'the hypothetical 1 mol dm^{-3} solution of the solute in which it behaves as if it were actually present in infinitely dilute solution'.

PROPERTIES OF IDEAL SOLUTIONS

The Lowering of the Vapour Pressure of the Solvent

Raoult's Laws states

$$p = p^* \chi_1$$

$$\therefore \qquad \frac{p^* - p}{p^*} = 1 - \chi_1$$

The left hand side of this equation is the relative lowering of the vapour pressure of the solvent caused by addition of solute.

The term on the right hand side is equal to the mole fraction of solute (χ_2). χ_2 is defined as

$$\chi_2 = \frac{\text{moles solute}}{\text{moles solute + moles solvent}}.$$

In *dilute solutions* the number of moles of solvent is very much greater than the number of moles of solutes.

Thus

$$\chi_2 = \frac{\text{moles solute}}{\text{moles solvent}}.$$

If we have x g solute of molecular weight M in 1 kg of water (M.W. = 18)

$$\chi_2 = \frac{x / M}{1000/18} = \frac{x}{55.5\, M}$$

Now, from above $\qquad \chi_2 = \dfrac{p^* - p}{p^*}$

$$\therefore \quad \boxed{\frac{p^* - p}{p^*} = \frac{x}{55.5\, M}}$$

So that if we measure the lowering of the vapour pressure of the solvent by addition of a known amount of solute, we can calculate the molecular weight of the solute.

Depression of the Freezing Point of the Solvent

It follows from the definition of an ideal solution

$$G_{\text{solvent}} = G^{\circ}_{\text{solvent}} + RT \ln \chi_1$$

that the addition of solutes has lowered the free energy of the solvent by an amount $RT \ln X_1$.[†] We shall now investigate the effect of addition of solute on the freezing point of the solvent.

Consider the freezing of the solution

$$\text{solvent (liquid)} \rightleftharpoons \text{solvent (solid)}.$$

At the freezing point of the pure solvent (T_f), of course, $\Delta G = 0$ (this is an equilibrium situation). However on addition of solute, the free energy of the solvent (liquid) has been lowered by an amount $RT \ln \chi_1$. What we wish to find is the new temperature at which $\Delta G = 0$, when the solute is present. (This is the new freezing point of the solution.) Thus we need to use an equation describing the variation ΔG with temperature. From eqn (3.6), we recall that this is

$$\frac{d(\Delta G/T)}{dT} = -\frac{\Delta H}{T^2}$$

Integrating between T_f, the freezing point of pure solvent (where $\Delta G = RT \ln X_1$) and T'_f, the new freezing point (where $\Delta G = 0$), we have

$$\int_{\Delta G = RT \ln X_1}^{\Delta G = 0} \mathrm{d}\left(\frac{\Delta G}{T}\right) = -\int_{T_f}^{T'_f} \frac{\Delta H}{T^2} \mathrm{d}T$$

$$\therefore \quad -R \ln X_1 = +\Delta H \left(\frac{1}{T'_f} - \frac{1}{T_f}\right)$$

† X_1, the mole fraction of solute is less than one, and so $\ln X_1$ is negative.

$$\therefore \qquad -\ln X_1 = \frac{\Delta H}{R}\left(\frac{T_f - T_f'}{T_f' \times T_f}\right).$$

If $T_f \approx T_f'$ then we can set $T_f' \times T_1 = (T_f)^2$ and X_1 can be replaced by $(1 - X_2)$ (where X_2 is the mole fraction of solute). We obtain

$$-\ln(1 - X_2) = \frac{\Delta H}{R} \cdot \frac{\Delta T_f}{T_f^2},$$

(where ΔT_f is the freezing point depression). If X_2 is small, $\ln(1 - X_2) \approx -X_2$

$$\therefore \qquad X_2 = \frac{\Delta H}{R} \times \frac{\Delta T_f}{T_f^2}.$$

Now ΔH = the latent heat of fusion, L_f.

$$\therefore \qquad \boxed{X_2 = \frac{L_f}{R} \times \frac{\Delta T_f}{T_f^2}}. \qquad (4.6)$$

This is the equation linking the mole fraction of solute (X_2) with the freezing point depression (ΔT_f). Note that we have assumed that (*a*) the solution is ideal and (*b*) that X_2 is small, i.e. the solution is dilute.

By considering the effect of a solute on the free energy of the solvent in the equilibrium

$$\text{Solvent} \rightleftharpoons \text{vapour}$$

we can obtain an analogous expression for the boiling point elevation caused by a solute

$$\boxed{X_2 = \frac{L_{vap}}{R} \times \frac{\Delta T_b}{T_b^2}} \qquad (4.7)$$

where L_{vap} is the latent heat of vaporization an T_b is the boiling point of pure solvent.

Now for the freezing point depression (eqn (4.6))

$$X_2 = \frac{L_f}{R} \times \frac{\Delta T_f}{T_f^2}$$

so that

$$\Delta T_f = \frac{RT_f^2}{L_f} \times X_2$$

$$X_2 = \frac{\text{moles of solute}}{\text{moles of solvent}}$$

If we have x g solute of molecular weight M in 1 kg water (M. wt = 18), then

$$X_2 = \frac{x/M}{1000/18} = \frac{x}{55.5\,M}.$$

substituting for X_2,

$$\Delta T_f = \frac{RT_f^2}{L_f} \times \frac{x}{55.5\,M}.$$

For water $L_f = 6.02$ kJ mol^{-1}, $T_f = 273.15$ K

$$\Delta T_f = 1.86\left(\frac{x}{M}\right)$$

so that if we had a 1 *molal* solution (1 mol solute per kg water) then

$$\Delta T_f = 1.86 \text{ K}$$

This is called the *molal freezing point depression constant* for water. For other solvents its value is different.

EXAMPLE 4.1 *A solution of 39 g glucose in 1 kg water has a freezing point 0.4 K below that of pure water. What is the molecular weight of glucose? The molal freezing point depression constant of water is 1.86 K.*

Solution

Let the molecular weight of glucose = M. $x = 39$ g

then the solution is (39/M) molal.

$$\Delta T_f = 0.4$$

$$0.4 = 1.86\left(\frac{39}{M}\right)$$

$$\therefore \qquad \text{Molecular weight} \quad M = \frac{1.86 \times 39}{0.4}$$

$$= 181$$

i.e. the molecular weight of glucose is 181 (actual value = 180).

Osmosis

Osmosis is the process whereby solvent spontaneously moves from one region in a solution where its activity is high, to another region where its activity is lower. In practice, osmosis takes place only when a semi-permeable membrane separates a solution, either from pure solvent or from a solution with the same solvent but having a different concentration of solute. The membrane that is used is in theory perfectly semi-permeable, which means that it is freely permeable to solvent but impermeable to all solutes. But such a completely semi-permeable membrane is very rare, and the membranes employed in practice to demonstrate osmosis between aqueous solutions are only *selectively* impermeable; they generally allow free passage of water and of solutes of small molecular weight, but are completely impermeable to substances of comparatively high molecular weight.

The osmotic pressure of a solution

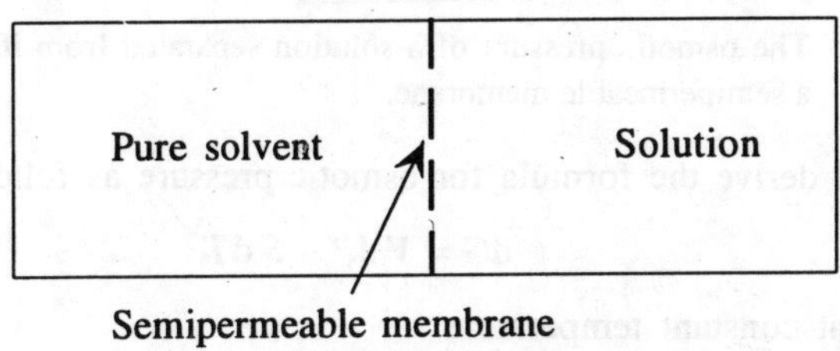

Fig. 4.2 Separation of solvent and solution by semipermeable membrane.

Consider a system in which a solution is separated from its pure solvent by a semipermeable membrane (i.e. the membrane is permeable to solvent, but not to solute) (Fig. 4.2).

Clearly the free energy of the solvent is different on the two sides of the membrane. On the solution side, that free energy of the solvent is given by eqn (4.5)

$$G_{\text{solvent}} = G^{\circ}_{\text{solvent}} + RT \ln X_1$$

whereas, on the solvent side, the free energy is $G^\circ_{solvent}$ (since we have pure solvent).

Since the free energies of the solvent on either side of the membrane are different, this is not an equilibrium situation, and there will be a tendency for solvent to pass into the solution.†

We can bring the system to equilibrium using various external factors. In the previous section we saw that changes in temperature could affect the value of $G_{solvent}$. Here we are concerned with the effects of pressure on the solution. The pressure required to bring the solution into equilibrium with the solvent is known as the *osmotic pressure* of the solution. This is illustrated in Fig. 4.3.

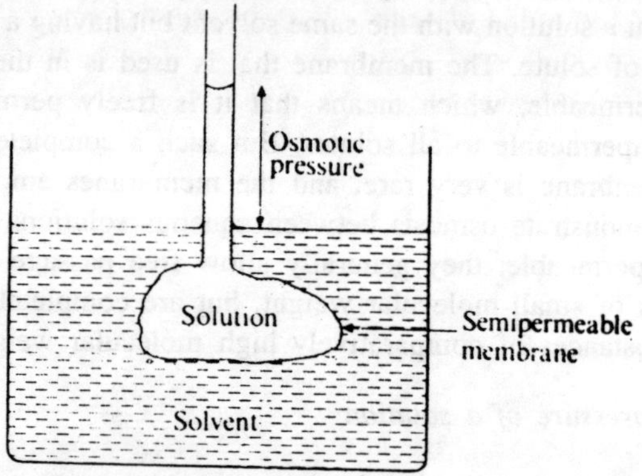

Fig. 4.3 The osmotic pressure of a solution separated from its solvent by a semipermeable membrane.

We shall derive the formula for osmotic pressure as follows:

$$dG = V\,dP - S\,dT$$

so that at constant temperature

$$dG = V\,dP.$$

Now the solvent on the solution side has a free energy $RT \ln X_1$ lower than that of pure solvent (this is because of dilution). The effect of applying the osmotic pressure must be to increase $G_{solvent}$ in the solution (G_A) so as to counteract this dilution effect. At equilibrium (change in G from applying osmotic pressure) + (change in G from dilution) = 0.

† As solvent passes into the solution, the mole fraction of solvent (X_1) in the solution tends towards 1 and hence $G_{solvent}$ tends towards $G^\circ_{solvent}$.

Let the osmotic pressure (in excess of 1 atm) = π atm, then

$$\int_0^{\pi} \mathrm{d}G_A + RT \ln X_1 = 0$$

$$\int_0^{\pi} V_A \, \mathrm{d}P = -RT \ln X_1.$$

V_A is the molar volume of solvent, and can be assumed to be independent of pressure (i.e. the solution is incompressible).

$$\therefore \qquad V_A \pi = -RT \ln X_1.$$

As before X_1 is replaced by $(1 - X_2)$ and if X_2 is small (i.e. if the solution is dilute) we can write:

$$-RT \ln X_1 = -RT \ln (1 - X_2)$$
$$= RT . X_2$$

$$\therefore \qquad V_A \pi = RT . X_2.$$

For *dilute aqueous* solutions, V_A is 0.018 $dm^3\ mol^{-1}$ (since the specific gravity is close to unity) and the above equation becomes

$$\boxed{\pi = RT . c} \qquad (4.8)$$

where c is the *molarity* of the solute (mol solute per dm^3 solution).

EXAMPLE 4.2 *What is the osmotic pressure of a solution containing 39 g glucose in 1 dm^3 solution? (T = 298 K)*

Solution

For dilute solutions

$$\pi = RTc.$$

If we measure π in atmospheres, we must use the appropriate value of R, i.e. $R = 0.082\ dm^3\ atm\ mol^{-1}\ K^{-1}$.

$$\therefore \qquad \pi = (0.082)(298)(c)$$

but $c = 39/180\ mol\ dm^{-3}$

$$\therefore \qquad \pi = (0.082)(298)(39/180)\ atm$$
$$= 5.3\ atm.$$

The osmotic pressure is equivalent to a column of water 55 metres high!

This example shows that measurement of the osmotic pressure of a dilute solution is much easier than the measurement of the rather small freezing point depression (0.4 K in this case).

The four properties, vapour pressure lowering, freezing point depression, boiling point elevation, and osmotic pressure, are known as the *colligative properties* of solutions. They have been used for determination of the molecular weight of small molecules (solutes), but in dealing with solutions containing macromolecules, we shall see later that in general only osmotic pressure measurements are feasible.

Anomalous Molecular Weights

Sometimes the molecular weight appears to be 'anomalous'. This could arise from a number of factors.

(1) An electrolyte, e.g. NaC1, will be *dissociated* in solution. This increases the number of species present in the solution and the colligative properties will be greater than those predicted on the basis of the molar concentration of solute. The ratio (observed effect/predicted effect) is given the symbol *i*, and is known as the *van't Hoff Factor* for the solution.

(2) In some cases the solute can *associate* in solution, eg. benzoic acid can form dimers in aqueous solution.

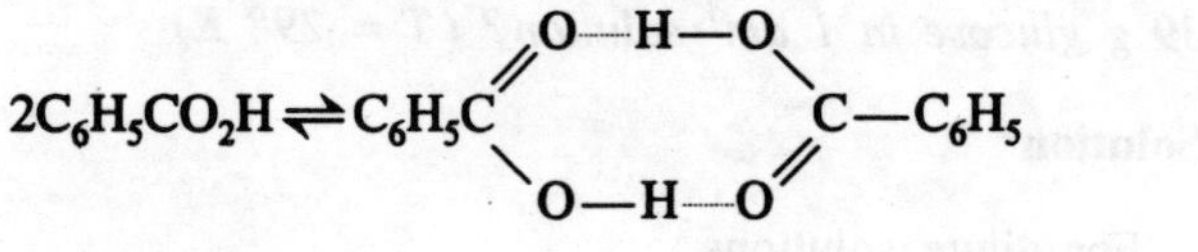

$$2C_6H_5CO_2H \rightleftharpoons C_6H_5C\begin{matrix} \diagup\!\!\!\!= O \cdots H - O \diagdown \\ \diagdown O - H \cdots O =\!\!\!\!\diagup \end{matrix}C - C_6H_5$$

Mol. wt.122 Mol. wt. 244

This reduces the number of species present in the solution, with a concomitant decrease in the colligative properties. The apparent molecular weight is between the monomer and dimer molecular weights depending on their relative numbers (concentrations).

(3) The solution may be non-ideal. This is likely to be particularly true when the solute and solvent are chemically dissimilar, such as in most of the solutions of interest to biochemists.

Non-Ideal Solutions—The Concept of Activity

From Raoult's Law we have been able to derive some simple and convenient equations to describe the properties of solutions. In practice the forces between molecules in a solution are not all uniform and there is a consequent departure from ideality. We would expect this departure to be small in dilute solutions, as indeed is the case. However, it remains convenient to keep the same form of the thermodynamic equations even for non-ideal solutions. To do this the term 'activity' is introduced as follows:

For ideal solutions Raoult's Law is

$$p = p^* X_1$$

while for non-ideal solutions this becomes

$$\boxed{p = p^* a_1} \tag{4.9}$$

where a_1 is the *activity* of the solvent. The thermodynamic equations are then exactly analogous if a_1 is substituted for X_1; a_1 is thus the mole fraction 'corrected' for the effects of non-ideality so that thermodynamic equations still hold.[†] The ratio of activity to mole fraction is termed the *activity coefficient* γ.

$$\gamma = \frac{a}{X} = \frac{\text{effective concentration}}{\text{true concentration}} \tag{4.10}$$

when $\gamma = 1$, the solution is obviously ideal. Eqn (4.10) enables us to calculate γ from a known value of X is a is measured.

EXAMPLE 4.3 *At 323 K, the vapour pressure of pure water is 0.122 atm. In a sucrose solution, where the mole fraction of water is 0.96, the vapour pressure is measured as 0.1165 atm (at 323 K). Calculate the activity coefficient of the water.*

Solution

From Raoult's Law—for non-ideal solution (eqn 4.9)

$$a_1 = \frac{p}{p^*} = \frac{0.1165}{0.122} = 0.955$$

[†] Thus eqn (2.1) $G = G^\circ + nRT \ln (p/p^0)$ is replaced by $G = G^0 + RT \ln a$ and this equation is then assumed to apply to each component (gaseous, liquid, solute, etc.).

since

$$X_1 = 0.96$$

we have

$$\gamma = \frac{0.955}{0.960} = 0.995$$

i.e. in this dilute solution the solvent behaves almost ideally.

Electrolyte Solutions

Clearly one case where the interactions between the species in solution are not going to be identical is that of solutions of electrolytes. By making the assumption that the departures from ideal behaviour in dilute solutions of electrolytes are a consequence of electrical interactions Debye and Hückel were able to derive the following expression for aqueous solutions:

$$\boxed{\log \gamma_\pm = -0.5\, Z_+ Z_- \sqrt{I},} \; ;^\dagger \qquad (4.11)$$

$\gamma_\pm$ is the mean activity coefficient of the ions. (The activity coefficient of the ions cannot be measured separately.) Z_+, Z_- are the charges carried by the positive and negative ions respectively. No account is taken of the sign of the charges. I is the *ionic strength*—defined as

$$\boxed{I = \tfrac{1}{2} \sum c_i Z_i^2 .} \qquad (4.12)$$

EXAMPLE 4.4 *Calculate the mean activity coefficient in solutions of (a) 0.05 mol dm^{-3} $MgCl_2$ and (b) 0.1 mol dm^{-3} Na_3PO_4.*

Solution

Both these salts are fully dissociated.

In (*a*) $Z_+ = 2$, $Z_- = 1$, $c_+ = 0.05$ mol dm^{-3} $c_- = (2 \times 0.05)$ mol dm^{-3} since there are two chloride ions per magnesium ion.

Thus

$$I = \tfrac{1}{2}(0.05 \times 2^2 + 2 \times 0.05 \times 1^2)$$
$$= 0.15.$$

† Eqn (4.11) is valid at 298 K. At other temperatures the constant (0.5) will be slightly different.

From the Debye–Hückel equation

$$\log \gamma_\pm = -0.5 \times 2 \times 1 \times \sqrt{(0.15)}$$
$$\gamma_\pm = 0.41$$

In (*b*) $Z_+ = 1$, $Z_- = 3$, $c_+ = (3 \times 0.1)$ mol dm^{-3}, $c_- = 0.1$ mol dm^{-3}

Thus

$$I = \frac{1}{2}(0.3 \times 1^2 + 0.1 \times 3^2)$$
$$= 0.6.$$

Hence

$$\log \gamma_\pm = -0.5 \times 3 \times 1 \times \sqrt{(0.6)}$$
$$\gamma_\pm = 0.07.$$

These mean ionic activity coefficients are related to the individual activity coefficients of the ions by the formula

$$(\gamma_\pm)^n = (\gamma_+)^{n+} (\gamma_-)^{n-}$$

where *n* is the *total* number of ions from one mole and *n*+ and *n*− are the numbers of the individual ions. Thus for $MgCl_2$ this would become

$$(\gamma_\pm)^3 = (\gamma_{Mg^{2+}})(\gamma_{Cl^-})^2$$

i.e.

$$\gamma_\pm = \{(\gamma_{Mg^{2+}})(\gamma_{Cl^-})^2\}^{\frac{1}{3}}$$

Similar expressions for the mean activities (i.e. $a_\pm$) and concentrations can also be written so that in the case of *single salt*

$$\gamma_\pm = a_\pm / c_\pm$$

for the example of 0.05 mol dm^{-3} $MgCl_2$ $c_\pm$ is given by

$$c_\pm = \{(0.05)^1 (2 \times 0.05)^2\}^{\frac{1}{3}}$$
$$c_\pm = 0.079 \text{ mol dm}^{-3}$$

Thus

$$a_\pm = y_\pm \cdot c_\pm$$
$$= 0.41 \times 0.079$$
$$a_\pm = 0.032 \text{ (mol dm}^{-3}\text{)}.$$

This is the mean ionic activity of the solute that would be obtained from measurements of the solution properties.

Problem

What is the mean activity of 0.1 mol dm^{-3} Na_3PO_4?

Solution

$$a_{\pm} = 0.016 \text{ (mol dm}^{-3}).$$

The conclusion from the preceding discussions is that *activities* ather than *concentrations* should always be used in the expressions for the equilibrium constants. This is particularly true for solutions of electrolytes.† One example of this is afforded by the solubility of a sparingly soluble salt.

Sparingly Soluble Salts

Suppose a sparingly soluble salt such as $CaCO_3$ is added to water to give a saturated solution. For the equilibrium

$$CaCO_3 \text{ (solid)} \rightleftharpoons Ca^{2+} \text{ (aqueous)} + CO_3^{2-} \text{(aqueous)}$$

$$K_s = \frac{a_{Ca^{2+}} \times a_{CO_3^{2-}}}{a_{CaCO_3}}$$

Since solid $CaCO_3$ is in its standard state, $a_{CaCo_3} = 1$.

Thus

$$K_s = a_{Ca^{2+}} \times a_{CO_3^{2-}}.$$

K_s is known as the *solubility product* of the salt. Thus

$$K_s = (\gamma_+ c_+)(\gamma_- c_-)$$
$$= \gamma_{\pm}^2 c_+ \cdot c_- .$$

If $\gamma_{\pm}$ can be obtained from the Debye–Hückel law then K_s can be calculated if c_+ and c_- are known.

EXAMPLE 4.5 *AgCl has a solubility product of 1.6 × 10^{-10} (mol dm^{-3}) at 298 K. Calculate the concentration of Ag^+ ions in solution if 0.01 mol dm^{-3} NaCl is present.*

---

† In accurate work, extrapolations of data (e.g. equilibrium constants) are made to a solution of zero ionic strength. When $I = 0$, the solution behaves ideally, and thus 'true' thermodynamic constants can be obtained.

Solution

The ionic strength of the solution is obviously entirely determined by the NaCl which is present in large excess over the dissolved AgCl.

Now

$$I = \tfrac{1}{2}\sum c_i Z_i^2$$

$$c_{Na^+} = c_{Cl} = 0.01 \text{ mol dm}^{-3}$$

$$\therefore \quad I = \tfrac{1}{2}\{(0.01)1^2 + (0.01)1^2\}$$

$$= 0.01.$$

From the Debye–Hückel expression

$$\log \gamma_\pm = -0.5\, Z_+ Z_- \sqrt{I}$$

$$\therefore \quad \gamma_\pm = 0.891^\dagger$$

Thus

$$K_s = a_{Ag^+} \cdot a_{Cl^-} = 1.6 \times 10^{-10}$$

$$a_{Cl^-} = \gamma_{Cl^-} \cdot c_{Cl^-}$$

$$= 0.891 \times 0.01$$

$$= 8.91 \times 10^{-3}$$

since $[Cl^-]$ is essentially that of the NaCl.

$$\therefore \quad a_{Ag^+} = \frac{K_s}{a_{Cl^-}} = \frac{1.6 \times 10^{-10}}{8.91 \times 10^{-3}} = 1.80 \times 10^{-8}$$

Hence

$$\gamma_{Ag^+} \cdot c_{Ag^+} = 1.80 \times 10^{-8}$$

and

$$c_{Ag^+} = \frac{1.80}{0.891} \times 10^{-8}$$

$$= 2.01 \times 10^{-8} \text{ (mol dm}^{-3})$$

i.e. 2.01×10^{-8} mol dm^{-3} AgCl dissolves under these conditions.

---

† Note that for this solution, where all the ions carry unit charge, $\gamma_\pm = \gamma_+ = \gamma_-$.

In this example the effect of adding NaC1 has been to depress the solubility of AgC1. This is because we have added a large excess of chloride ions. If we added a salt, e.g. $NaNO_3$ with no common ion, we would find that the solubility of AgCl would be increased. This arises because the high ionic strength would lower $\gamma_\pm$ and so $c_\pm$ would have to increase to compensate, so that K_s remains constant.

The Concept of Chemical Potential

We have seen that reactions or systems at equilibrium are characterized by the condition that $\Delta G = 0$. If we consider a single component in two environments (as for example the solvent in an osmotic pressure experiment or the diffusible ions in the Donnan effect, p. 66) then the condition for equilibrium is that the free energy of this component is the same in the two environments. Clearly if they are not the same there will be a tendency to equalize them, e.g. by the 'flow' of solute from the high free energy environment to the low free energy environment (so that ΔG is negative). This can be likened to an electrical circuit where current flows from a region of *high* electrical potential to one of *low* electrical potential. The greater this difference in potential the greater this flow of current. In the same way, we can think of free energy as a *chemical potential, μ*. In fact the chemical potential of a component in a mixture is essentially equal to its molar Gibbs free energy. In describing solution properties many authors use 'chemical potential' rather than Gibbs free energy. We have not done so because the treatments are entirely equivalent and the introduction of chemical potential might lead to some confusion. However in the case of solutions of *ions,* it is more meaningful to use the term potential since we are dealing with charged species. We shall discuss the electrochemical potential of ions in Chapter 8.

PROBLEMS

1. On Pike's Peak (Colorado Springs) water boils at 364.1 K. Calculate the atmospheric pressure on Pike's Peak given that the latent heat of vapourization is 43.6 kJ mol^{-1} and that water boils at 373.15 K at 1 atm pressure.
2. At 295 K the vapour pressure of a solution of 10 g of a solute in 0.1 dm^3 water is 0.0247 atm. If the vapour pressure of pure water is 0.025 atm at the same temperature, calculate the molecular weight of the solute. What difference would result if the solute is 50 per cent ionized according to the equation below?

$$AB \rightleftharpoons A^+ + B^-$$

3. At 310 K the vapour pressure of a 60 per cent solution (w/w) of glycerol in water † is 0.0434 atm. If the vapour pressure of water is 0.0620 atm at this temperature calculate the activity and activity coefficient of water in this solution (assume the vapour pressure of glycerol is negligible).
4. Calculate the freezing point depression if: (1) 0.1 g of ammonium sulphate (molecular weight 132) is dissolved in 10 g of distilled water; (2) 0.1 g of a protein of molecular weight 20 000 is dissolved in 10 g of distilled water; (3) a mixture of 0.1 g protein and 0.1 g ammonium sulphate is dissolved in 10 g of water. (The freezing point depression constant for a solution of 1 mol solute per 1 kg water is 1.86 K.) In the light of this result *comment* on the possibility of using this method to determine the molecular weight of a macromolecule.
5. Several insects are able to withstand prolonged exposure to low temperatures. In many cases the haemolymph (blood) of these insects contains large amounts of glycerol. What protection against freezing would be offered by the glycerol in the haemolymph of the parasitic wasp *Bracon cephi* which occurs at a concentration of approximately 30 per cent (w/w)? (The latent heat of fusion of water is 6.01 kJ mol^{-1}.)
6. An aqueous solution of glucose (molecular weight 180) has an osmotic pressure of 1.55 atm at 298 K. What is the freezing point depression of this solution? (Freezing point depression constant = 1.86 K for 1 mol solute per kg. water; R = 0.082 dm^3 atm mol^{-1} K^{-1}.)
7. Calculate the ionic strength of
   (a) 0.1 mol dm^{-3} $MgCl_2$.
   (b) 0.05 mol dm^{-3} Na_2HPO_4.
   (c) A 1 mmol dm^{-3} protein solution (as its sodium salt) at a pH at which the protein carries a charge of −10 units.
   (d) A 0.1 mol dm^{-3} solution of acetic acid ($K_a = 1.75 \times 10^{-5}$ (mol dm^{-3})).
8. State the Debye–Hückel Law for a dilute aqueous solution of a strong electrolyte at 298 K.

   Calculate the activities of sodium and chloride ions in an aqueous solution of 1 mmol dm^{-3} sodium chloride, and also in an aqueous solution containing 1 mmol dm^{-3} sodium chloride and 3 mmol dm^{-3} potassium bicarbonate.
9. The total concentration of ferric ion in the plasma is estimated to be 50 μ mol dm^{-3}. Calculate the pH at which 99 per cent of the Fe^{3+} would be

† i.e. the mixture contains 60 per cent (by weight) glycerol and 40 per cent water.

precipitated, given that the solubility product for $Fe(OH)_3$ is 10^{-36} at 310 K. Assume that the ionic product, K_w, for water is 10^{-14} (K_w is the product of $[H^+]$ and $[OH^-]$, both expressed as mol dm^{-3}).

Comment on the state of Fe^{3+} in the plasma where the $[H^+]$ is about 10^{-7} mol dm^{-3}.

10. The mean standard enthalpy and entropy changes in the range 275 K to 298 K for the dissolution of $Ca(OH)_2$ in water to give an ideal solution of its constituents are –16.3 kJ mol^{-1} and –153 J K^{-1} mol^{-1} respectively. Calculate the ideal solubility of $Ca(OH)_2$ at 298 K and 275 K (atomic weights Ca = 40, H = 1, O = 16).

Solution to Problems

1. Atmospheric pressure = 0.705 atm.

   The Clapeyron–Clausius equation is used to find the vapour pressure of water at 364.1 K.

2. Molecular weight = 150.

   If the solute is 50 per cent dissociated we have a 1.5-fold increase in the number of particles present (i = 1.5), i.e. the vapour pressure lowering will be 0.00045 atm (new vapour pressure = 0.024 55 atm).

3. Activity (a) of water (i.e. P_A / P_A^*) = 0.701.

   Now the mole fraction of water = 0.773

   so that the activity coefficient = a / X = 0.906.

4. (1) Freezing point depression = 0.42 K (note that $(NH_4)_2SO_4$ is completely dissociated to give 3 ions).

   (2) Freezing point depression = 0.00093 K.

   (3) Freezing point depression = 0.42093 K (i.e. the sum of the contributions from the individual species).

   This result shows that the freezing point depression is too insensitive for the determination of molecular weights of macromolecules.

5. Freezing point depression = 8.3 K. This result is derived by using the full expression in the derivation of eqn (5.6). If the approximation for dilute solutions is used ($\Delta T_f = 1.86\,(x/M)$) the result is 8.7 K. Both methods assume that the solution is ideal.

   The insects can actually survive exposure to temperatures considerably lower than this, because the haemolymph can be supercooled without freezing.

6. Freezing point depression = 0.12 K (the concentration of glucose is so low we can equate concentrations expressed in terms of mol kg^{-1} and mol dm^{-3}).

7. (a) $I = 0.3$
   (b) $I = 0.15$
   (c) $I = 0.055$
   (d) $I = 0.00131$ (the concentration of H^+ and acetate$^-$ are both 1.31 mmol dm^{-3})
8. $\log \gamma_\pm = -0.5\, Z_A Z_B \sqrt{(I)}$.
   (a) In 1 mmol dm^{-3} NaCl, $\gamma_\pm = 0.966$, hence $a_{Na^+} = a_{Cl^-} = -9.66 \times 10^{-4}$.
   (b) In 1 mmol dm^{-3} NaCl and 3 mmol dm^{-3}. $KHCO_3$, $\gamma_\pm = 0.93$, hence $a_{Na^+} = a_{Cl^-} = 9.3 \times 10^{-4}$.
9. When 99 per cent of the Fe^{3+} is precipitated, $[Fe^{3+}] = 0.5 \times 10^{-6}$ mol dm^{-3}. Hence $[OH^-] = 1.26 \times 10^{-10}$, i.e. pH = 4.1.

   Thus above pH 4.1 more than 99 per cent of the Fe^{3+} would be precipitated. At physiological pH only a very small concentration of Fe^{3+} ($\approx 10^{-15}$ mol dm^{-3}) can remain in solution. This suggests that Fe^{3+} in the plasma must be complexed with other species (in this case, the protein ferritin).
10. At 298 K, the solubility product is 7.84×10^{-6}, hence the solubility of $Ca(OH)_2$ is 0.93 g dm^{-3}.

    At 275 K, the solubility product is 1.355×10^{-5}, hence the solubility is 1.11 g dm^{-3}.

5

Electrochemistry

Oxidation–Reduction Processes

We are devoting a separate chapter to a discussion of the thermodynamics of oxidation–reduction processes, not only to emphasize their importance in biochemistry, but also because many of these processes can be studied in electrochemical cells where true *reversible* conditions can be achieved.

Oxidation–reduction processes can be most easily understood in terms of the transfer of electrons, e.g. the oxidation of Fe^{2+} to Fe^{3+}. One of the most important biochemical examples is the so-called 'mitochondrial electron transport chain', consisting of various cytochromes, flavoproteins etc. each of which can undergo oxidation and reduction. Electrons from reducing agents (such as NADH or succinate) are passed along the chain via these various cytochromes etc. to the oxidizing agent (in this case, O_2). In this way the free energy available from the oxidation of say, NADH by O_2 ($\Delta G^{\circ\prime} = -214.4$ kJ mol^{-1}) is utilized in a series of steps to produce three moles of ATP from ADP and phosphate. The ATP thus formed is used in a whole range of ways in the cell (e.g. mechanical work, biosynthesis, transport of metabolites etc.).

The thermodynamics of oxidation–reduction processes can be described by reference to a simple example.

Consider the reaction in which Zn is added to a solution $CuSO_4$. Precipitation of Cu

$$Zn + Cu^{2+} \rightleftharpoons Zn^{2+} + Cu\downarrow,$$

occurs spontaneously, and under such conditions we would describe the reaction as being carried out irreversibly. For this reaction ΔG° is -213.2 kJ mol^{-1} at 298 K (corresponding to an equilibrium constant of 10^{37}). The free energy available from this reaction could be utilized in the form of electrical work, by setting up an appropriate *electrochemi-*

cal cell. In order to do this we divide the overall reaction into two separate oxidation–reduction processes:

$$Zn - 2e^- \rightarrow Zn^{2+} \qquad \text{oxidation (i.e. loss of electrons),}$$
$$Cu^{2+} + 2e^- \rightarrow Cu \qquad \text{reduction (i.e. gain of electrons).}$$

(The overall reaction is obviously the sum of these two processes.)

The separate reactions are known as the *half-cell reactions.* Neither half-cell-reaction could be studied in isolation, since the Zn would gain an enormous positive potential and the Cu a negative one. However, the difference between them can be studied if they can be suitably coupled to form an *electrochemical cell.*

The electrodes (Zn and Cu) dip into solutions of their ions (the sulphate ions being necessary to maintain electrical neutrality). To avoid the spontaneous precipitation of Cu it is necessary to keep the solutions of $ZnSO_4$ and $CuSO_4$ separate, while maintaining electrical contact between them. This is achieved by a 'salt bridge', which contains concentrated KCl† in a gel and dips into each of the half cells.

If the Zn and Cu electrodes are connected, electrons flow from the Zn to the Cu as Zn^{2+} ions are formed and Cu^{2+} ions are reduced. As the reaction proceeds towards equilibrium the driving force (ΔG) falls and so the amount of electrical work obtained from the cell decreases.

Suppose however that we *apply* a suitable potential difference across the electrodes so as to oppose the effect of the electrochemical cell.

As the applied potential difference is increased, the current flowing in the circuit will decrease. At a certain point (*the null point*), no current flows (i.e. the applied potential difference exactly equals that of the electrochemical cell). Under these conditions the potential difference of the cell is known as its *electromotive force (e.m.f.).* If the applied potential difference is increased beyond this value, the current will reverse its direction, (i.e. the cell reaction proceeds in the opposite direction $Cu + Zn^{2+} \rightarrow Cu^{2+} + Zn$). As the potential difference can be varied smoothly about the null point with no discontinuities in the

† KCl is chosen because K^+ and Cl^- have similar current carrying properties (mobilities), so that overall current can be carried in either direction with equal facility. The diffusion of Zn^{2+} or Cu^{2+} ions into the bridge is very slow and can effectively be ignored. However, it should be noted that a salt bridge (or any junction between liquids) contributes a small degree of irreversibility to the system. For most practical purposes, there is no need to consider this.

current flow, the electrochemical cell is said to be operating *reversibly* at the null point. We have therefore used the applied potential difference to prevent the tendency of Zn to reduce Cu^{2+} ions in the cell reaction. A slight increase or decrease in the applied potential difference would drive the reaction in either direction, i.e. we have satisfied the criterion of *reversibility*.

The Thermodynamics of Reversible Cells

The First Law of thermodynamics states that

$$\Delta E = \Delta q + \Delta w,$$

where Δw is the work done *on* the system and consists of work done *by* the surroundings $(-P\Delta V)$ and also in this case the electrical work done $(-\Delta w_{elec})$.

So

$$\Delta E = \Delta q - P\Delta V - \Delta w_{elec}.$$

Now at constant pressure

$$\Delta H = \Delta E + P\Delta V$$

So

$$\Delta H = \Delta q - \Delta w_{elec}.$$

Since the electrochemical cell process is made reversible, $\Delta q = T\Delta S$ (from the Second Law)

$$\therefore \qquad \Delta H = T\Delta S - \Delta w_{elec}.$$

Using the equation $\Delta G = \Delta H - T\Delta S$ we obtain $\Delta G = -\Delta w_{elec}$ (for electrochemical cell processes at constant pressure).

Thus ΔG measures the amount of 'useful' work obtainable from the electrochemical cell.

Suppose the reaction involves transfer of n electrons (e.g. $n = 2$ in the case of $Zn + Cu^{2+} \rightleftharpoons Zn^{2+} + Cu$). Then the electrical work done in transporting n electrons through a potential difference E is given by $\Delta w_{elec} = nFE$.

F is a conversion constant and is equal to 96.5 kJ V^{-1} mol^{-1}.

Thus we arrive at the equation

$$\boxed{\Delta G = -nFE}$$

for all reversible cell processes. Measurement of the e.m.f. E of an oxidation–reduction process gives ΔG directly, provided that the number of electrons n involved in this process is known. In the same way that ΔG° measures the free-energy change when reactants and products are in their standard states, E° refers to the *e.m.f of the cell under standard state conditions.*

EXAMPLE 5.1 *For the reaction*

$$Zn + Cu^{2+} \rightleftharpoons Zn^{2+} + Cu,$$
$$\Delta G^\circ = 213.2 \text{ kJ mol}^{-1}.$$

What is the value of E° for this reaction ?

Solution

$$\Delta G^\circ = -nFE^\circ,$$

$$\therefore \quad E^\circ = -\frac{\Delta G^\circ}{nF},$$

$$= +\frac{213.2}{2 \times 96.5} V \quad \text{since } n = 2,$$

$$= 1.1 \text{ V}.$$

Note that since ΔG° is negative for reactions to be favourable, E° must be positive.

Types of Half-Cells

As mentioned earlier all reactions in electrochemical cells can be broken down into two half cell reactions of the type

$$\text{Oxidized state} + ne \rightleftharpoons \text{Reduced state}.$$

(One of these reactions will be driven in reverse.)

We shall now illustrate the various types of half cell which are commonly used and indicate the nomenclature used to represent them.†

---

† The convention used here is that a vertical line indicate the physical separation of two phase, e.g. Zn (solid) and Zn^{2+} solution ($Zn^{2+}|Zn$). A comma is used to indicate that the two species are in the *same phase*—e.g. Fe^{2+} and Fe^{3+}(Fe^{2+}, Fe^{3+}) in solution or Ag and AgCl both in the solid phase (Ag, AgCl). A double vertical line (||) is used to denote a liquid–liquid junction (usually achieved by a salt bridge).

(1) *A metal electrode in a solution of its ions*

Many examples of this are known, e.g.

$$Zn^{2+} + 2e^- \rightleftharpoons Zn \quad \ldots\ Zn^{2+} | Zn$$

$$Ag^+ + e^- \rightleftharpoons Ag \quad \ldots\ Ag^+ | Ag.$$

(2) *Gas in contact with a solution of its ions*

The best known example of this is the hydrogen electrode

$$H^+ + e^- \rightleftharpoons \tfrac{1}{2}H_2 \quad \ldots\ H^+ | \tfrac{1}{2}H_2 \cdot Pt.$$

Hydrogen gas is bubbled over a Pt black electrode, which dissociates the H_2 molecules. Thus a monatomic layer of hydrogen is formed in contact with the solution of H^+ ions.

Cl_2 and O_2 electrodes can also be used in this way.

(3) *Two different oxidation states of the same species*

Here an inert electrode (such as Pt) is in contact with the solution, and thus provides a means of allowing electrons to enter or leave the solution.

For example

$$Fe^{3+} + e^- \rightleftharpoons Fe^{2+} \quad \ldots\ Pt | Fe^{3+}, Fe^{2+}$$

$$Sn^{4+} + 2e^- \rightleftharpoons Sn^{2+} \quad \ldots\ Pt | Sn^{4+}, Sn^{2+}$$

$$Fe(CN)_6^{3-} + e^- \rightleftharpoons Fe(CN)_6^{4-} \quad \ldots\ Pt | Fe\,(CN)_6^{3-}, Fe\,(CN)_6^{4-}$$

(4) *Metal in contact with its insoluble salt*

A metal is coated with a thin layer of one of its insoluble salts and this is in contact with a solution containing the anion of the insoluble salt. The two most common examples of this type are[†]

$$\underset{\text{(solid)}}{AgCl} + e^- \rightleftharpoons \underset{\text{(solid)}}{Ag} + Cl^- \quad \ldots\ AgCl, Ag | Cl^-$$

[†] The overall reaction in these half cells can be thought of as consisting of two simpler reaction, i.e.

$$AgCl \rightleftharpoons Ag^+ + Cl^-$$

and

$$Ag^+ + e^- \rightleftharpoons Ag.$$

and

$$\tfrac{1}{2}\underset{\text{(solid)}}{Hg_2Cl_2} + e^- \rightleftharpoons \underset{\text{(liquid)}}{Hg} + Cl^- \quad \ldots\ \tfrac{1}{2}Hg_2Cl_2, Hg\,|\,Cl^-$$

The latter is known as the *calomel electrode,* and is often used as a convenient reference electrode.

Half-Cell Electrode Potentials

Clearly individual half-cell potentials cannot be measured directly as there is no way of measuring the potential difference between an electrode and the surrounding electrolyte. We can only measure the *difference* between the potentials of two half cells when they are linked to form an electrochemical cell.

A series of the relative values of electrode potentials could be obtained if each half cell were combined with a standard half cell. The hydrogen electrode has been chosen as this standard and assigned a standard e.m.f. (E°) of 0 volts, i.e. in the reaction

$$H^+ + e^- \rightleftharpoons \tfrac{1}{2}H_2\,(Pt)$$

when each component is in its standard state (H_2, 1 atm; H^+, a 1 mol dm^{-3} ideal solution) the e.m.f. of this electrode = 0 volts. (This is the standard hydrogen electrode.)

Some representative *standard electrode potentials* E° for half cells (i.e. with each component in its standard state) are given in Table 5.1.

TABLE 5.1

Electrode	Reaction	E°(V)
$Na^+\,|\,Na$	$Na^+ + e^- \rightarrow Na$	– 2.71
$Zn^{2+}\,|\,Zn$	$Zn^{2+} + 2e^- \rightarrow Zn$	– 0.76
$H^+\,|\,\tfrac{1}{2}H_2 \cdot Pt$	$H^+ + e^- \rightarrow \tfrac{1}{2}H_2(Pt)$	0
$Cu^{2+}\,|\,Cu$	$Cu^{2+} + 2e^- \rightarrow Cu$	+ 0.34
$Pt\,|\,Fe^{3+}, Fe^{2+}$	$Fe^{3+} + e^- \rightarrow Fe$	+ 0.77
$Pt \cdot \tfrac{1}{2}Cl_2\,|\,Cl^-$	$(Pt)\tfrac{1}{2}Cl_2 + e^- \rightarrow Cl^-$	+ 1.36

The sign of E° refers to the reaction as written in Table 5.1, i.e.

$$\boxed{\text{Oxidized form} + ne^- \rightarrow \text{Reduced form}}$$

Thus a negative E° value (i.e. ΔG° is positive) implies that the oxidized form is favoured. For instance H_2 will not reduce Zn^{2+} to Zn. Correspondingly a positive E° (i.e. ΔG° negative) means that the reduced state is favoured. Thus H_2 would reduce Cu^{2+} to Cu.

Electrochemical Cells

Consider the linking of two half cells, for instance the Cu and Zn half cells, to form an electrochemical cell. We can represent this electrochemical cell as:†

$$Zn^{2+}|Zn||Cu^{2+}|Cu.$$

It is *essential* to note that when the electrochemical cell is written in this way, it refers to the reaction as written below

$$Zn + Cu^{2+} \rightleftharpoons Zn^{2+} + Cu.$$

If the electrochemical cell were written

$$Cu^{2+}|Cu||\ Zn^{2+}|Zn$$

this would refer to the reaction

$$Zn^{2+} + Cu \rightleftharpoons Zn + Cu^{2+}.$$

The *convention* is that the reaction corresponding to a given electrochemical cell,

$$\underset{\text{Left}}{(Ox)|(Red)}||\underset{\text{Right}}{(Ox)|(Red)}$$

is

$$\boxed{\text{Left (Red)} + \text{Right (Ox)} \rightleftharpoons \text{Left (Ox)} + \text{Right (Red).}}$$

EXAMPLE 5.2 *What is the reaction corresponding to the electrochemical cell below?*

$$Pt|Fe^{3+}, Fe^{2+}||Sn^{4+}, Sn^{2+}|Pt.$$

What are the individual half cell reactions?

Solution

Using the above convention we see that the reaction is

---

† Sometimes this type of cell is written with Zn on the extreme left hand side, i.e. $Zn|Zn^{2+}||Cu^{2+}|Cu$. However the two representation refer to the same electrochemical cell. (See the convention referred to above.)

$$2Fe^{2+} + Sn^{4+} \rightleftharpoons 2Fe^{3+} + Sn^{2+}.$$

Note that two electrons are involved in the reduction of Sn^{4+} to Sn^{2}. Hence two equivalents of Fe^{2+} are required to maintain the electron balance.

The individual half-cell reactions are

Right $\quad Sn^{4+} + 2e^- \rightleftharpoons Sn^{2+}$

Left $\quad 2Fe^{3+} + 2e^- \rightleftharpoons 2Fe^{2+}$

And the overall reaction is obtained by subtraction (Right–Left).

EXAMPLE 5.3 *Devise a cell to study the reduction Co^{3+} to Co^{2+} by H_2 gas.*

Solution

The reaction is

$$Co^{3+} + \tfrac{1}{2}H_2 \rightleftharpoons Co^{2+} + H^+.$$

Using the convention formula, the electrochemical cell would be

$$H^+ | \tfrac{1}{2}H_2, Pt || Co^{3+}, Co^{2+} | Pt.$$

Left Right

The Sign Convention for Electrochemical Cells

The above examples illustrate that the overall reaction in an electrochemical cell is obtained by substraction of the reaction at the *left-hand half-cell* from that at the *right-hand half-cell.*

Thus E° for the electrochemical cell is given by

$$\boxed{E^\circ\,(\text{cell}) = E^\circ\,(\text{right}) - E^\circ\,(\text{left})}$$

EXAMPLE 5.4 *What is the E° for the electrochemical cell?*

$$H^+ | \tfrac{1}{2}H_2, Pt || Pt | Fe^{3+}, Fe^{2+}$$

given that

$$E^\circ \text{ for } Pt | Fe^{3+}, Fe^{2+} \text{ is } +0.77\text{ V}$$

and

$$E^\circ \text{ for } H^+ | \tfrac{1}{2}H_2, Pt \text{ is } 0\text{ V}.$$

Solution

From the equation

$$E^\circ \text{ (cell)} = E^\circ \text{ (right)} - E^\circ \text{ (left)}$$

we see that

$$E^\circ \text{ (cell) is } + 0.77 \text{ V.}^\dagger$$

The cell reaction is

$$\tfrac{1}{2}H_2 + Fe^{3+} \rightleftharpoons H^+ + Fe^{2+}$$

and since E° is positive (i.e. ΔG° negative) the above reaction would clearly proceed in the direction of Fe^{3+} reduction.

EXAMPLE 5.5 *What is the equilibrium constant K for the above reaction? (Assume T = 298 K.) (R = 8.31 J K^{-1} mol^{-1}.)*

Solution

The electrochemical cell reaction involves the transfer of *one* electron (i.e. $n = 1$).

Using $\Delta G^\circ = -n FE^\circ$, $n = 1$, $F = 96.5$ kJ V^{-1} mol^{-1}, and $E^\circ = +0.77$ V,

$$\therefore \qquad \Delta G^\circ = -1 \times 96.5 \times 0.77 \text{ kJ mol}^{-1}$$

i.e.

$$\Delta G^\circ = -74.31 \text{ kJ mol}^{-1}.$$

Now

$$\Delta G^\circ = -RT \ln K$$

$$\therefore \qquad -74310 = -8.31 \times 298 \times \ln K$$

i.e.

$$\ln K = 30.00$$

and

$$K = 1.08 \times 10^{13} \text{ (mol dm}^{-3} \text{ atm).}$$

This example illustrates how e.m.f. measurements can be used to determine ΔG° (and hence K) for a reaction. It is also possible to calculate ΔH° and ΔS° as illustrated below:

---

† This illustrates the general point that standard potentials are measured with respect to the hydrogen electrode when the latter is written to form the left hand half cell.

Calculation of Thermodynamic Parameters

Since $\Delta G^\circ = \Delta H^\circ - T\Delta S^\circ$

$$\frac{\mathrm{d}(\Delta G^\circ)}{\mathrm{d}T} = -\Delta S^\circ$$

(assuming ΔH° is constant over the temperature range).

Now $\Delta G^\circ = -nFE^\circ$

$$\therefore \quad \boxed{\frac{\mathrm{d}E^\circ}{\mathrm{d}T} = \frac{\Delta S^\circ}{nF}}$$

So by determining E° at two temperatures, ΔS° for the electrochemical cell reaction can be obtained. ΔH° is then readily calculated since

$$\Delta H^\circ = \Delta G^\circ + T\Delta S^\circ.$$

So far we have only discussed the e.m.f.s of electrochemical cells under standard state conditions. We now derive a more general equation for the e.m.f. of an electrochemical cell under any given set of conditions.

The Nernst Equation

Consider the equilibrium

$$a\mathrm{A} + b\mathrm{B} \rightleftharpoons c\mathrm{C} + d\mathrm{D}.$$

For this reaction, under standard state conditions ΔG° is given by

$$\Delta G^\circ = c(G_\mathrm{C}^\circ) + d(G_\mathrm{D}^\circ) - a(G_\mathrm{A}^\circ) - b(G_\mathrm{B}^\circ).$$

Under any arbitrary non-standard conditions

$$\Delta G = c(G_\mathrm{C}) + d(G_\mathrm{D}) - a(G_\mathrm{A}) - b(G_\mathrm{B}).$$

So by difference

$$\Delta G - \Delta G^\circ = c(G_\mathrm{C} - G_\mathrm{C}^\circ) + d(G_\mathrm{D} - G_\mathrm{D}^\circ) - a(G_\mathrm{A} - G_\mathrm{A}^\circ) - b(G_\mathrm{B} - G_\mathrm{B}^\circ)$$

Now we have to derive expressions for the terms $(G_\mathrm{C} - G_\mathrm{C}^\circ)$, etc., for each component. The exact expression obviously depends on whether the components are gases, solutes or solids (Chapter 3).

For n moles of a gas $\quad G - G^\circ = nRT \ln (p/p^\circ)$

For a solid $G = G^\circ$

For a solute $G - G^\circ = RT \ln (c/c^\circ)$

where p and c represent pressure and concentration respectively. Note we are assuming ideal behaviour: if a solute, for example, is not behaving ideally we set its activity, $a = c\gamma$ where γ is the activity coefficient (Chapter 5).

All the terms could be combined in the form of a quotient. If all the components were solutes this would be

$$\Delta G - \Delta G^\circ = RT \ln \left(\frac{(c_C/c_C^\circ)^c \, (c_D/c_D^\circ)^d}{(c_A/c_A^\circ)^a \, (c_B/c_B^\circ)^b} \right)$$

Using $\Delta G = -nFE$, and remembering that $c_C^\circ = c_D^\circ$ etc. = 1 mol dm^{-3} we have

$$\boxed{E = E^\circ - \frac{RT}{nF} \ln \frac{[C]^c \, [D]^d}{[A]^a \, [B]^b}}$$

where [ ] represents concentration expressed in terms of mol dm^{-3}.

This is the *Nernst equation* written in terms of concentrations.

Let us consider some examples of the application of this equation.

EXAMPLE 5.6 *The standard electrode potential for the Pt$|Fe^{3+}$, Fe^{2+} half cell is 0.77 V. What is the value of E for this half cell when [Fe^{3+}] = 0.2 mol dm^{-3} and [Fe^{2+}] = 0.05 mol dm^{-3}? (T = 298 K.)*

Solution

For the reaction below ($n = 1$)

$$Fe^{3+} + e^- \rightleftharpoons Fe^{2+}$$

the Nernst equation is

$$E = E^\circ - \frac{RT}{F} \ln \frac{[Fe^{2+}]}{[Fe^{3+}]}$$

assuming the solutions are ideal.

$$E = 0.77 - \frac{RT}{F} \ln \frac{0.05}{0.2}$$

$$= 0.77 - \frac{8.31\,(298)}{96\,500} \ln (0.25)$$

$$= 0.77 + 0.036 \text{ V}$$

$$= 0.806 \text{ V}.$$

EXAMPLE 5.7 *What is the e.m.f. of the electrochemical cell*

$$Pt,\ \tfrac{1}{2}H_2 | H^+Cl^- | AgCl,\ Ag$$

if the concentration of HCl is 10^{-3} mol dm^{-3} and the partial pressure of H_2 is 1 atm? E° for the cell is 0.2225 V? (T = 298 K.)

Solution

The cell reaction is

$$\tfrac{1}{2}H_2 + \underset{\text{(solid)}}{AgCl} \rightleftharpoons H^+ + Cl^- + \underset{\text{(solid)}}{Ag}.$$

For this reaction, n = 1.

The Nernst equation becomes

$$E = E^\circ - \frac{RT}{F} \ln \frac{[H^+][Cl^-]}{(p_{H_2})^{\frac{1}{2}}}$$

since Ag and AgCl are both solids and hence in their standard state they do not appear in the expression above.

Now $[H^+] = [Cl^-] = 10^{-3}$ mol dm^{-3} and p_{H_2} = 1 atm.

$$E = 0.2225 - \frac{8.31\,(298)}{96\,500} \ln 10^{-6}$$

$$= 0.2225 + 0.3545 \text{ V}$$

$$E = 0.577 \text{ V}.$$

EXAMPLE 5.8 *What is the value of E for the hydrogen electrode if the partial pressure of H_2 = 0.1 atm? The concentration of H^+ is 1 mol dm^{-3}. (T = 298 K.)*

Solution

For the reaction below, n = 1

$$H^+ + e^- \rightleftharpoons \tfrac{1}{2}H_2\,(g).$$

Now for the gas (H_2).

$$G - G^\circ = nRT \ln p_{H_2}$$

and for the solute (H^+)

$$G - G^\circ = RT \ln [H^+]$$

assuming the solution is ideal.

So the Nernst equation becomes

$$E = E^\circ - \frac{RT}{F} \ln \frac{(p_{H_2})^{\frac{1}{2}}}{[H^+]}$$

In this case

$$p_{H_2} = 0.1, \qquad [H^+] = 1 \text{ mol dm}^{-3}$$

$$E = 0 - \frac{8.31\,(298)}{96500} \ln (0.1)^{\frac{1}{2}}$$

$$= 0 + 0.0295 \text{ V}$$

$$E = 0.0295 \text{ V}.$$

EXAMPLE 5.9 *What is the value of E for the half cell Pt|NAD$^+$ + H$^+$, NADH at pH 7, given that E° is – 0.11 V? (T = 298 K). Assume that NAD$^+$ and NADH are both present at 1 mol dm^{-3} concentration.*

Solution

The half-cell reaction for this most important biochemical oxidation–reduction system is:

$$\boxed{NAD^+ + H^+ + 2e^- \rightleftharpoons NADH}$$

The Nernst equation is then

$$E = E^\circ - \frac{RT}{2F} \ln \frac{[NADH]}{[NAD^+]\,[H^+]}$$

Now $[H^+] = 10^{-7}$ mol dm^{-3} at pH 7.

$$\therefore \qquad E = -0.11 - \frac{RT}{2F}\ln\frac{1}{10^{-7}}$$

$$= -0.11 - \frac{8.31(298)}{2(96500)}\ln 10^{7}$$

$$= -0.11 - 0.21 \text{ V}$$

$$= -0.32 \text{ V.}$$

Thus at pH 7 the value of E for this half cell is – 0.32 V.

Biochemical Standard States

We introduced the concept of a 'biochemical standard state' where all the components of a reaction are present in their standard states, except H^+ which is present at 10^{-7} mol dm^{-3}, corresponding to pH 7. In the same way that the standard free energy change under these conditions is denoted by $\Delta G°'$, we shall use the symbol $E°'$ to refer to standard electrode potentials at pH 7. Thus in the example above (Pt | $NAD^+ + H^+$, NADH) $E°' = -0.32$ V.

EXAMPLE 5.10 *The $E°'$ values for the reactions*

$$\text{acetaldehyde} + 2H^+ + 2e^- \rightleftharpoons \text{ethanol} \qquad (i)$$

and

$$NAD^+ + H^+ + 2e^- \rightleftharpoons NADH \qquad (ii)$$

are – 0.197 V and – 0.32 V respectively.

What is the equilibrium constant for the coupled reaction?

$$\text{acetaldehyde} + NADH + H^+ \rightleftharpoons \text{ethanol} + NAD^+ \quad (T = 298 \text{ K}).$$

Solution

$E°'$ for the overall reaction is obtained by subtracting the $E°'$ for reaction (*ii*) from that for reaction (*i*):

$$E°' = -0.197 - (-0.32) \text{ V}$$

$$= +0.123 \text{ V.}$$

Since

$$\Delta G = -nFE$$

$$\Delta G°' = -2(96.5)(0.123) \text{ kJ mol}^{-1}$$

$$= -23.7 \text{ kJ mol}^{-1}.$$

From

$$-\Delta G^{\circ\prime} = RT \ln K'$$
$$K' = 1.43 \times 10^4 \text{ (mol dm}^{-3}\text{)}.$$

Thus the equilibrium constant at pH 7 is 1.43×10^4 (mol dm^{-3}).

The Nernst Equation and Chemical Equilibrium

The above examples illustrate the use of the *Nernst equation* in relating the value of E° to the e.m.f. E under any arbitrary known set of conditions (e.g. concentrations). Values of E° for different electrochemical cells can be obtained and tabulated in the same way as values of ΔG°, etc.

It is a useful exercise to compare the Nernst equation with the equation linking ΔG° and equilibrium constant (i.e. $-\Delta G^\circ = RT \ln K$). It will be recalled that this latter equation was obtained as a special case of the more general relationship.[†]

$$\Delta G - \Delta G^\circ = RT \ln \frac{[\mathrm{C}]^c\,[\mathrm{D}]^d}{[\mathrm{A}]^a\,[\mathrm{B}]^b}$$

since at equilibrium $\Delta G = 0$ and the values of [A], [B], etc., are their *equilibrium concentrations.*

However the above general equation can be used to calculate ΔG under any set of conditions. In this respect, the general equation is entirely equivalent to the Nernst equation, which for solute components is:

$$E = E^\circ - \frac{RT}{nF} \ln \frac{[\mathrm{C}]^c\,[\mathrm{D}]^d}{[\mathrm{A}]^a\,[\mathrm{B}]^b}$$

(noting that $\Delta G = -nFE$).

In practice we usually set up an electrochemical cell with the components not in their standard states and measure E. The reaction is prevented from proceeding towards equilibrium by application of an opposing e.m.f. of magnitude E. If, however, we allowed the reaction to proceed (i.e. current to flow) the value of E would fall until, at equilibrium, E would equal zero (since at equilibrium $\Delta G = 0$).

[†] This is written in terms of concentrations, rather than activities.

The Effect of Non-Ideality

So far we have assumed that all components in the electrochemical cells behave ideally. This assumption can lead to errors especially in the case of electrolyte solutions. As an example consider the cell

$$\text{Pt}, \tfrac{1}{2}\text{H}_2\,(1\ \text{atm})\,|\,\text{H}^+\text{Cl}^-\,(\text{conc.} = c)\,|\,\tfrac{1}{2}\text{Hg}_2\text{Cl}_2,\ \text{Hg}.$$

The cell reaction is

$$\tfrac{1}{2}\text{H}_2 + \tfrac{1}{2}\text{Hg}_2\text{Cl}_2 \rightleftharpoons \text{Hg} + \text{H}^+ + \text{Cl}^-$$

and the Nernst equation,

$$E = E^\circ - \frac{RT}{nF}\ln\,[\text{H}^+]\,[\text{Cl}^-] \qquad (n = 1)$$

can be used to calculate the apparent E° as a function of the HCl concentration (c).

The following results are obtained.

c(M)	E(V)	E°(V)
10^{-1}	0.4046	0.2866
10^{-2}	0.5099	0.2739
10^{-3}	0.6239	0.2699
10^{-4}	0.7406	0.2686

E° is not constant because of the non-ideality of the solution. It is possible to account for this departure from ideality by substituting activities ($a = c\gamma$) for concentrations in the Nernst equation and calculating γ from the Debye-Hückel theory for dilute electrolyte solutions. When this is done E° has a constant value (0.2680 V), and this would be the true E° obtainable by extrapolation of the above E° values to zero ionic strength (i.e. infinite dilution).

Coupled Oxidation–Reduction Processes

From our discussion of the thermodynamics of oxidation-reduction processes it is clear that e.m.f. values can be thought of as entirely analogous to free energy changes. Now we can use ΔG° values for individual reactions to predict the equilibrium position in a 'coupled' reaction and similarly in a 'coupled' oxidation–reduction process,

$$(\mathrm{Ox})_1 + (\mathrm{Red})_2 \rightleftharpoons (\mathrm{Ox})_2 + (\mathrm{Red})_1$$

we can use E° values to predict the position of equilibrium ($\Delta G^\circ = -nFE^\circ$) provided of course n is known. One of the most important examples of such processes is the mitochondrial 'electron transport chain', where a series of 'half-cell reactions' are arranged in order of increasing $E^{\circ\prime}$ (i.e. stronger oxidizing power). Electrons are passed along the chain, and the total energy available from oxidation of NADH is utilized (by a process not yet fully understood) in the production of ATP from ADP and phosphate. Two electrons are assumed to be passed along the chain so that from $\Delta G^{\circ\prime} = -nFE^{\circ\prime}$, we see that for each 0.1 V change in $E^{\circ\prime}$ the $\Delta G^{\circ\prime}$ change is 19.3 kJ mol^{-1}.

There are three places indicated in the chain where the difference in $\Delta G^{\circ\prime}$ is sufficient to drive the phosphorylation of ADP to produce ATP ($\Delta G^{\circ\prime}$ for this process is 30.5 kJ mol^{-1}), and these three places have been tentatively described as the 'phosphorylation' sites. It should be remembered, however, that this assumes that the various components are all in their standard states, and clearly the *in vivo* situation (where the components are not present in their standard states, and the $\Delta G'$ values differ considerably from the $\Delta G^{\circ\prime}$ values) will be considerably more complicated than this.

Oxidative Phosphorylation in The Mitochondria

The mechanism of coupling of electron transport to ATP synthesis has been difficult to unravel. There is now a general, but not universal, consensus that protons play a key role in the coupling mechanism and that a *chemiosmotic hypothesis* gives a very useful description of the mechanism. In this hypothesis, the electron flow from one electron carrier to another at a higher redox potential is coupled to the movement of protons from the inside (matrix) to the outside of the mitochondrion. This hypothesis is the only one which is at present amenable to a quantitative treatment and we therefore develop the necessary thermodynamic treatment.

We first have to introduce a new thermodynamic function, the *electrochemical potential* of an ion. We have already briefly mentioned the concept of *chemical potential* of a solute, μ, which is essentially equivalent to its molar Gibbs free energy. For an ideal solution the *chemical* potential of an ion, μ_i, can be defined as

$$\mu_i = \mu_i^\circ + RT \ln X_i$$

where μ_i° is the chemical potential of the pure component in its standard state. For an ion, however, the chemical potential, μ_i, alone is not the correct term to describe its thermodynamic behaviour since we also have to take account of the interaction of the ion with the electrostatic field, ψ, in which it is placed. This field is provided by the neighbouring charged species. This additional term equals $zF\psi$ where z is the charge on the ion and F is the conversion factor (96.5 kJ V^{-1} mol^{-1}).

The *electrochemical potential*, $\tilde{\mu}_i$, is therefore defined as

$$\tilde{\mu}_i = zF\psi + \mu_i$$

$$= zF\psi + \mu_i^\circ + RT \ln X_i. \qquad (5.1)$$

In the chemiosmotic hypothesis, the movement of protons from the matrix to the outside of the mitochondria will cause[†] the inside (matrix) of the mitochondria to become negatively charged and alkaline and the outside to become positively charged and acidic. There is now energy stored as a gradient across the membrane (which is highly impermeable to protons) and this energy *is* the difference in electrochemical potential for the hydrogen ion. Therefore, from eqn (5.1)

$$\tilde{\mu}_{i\,in} - \tilde{\mu}_{i\,out} = zF\psi_{in} - zF\psi_{out} + RT \ln\left(\frac{[H^+]_{in}}{[H^+]_{out}}\right)$$

or

$$\Delta\tilde{\mu}_i = zF\Delta\psi - 2.3\,RT\,\Delta\text{pH}$$

where ΔpH is the pH difference between the inside an outside and $\Delta\psi$ is the membrane potential across the membrane (inside minus outside). It is convenient to use units of mV for the electrochemical potential and the value of $\Delta\tilde{\mu}_i$ in mV is termed the *protonmotive force* Δp. To convert the terms in the above equation to electrical units (i.e. volts) we divide through by the conversion factor F. Putting $\Delta\tilde{\mu}_i / F = \Delta p$ and noting that $z = 1$ for the proton, we obtain

$$\boxed{\Delta p = \Delta\psi - \frac{2.3RT}{F}\Delta\text{pH}}\ ^{\ddagger} \qquad (5.2)$$

[†] Neglecting any buffering effects and provided that no other ions are moving.
[‡] To obtain the quantities in mV, we should note that the value of F is 96.5 J mV^{-1} mol^{-1}

$\Delta\psi$ is the membrane potential (in mV).

We now consider two very important questions.

(1) How large does the protonmotive force have to be to drive oxidative phosphorylation?

(2) How is the protonmotive force measured ?

We shall consider each in turn.

(1) *The magnitude of the protonmotive force*

Nothing has been said so far on the question of *how many* protons are translocated as electrons pass down the respiratory chain. This is, at present, a controversial topic. However, it should be noted that the driving force required to move the proton(s) (which is provided by the redox potential difference between two adjacent carriers) is directly proportional to the number of protons being moved.

How does this protonmotive force drive ATP synthesis? The proposal is that ATP synthesis occurs when a defined number of protons pass down their electrochemical gradient via mitochondrial ATPase.† It follows that the protonmotive force must be large enough to favour ATP synthesis from ADP and P_i. The required magnitude of Δp could be assessed in the following experiment. A suspension of mitochondria is supplied with oxidisable substrate, ADP and P_i, and allowed to respire until the ATP concentration reaches a steady value. The final concentrations of ADP, P_i, and ATP are then measured. This enables the $\Delta G'$ for the reaction ADP + P_i = ATP + H_2O to be calculated provided that $\Delta G^{\circ\prime}$ is also known since

$$\Delta G' = \Delta G^{\circ\prime} + RT \ln \frac{[\text{ATP}]}{[\text{ADP}]\,[\text{P}_i]} \tag{5.3}$$

$\Delta G'$ may then be related to the protonmotive force by the equation we have already met, $\Delta G' = nFE'$. Thus

$$\boxed{\Delta G' = -\, n \,.\, F \,.\, \Delta p} \tag{5.4}$$

where *n* is the number of protons that pass through the ATPase for each ATP molecule synthesized. In the *chemiosmotic hypothesis the value of* $n = 2$ and the required value of Δp can be calculated.

† The membrane is highly impermeable to protons and it is only via this ATPase the protons can move back across the membrane into the matrix.

(2) *Measurement of the protonmotive force*

We now require a method of estimating the magnitude of the protonmotive force so that the thermodynamic feasibility of the chemiosmotic hypothesis can be tested. Eqn (5.2) shows that we need to know ΔpH and the membrane potential $\Delta\psi$ to obtain Δp. We shall consider each in turn.

(*i*) *Measurement of ΔpH*

The most widely used method for estimating ΔpH involves the use of a metabolically inert weak base or acid. Biological membranes are generally not permeable to charged molecules and thus the underlying assumption made is that only the uncharged form, of the acid or base, can cross the membrane.

Determination of ΔpH

EXAMPLE 5.11 *A 1 cm^3 suspension of rat liver mitochondria in 0.25 mol dm^{-3} sucrose pH 6.0 is incubated with succinate and [^3H]-acetate for 2 minutes before the mitochondria were very rapidly separated from the incubation medium. Analysis showed that one sixth of the total acetate had been taken up by the mitochondria. If the concentration of mitochondria was 5 g protein dm^{-3}, what was the magnitude of the pH gradient ? Assume that the internal volume of mitochondria is 0.4 cm^3 (g protein)$^{-1}$.*

Solution

Assume that only CH_3COOH (denoted HA) but not CH_3COO^- (denoted A^-) can permeate the membrane. Then at equilibrium

$$[HA]_{in} = [HA]_{out}$$

but

$$K_a = \frac{[H^+]_{in}\,[A^-]_{in}}{[HA]_{in}} = \frac{[H^+]_{out}\,[A^-]_{out}}{[HA]_{out}}$$

so that

$$[H^+]_{in}/[H^+]_{out} = [A^-]_{out}/[A^-]_{in}$$

However the measurements only tell us the total acetate [A^-_{in} + AH_{out}] inside the mitochondria. Fortunately we only need to know the total concentration as the following will show:

$$\frac{[\text{Acetate}]_{\text{total in}}}{[\text{Acetate}]_{\text{total out}}} = \frac{[\text{AH}]_{\text{in}} + [\text{A}^-]_{\text{in}}}{[\text{AH}]_{\text{out}} + [\text{A}^-]_{\text{out}}}$$

$$= \frac{[\text{AH}]_{\text{in}} + (\text{K}_\text{a}\,([\text{AH}]_{\text{in}}/[\text{H}^+]_{\text{in}}))}{[\text{AH}]_{\text{out}} + (\text{K}_\text{a}\,([\text{AH}]_{\text{out}}/[\text{H}^+]_{\text{out}}))}$$

But we have already noted that $[\text{AH}]_{\text{in}} = [\text{AH}]_{\text{out}}$, so

$$\frac{[\text{Acetate}]_{\text{total in}}}{[\text{Acetate}]_{\text{total out}}} = \frac{1 + K_\text{a}/[\text{H}^+]_{\text{in}}}{1 + K_\text{a}/[\text{H}^+]_{\text{out}}}$$

Dividing by K_a on the right hand side we obtain

$$\frac{[\text{Acetate}]_{\text{total in}}}{[\text{Acetate}]_{\text{total out}}} = \frac{1/K_\text{a} + 1/[\text{H}^+]_{\text{in}}}{1/K_\text{a} + 1/[\text{H}^+]_{\text{out}}}.$$

If the pK_a of the acid is << pH_{in} or pH_{out} (i.e. if K_a for the acid is greater than $[\text{H}^+]_{\text{out}}$ or $[\text{H}^+]_{\text{in}}$) then:

$$\frac{[\text{Acetate}]_{\text{total in}}}{[\text{Acetate}]_{\text{total out}}} = \frac{[\text{H}^+]_{\text{out}}}{[\text{H}^+]_{\text{in}}}.$$

Since 1/6th of the acetate was taken up into the mitochondria 5/6th remain on the outside in the suspending medium. The relative volumes of the mitochondria and the suspending medium are known so that the acetate concentration ratio in the above equation can be obtained as follows. The internal volume is 0.002 cm^3 (0.0004×5) while the total volume is 1 cm^3. The relative concentrations of acetate are then obtained by dividing the relative amounts of acetate in each compartment by the corresponding volume. Thus

$$\frac{[\text{Acetate}]_{\text{total in}}}{[\text{Acetate}]_{\text{total out}}} = \frac{\frac{1}{6} \div 0.002}{\frac{5}{6} \div 1} = \frac{100}{1}$$

$$\therefore \qquad [\text{H}^+]_{\text{out}}/[\text{H}^+]_{\text{in}} = 100.$$

Since the external pH is 6.0, the internal pH is estimated to be 8.0, i.e. the inside is more alkaline with $\Delta\text{pH} = 2.0$

The distribution of a weak acid or base is a method which has found wide application in biology for the estimation of the pH in cellular spaces. The reader should note that the method involves a number of assumptions which cannot always be justified by experiment. Amongst these are the assumptions of equal activity coefficients on each side of the membrane, no binding or metabolism of the weak acid or base, and that complete equilibration via the uncharged form has occurred. It should also be noted that the accumulation of a species is always easier to measure than its exclusion and so if the organelle under investigation is expected to have a high pH relative to the medium then a weak acid probe should be used. Conversely, when a more acidic internal pH is suspected the distribution of a weak base should he measured.

(ii) Measurement of membrane potential

The membrane potential, $\Delta\psi$ is estimated on the basis of an ion distribution across a membrane. To understand the method we recall the definition of the electrochemical potential of an ion, eqn (5.1). The expressions for the difference $\Delta\tilde{\mu}_i$ in electrochemical potentials that can occur across a membrane boundary is

$$\Delta\tilde{\mu}_i = zF\Delta\psi + RT \ln \frac{[X_i]_{in}}{[X_i]_{out}}. \tag{5.5}$$

Let us consider how this equation can be applied to the behaviour of a positively charged ion to which the mitochondrial membrane is permeable. (The membrane becomes permeable to certain hydrophobic organic cations and to certain inorganic ions such as K^+ or Rb^+ when an ionophore, which is an 'ion carrier' molecule, is present.) From the earlier description of the chemiosmotic hypothesis we expect respiratory chain-linked redox reactions to make the outside of the mitochondria positive, relative to the inside, following the translocation of H^+ ions. The positively charged ion (e.g. K^+ or Rb^+) will enter the mitochondria until it reaches a distribution across the mitochondrial membrane which is in electrochemical equilibrium, i.e. $\Delta\tilde{\mu}_i = 0$. When this condition is satisfied, then from eqn (5.5)

$$\Delta\psi = -\frac{RT}{zF} \ln\left(\frac{[X]_{in}}{[X]_{out}}\right). \tag{5.6}$$

Thus if $[X]_{in}$ and $[X]_{out}$ can be measured, $\Delta\psi$ can be estimated.

Knowing $\Delta\psi$ and ΔpH, Δp can then be determined from eqn (5.2).

We have attempted to provide the necessary thermodynamics for a basic understanding of the chemiosmotic hypothesis as applied to mitochondria. The same hypothesis is in principle applicable to energy linked transport processes in chloroplasts and bacteria. A cautionary note is apposite at this point. A large amount of qualitative experimental evidence has been used to support the chemiosmotic hypothesis— but a mechanism can only be established by thermodynamic measurements. We have outlined some of the thermodynamic measurements which involve a number of assumptions which may not be satisfied. Research workers are still actively concerned with the mechanism of oxidative phosphorylation and it may yet transpire that the protonmotive force has been either over- or under-estimated by these measurements. At present however, the chemiosmotic hypothesis provides the only framework for a quantitative thermodynamic analysis of the coupling of electron transport to ATP synthesis.

PROBLEMS

1. What are the reactions in the following half cells?
   (*a*) $Zn^{2+}|Zn$
   (*b*) $H^+|\frac{1}{2}H_2 . Pt$
   (*c*) $Pt|Co^{3+}, Co^{2+}$
   (*d*) $AgBr, Ag|Br$
   (*e*) $\frac{1}{2}Hg_2Cl_2, Hg|Cl^-$ (the calomel electrode).
   (*f*) Pt|Fumarate^{2-} + 2H$^+$, succinate^{2-}
   (*g*) Pt|Cytochrome c (Fe^{3+}), Cytochrome c (Fe^{2+})
   (*h*) Pt|CO_2 + H^+, Formate$^-$.
   (*i*) Pt|NAD^+ + H^+, NADH.
2. Write down the cell reaction for each of the following cells:
   (*a*) $Cu^{2+}|Cu||Zn^{2+}|Zn$
   (*b*) $H^+|\frac{1}{2}H_2 . Pt||Ag^+|Ag$
   (*c*) $Pt . \frac{1}{2}H_2|HCl||AgCl, Ag$
   (*d*) $H^+|Pt . \frac{1}{2}H_2||Fe^{3+}, Fe^{2+}|Pt$

(*e*) $Pt|NAD^+ + H^+, NADH||oxaloacetate^{2-} + 2H^+, malate^{2-}|Pt$.

Comment on the involvement of H^+ ions in (*e*). Would you write this cell differently?

Calculate the standard e.m.f.s of the above cell (*a*)–(*d*) given the following standard electrode potentials (in volts) all at pH = 0: $Cu^{2+}|Cu$, 0.34; $Zn^{2+}|Zn$, – 0.76; $H^+|\frac{1}{2}H_2$. Pt = 0; $Ag^+|Ag$ = 0.80; AgCl, $Ag|Cl^-$ = 0.22; $Pt|Fe^{3+}, Fe^{2+}$, + 0.77.

For cell (*e*) E° values at pH 7 (known as $E^{\circ\prime}$—by analogy with $\Delta G^{\circ\prime}$) are: $Pt|NAD^+ + H^+$, NADH, – 0.32 V; $Pt|Oxaloacetate^{2-} + 2H^+$, $malate^{2-}$, – 0.17 V.

Calculate $E^{\circ\prime}$ for the cell.

3. Write down the cells equivalents to the following reactions:

   (*a*) $Sn^{2+} + Pb \rightleftharpoons Pb^{2+} + Sn$

   (*b*) Lactate + NAD $\rightleftharpoons$ pyruvate + NADH + H^+.

4. Discuss the meaning of the term ***reversible*** as used in the second law of thermodynamics.

   In the reaction

$$Fe + Cu^{2+} \rightleftharpoons Fe^{2+} + Cu,$$

$$\Delta H^\circ_{298} = -148.8 \text{ kJmol}^{-1}$$

   and

$$\Delta S^\circ_{298} = 8.8 \text{ J K}^{-1} \text{ mol}^{-1}.$$

   Under standard state conditions:

   (*a*) In which direction would the reaction proceed?

   (*b*) In a cell an e.m.f. is applied to prevent this reaction occurring—calculate this e.m.f. (298 K).

   (*c*) Discuss ***briefly*** the situation when the applied e.m.f. is different from this value.

5. The standard e.m.f. of the cell

$$Zn^{2+}|Zn||Fe^{3+}, Fe^{2+}|Pt$$

   is 1.53 V at 298 K and 1.55 V at 323 K.

   What is the cell reaction?

   Calculate ΔG°, ΔH°, and ΔS° and state any assumptions you make.

6. Derive the Nernst equation for a half cell. Using this calculate the ratio of oxidized to reduced forms of cytochrome c_1 at the following potentials: 0.3, 0.25, 0.2, 0.15 and 0.1 volts (T = 298 K)

$$Pt|Cyt\text{-}c_1\ (Fe^{3+}),\ Cyt\text{-}c_1\ (Fe^{2+}) \qquad (E^{\circ\prime} = 0.21\text{V})$$

7. (*a*) For the $NAD^+ + H^+$, NADH couple, the value of $E^{\circ\prime}$ (i.e. at pH = 7) is – 0.32 V. What is the value of the e.m.f. at pH = 0 (i.e. E°)?

   If both NAD^+ and NADH are in their standard state (i.e. 1 mol dm^{-3}) but the pH = 6, what is the new e.m.f.?

   (*b*) For the oxaloacetate^{2-} + $2H^+$, malate^2 couple, $E^{\circ\prime} = -0.175$ V (i.e. pH = 7). What is the e.m.f. at pH = 6, if both anions are in their standard state?

   (*c*) From the data in (*a*) and (*b*) calculate the value of the equilibrium constant for the oxidation of malate^2 by NAD^+ at pH = 6 and pH = 7. Comment on your answer. Assume the temperature is 298 K.

8. The pH of a solution could be measured by using a standard hydrogen electrode as in the following cell:

$$\text{Pt}\,.\,\tfrac{1}{2}H_2(1\text{ atm})|HCl(x\text{ mol dm}^{-3})|\tfrac{1}{2}Hg_2Cl_2,\ Hg.$$

   This cell has an e.m.f. of 0.48 V at 298 K. What is the pH of the solution? (E° for the calomel electrode is 0.24 V.)

9. In the photosynthetic chain the following oxidation-reduction couples are involved:

   Cyt-b (Fe^{3+}), Cyt-b (Fe^{2+}), $E^{\circ\prime} = 0.06$ V

   Cyt-f (Fe^{3+}), Cyt-f (Fe^{2+}) $E^{\circ\prime} = 0.36$ V

   What is the $\Delta G^{\circ\prime}$ of the overall reaction in which these are coupled? Is this energy sufficient to synthesize ATP from ADP and P_i, if $\Delta G^{\circ\prime}$ for this latter reaction is + 30.5 kJ mol^{-1}?

   Would the same conclusion be reached if we considered the flow of two electrons along the redox chain, i.e.

$$2\text{ Cyt-b }(Fe^{3+}) + 2e^- \rightleftharpoons 2\text{ Cyt-b }(Fe^{2+})$$

$$2\text{ Cyt-f }(Fe^{3+}) + 2e^- \rightleftharpoons 2\text{ Cyt-f }(Fe^{2+})?$$

10. A suspension of mitochondria (5 g protein dm^{-3}) was incubated with [^{3}H]-acetate, $^{86}Rb^+$, valinomycin (ionophore which makes the membrane permeable to Rb^+), ADP, P_i and succinate, for a sufficient period to permit the ATP level to reach a constant value. At the end of this incubation the mitochondria were rapidly filtered from part of the suspension and counted for radioactivity from [^{3}H]-acetate and $^{86}Rb^+$. It was found that 1/11th of the total [^{3}H]-acetate added and 1/6th of the total $^{86}Rb^+$ added was also found in the mitochondria. The remaining mitochondrial suspension was analysed for ATP, ADP, and P_i. These concentrations were 1 mmol dm^{-3}, 0.02 mmol dm^{-3}, and 1 mmol dm^{-3} respectively. Are these data compatible with the view that 2 protons must past across the mitochondrial membrane for each ATP molecule synthesised? Assume that the internal volume of mitochondria is 0.4 $cm^3\ g^{-1}$ protein. (T = 300 K, F = 96.5k

kJ V^{-1} mol^{-1}. $R = 8.31$ J K^{-1} mol^{-1}. $\Delta G^{\circ\prime}$ for the reaction ADP + P_i = ATP is 30.5 kJ mol^{-1}.)

Answers

1. (*a*) $Zn^{2+} + 2e^- \rightarrow Zn$.

   (*b*) $H^+ + e^- \rightarrow \frac{1}{2} H_2$.

   (*c*) $Co^{3+} + e^- \rightarrow Co^{2+}$.

   (*d*) $AgBr + e \rightarrow Ag + Br^-$.

   (*e*) $\frac{1}{2} Hg_2Cl_2 + e^- \rightarrow Hg + Cl^-$.

   Note **that in (*d*) and (*e*) we can obtain the reaction by adding two reactions e.g. Ag Br(s) → $Ag^+ + Br^-$ and $Ag^+ + e^- \rightarrow Ag(s)$.**

   (*f*)
   $$\begin{matrix} CHCO_2^- \\ || \\ CHCO_2^- \\ \text{fumarate} \end{matrix} + 2H^+ + 2e^- \rightarrow \begin{matrix} CH_2CO_2^- \\ | \\ CH_2CO_2^- \\ \text{succinate} \end{matrix}$$

   Note that H^+ acts as one of the reactants (i.e. reduction by H_2 can be thought of as reduction by $2H^+ + 2e^-$).

   (*g*) Cyt c (Fe^{3+}) + e^- → Cyt c (Fe^{2+}).

   (*h*) $CO_2 + H^+ + 2e^- \rightarrow HCO_2^-$.

   (*i*) $NAD^+ + H^+ + 2e^- \rightarrow NADH$.

   This is equivalent to reduction of NAD^+ by the hydride ion ($H^- = H^+ + 2e^-$).

2. The electrochemical cell reaction is given by

   Left (reduced) + Right (oxidized) → Left (oxidized) + Right (reduced)

   and E° by E°(right) – E° (left).

   (*a*) $Cu + Zn^{2+} \rightarrow Cu^{2+} + Zn$, $E^\circ = -1.1$ V.

   (*b*) $\frac{1}{2}H_2 + Ag^+ \rightarrow H^+ + Ag$, $E^\circ = 0.8$ V.

   (*c*) $\frac{1}{2}H_2 + AgCl \rightarrow H^+ + Cl^- + Ag$, $E^\circ = 0.22$ V.

   (*d*) $\frac{1}{2}H_2 + Fe^{3+} \rightarrow H^+ + Fe^{2+}$, $E^\circ = 0.77$ V.

   (*e*) NADH + $oxaloacetate^{2-}$ + $2H^+$ → NAD^+ + H^+ + $malate^{2-}$,
   $E^{\circ\prime} = 0.15$ V.

   We have written the electrochemical cell this way to emphasize that $2H^+$ are involved in the $oxaloacetate^{2-}$, $malate^{2-}$ half cell and one H^+ in the

NAD^+, NADH half cell. *Note* that $2e^-$ are involved in both half cells (see the answer to Question 1).

3. The electrochemical cell reaction is given by

   Left (reduced) + Right (oxidized) → Left (oxidized) + Right (reduced).

   (*a*) The left-hand half cell is $Pb^{2+}|Pb$ and the electrochemical cell is thus:

   $$Pb^{2+}|Pb||Sn^{2+}|Sn.$$

   (*b*) Pt|pyruvate$^-$ + $2H^+$, lactate$^-$||NAD^+ + H^+, NADH|Pt.

   See the comment on Question 2, part (*e*), regarding the involvement of H^+ ions in this reaction.

4. The electrochemical cell is said to be *reversible* since the reaction can be made to proceed in either direction.

   (*i*)
   $$\Delta G^\circ = \Delta H^\circ - T\Delta S^\circ$$
   $$= -151.4 \text{ kJ mol}^{-1}$$

   Hence the reaction would *proceed in the direction as written* (i.e. left → right).

   (*ii*) Since $\Delta G^\circ = -nFE^\circ$ and $n = 2$,
   $$E^\circ = 0.78 \text{ V}.$$

   Thus an e.m.f. of this magnitude must be applied to prevent this reaction occurring (when the components are in their standard states).

   (*iii*) If the applied e.m.f. is *greater* than 0.78 V the reaction proceeds from *right to left*, and vice versa.

5. The cell reaction is

   $$Zn + 2Fe^{3+} \rightarrow Zn^{2+} + 2Fe^{2+}.$$

   Since $\Delta G^\circ = -nFE^\circ$ and $n = 2$

   $$\Delta G^\circ_{298} = -295.3 \text{ kJ mol}^{-1}$$

   Now

   $$\Delta S^\circ = -\frac{d(\Delta G^\circ)}{dT}$$

   $$= nF\frac{d(E^\circ)}{dT}$$

   $$= 154.4 \text{ J K}^{-1} \text{ mol}^{-1}$$

   $$\therefore \quad \Delta H^\circ = -249.3 \text{ KJ mol}^{-1}$$

The assumptions made are that ΔH° and ΔS° are *both independent of temperature* over this range.

6. For a half cell, the Nernst equation becomes

$$E = E^\circ + \frac{RT}{nF} \ln \frac{[\text{Oxidised}]}{[\text{Reduced}]} \quad \text{(see text)}$$

or at pH 7

$$E' = E^{\circ\prime} + \frac{RT}{nF} \ln \frac{[\text{Oxidised}]}{[\text{Reduced}]}. \quad (n = 1 \text{ in this case})$$

E	0.3	0.25	0.2	0.15	0.1
(Oxidized/Reduced)	33.4	4.75	0.68	0.097	0.014

Because of the logarithmic nature of the Nernst equation, even quite large errors in the ratio [oxidized]/[reduced] lead to relatively small changes in E.

7. E° (at pH 0) is related to $E^{\circ\prime}$ (at pH 7) by the equation

$$E^{\circ\prime} = E^\circ + \frac{RT}{2F} \ln \frac{[\text{NAD}^+][\text{H}^+]}{[\text{NADH}]}$$

(*i*) Thus at pH 0, $E^\circ = -0.11$ V.

Using this value of E° (pH 0) we can then compute the e.m.f. at pH 6.

e.m.f. (pH 6) = – 0.29 V.

(*ii*) A similar approach gives E° (pH 0) for this half cell as 0.239 V. *Note* that there are *two* protons involved here, so the Nernst Equation has a term in $[\text{H}^+]$. Thus the e.m.f. at pH 6 is – 0.116 V.

(*iii*) The reaction is

$$\text{NAD}^+ + \text{H}^+ + \text{malate}^{2-} \rightarrow \text{NADH} + \text{oxaloacetate}^{2-} + 2\text{H}^+.$$

At pH 7, $E^{\circ\prime}$ for this reaction is – 0.145 V.

Now $\ln K' = -\Delta G^{\circ\prime}/RT = nFE^{\circ\prime}/RT$, hence

$$K'(\text{pH } 7) = 1.24 \times 10^{-5} \text{ (mol dm}^{-3}\text{, pH 7)}$$

At pH 6, the e.m.f. is – 0.174 V hence $\ln K$ (pH 6) = nFE/RT

$$K \text{ (pH 6)} = 1.29 \times 10^{-6} \text{ (mol dm}^{-3}\text{, pH 6)}$$

Thus by increasing $[\text{H}^+]$ (i.e. lowering the pH from 7 to 6) the equilibrium is driven towards the left. The K referred to here (at any given pH) does not include the term in $[\text{H}^+]$ since this is already taken into account in converting E° (pH 0) to E at any other pH.

i.e. this $$K = \frac{[\text{NADH}]\,[\text{oxaloacetate}^{2-}]}{[\text{NAD}^+]\,[\text{malate}^{2-}]}.$$

If we included the term in $[H^+]$ in the expression for K, K would be independent of pH. Thus the true equilibrium constant in this case would be 1.25×10^{-12} (mol dm^{-3}).

8. The cell reaction is

$$\tfrac{1}{2}H_2 + \tfrac{1}{2}HgCl_2 \rightleftharpoons Hg + H^+ + Cl^-.$$

From the Nernst equation (since $n = 1$),

$$E = E^\circ - \frac{RT}{F} \ln \frac{a_{H^+}\, a_{Cl^-}}{(p_{H_2})^{\frac{1}{2}}}$$

If we assume the solution to be ideal and replace activity by concentration for H^+ and Cl^-, we obtain (since $p_{H_2} = 1$):

$$c_{H^+}.c_{Cl^-} = 8.68 \times 10^{-5}$$

$$c_{H^+} = 9.31 \times 10^{-3} \text{ (mol dm}^{-3}\text{) } (= c_{Cl^-}).$$

Thus the pH of the solution is 2.03.

9. The coupled reaction

$$\text{Cyt f } (Fe^{3+}) + \text{Cyt b } (Fe^{2+}) \rightleftharpoons \text{Cyt f } (Fe^{2+}) + \text{Cyt b } (Fe^{3+})$$

has an $E^{\circ\prime}$ of 0.3 V.

For a *one* electron transfer (i.e. one mole of each component involved), $\Delta G^{\circ\prime} = -29$ kJ mol^{-1}.

This *would not be sufficient* to drive the synthesis of ATP from ADP and P_i.

However, if *two* electrons were passed along the redox chain (i.e. two moles of each component involved), $\Delta G^{\circ\prime}$ would be -57.9 kJ and this would clearly *be sufficient* to drive the synthesis of one mole of ATP from ADP and P_i.

10. We can use eqn (5.3) to calculate $\Delta G' = 57.2$ kJ mol^{-1}. From eqn (5.4) $\Delta p = -592/n$ mV.

We can also estimate Δp from eqn (5.2) if we know $\Delta\psi$ and ΔpH. The calculation of ΔpH is analogous to the worked example on p. 126 and $\Delta\text{pH} = +1.66$ with the inside being more alkaline. $\Delta\psi$ can be obtained from $^{86}Rb^+$ distribution data using eqn (5.6) and $\Delta\psi = -118$ mV.

From eqn (5.2) $\Delta p = -216$ mV.

Comparison with the first value of Δp gives $n = 2.7$. This means that either the chemiosmotic theory is not an adequate description or that the stoichiometry of proton movement is actually higher than the expected value of 2.0.

These data represent typical experimental results and it is not difficult to see why the mechanism of oxidative phosphorylation remains such a controversial topic.

6

Acids and Bases

In this chapter we shall consider the hydrogen ion concentrations of aqueous solutions and the practice of reporting them as pH values. We shall discuss the nature of acids and bases, and the composition of buffer mixtures, which by acting as 'reservoirs' of acid and base, can stabilize the hydrogen ion concentration of an aqueous solution. The biological relevance of this study will also be considered.

The Ion Product of Water

All aqueous solutions contain positively charged hydrogen ions (or protons, H^+), and negatively charged hydroxyl ions (OH^-). In pure water these are entirely derived from the ionization of water molecules,

$$H_2O \rightleftharpoons H^+ + OH^-$$

a process that is also referred to as the *dissociation* of water (into its component ions), or as the *protolysis* of water (to emphasize the liberation of H^+ ions). In fact, it is not strictly true to speak of the liberation of H^+ ions in an aqueous medium, for in this situation the H^+ ion is always hydrated and is predominantly present as H_3O^+, the hydronium ion. Thus the protolysis of water might better be represented by the following equation,

$$H_2O + H_2O \rightleftharpoons H_3O^+ + OH^-$$

Yet, except when it is necessary to emphasize the part played by water in another protolytic reaction, we can overlook the fact that H^+ actually exists in an aqueous medium in the form of H_3O^+ ions, for this need not affect our appraisal of the origin and ultimate fate of these protons; we may therefore continue to think of water dissociating to yield H^+ plus OH^- ions.

If $H_2O \rightleftharpoons H^+ + OH^-$, then at equilibrium, $(H^+)(OH^-)/(H_2O) = K_a$, where K_a is the temperature-dependent, *acid dissociation constant* of water and the bracketed terms represent the equilibrium activities of H^+, OH^- and H_2O. At 298.15 K (i.e. 25°C), K_a of water = 1.8×10^{-16}, and since the water is so little dissociated, $[H_2O]$ in the denominator of the equation (above) that defines the values of K_a of water can be assumed to be insignificantly less than the total activity of pure water at 298 K which is a constant. Thus,

$$[H^+][OH^-] = K_a[H_2O] = K_w$$

where K_w is the *ion product* (or ion product constant) of water.

In pure water, and in the dilute aqueous solutions that we shall consider in this chapter, we may assume that the activities of H^+ and OH^- ions will equal their concentrations. Therefore, $K_w = [H^+][OH^-]$, where $[H^+]$ and $[OH^-]$ are the equilibrium concentrations of these ions in any aqueous solution expressed in mol dm^{-3}. In pure water at 298 K, $K_w = 10^{-14}$, $[H^+]$ equals $[OH^-]$, and so both equal 10^{-7} mol dm^{-3} (the value of K_w increases by about 8% for every K rise in temperature, until at 310 K it is 2.4×10^{-14}).

The equation derived above to define K_w states that the product of the concentrations of hydrogen and hydroxyl ions in an aqueous solution is constant at a given temperature, and that any change in its hydrogen ion concentration will be reflected as a converse change in its hydroxyl ion concentration. Thus in any aqueous solution at 298 K, if the hydrogen ion concentration is 10^{-3} mol dm^{-3}, the hydroxyl ion concentration must be 10^{-11} mol dm^{-3}. Since values involving such negative powers of 10 are inconvenient to write and are troublesome to use in calculations, Sorensen suggested that the hydrogen ion concentrations of dilute aqueous solutions would be better expressed as pH values.

Meaning of the Term pH

The relationship between a pH value and the hydrogen ion concentration (in mol dm^{-3}) that it represents, can be explained in various ways; e.g.

The pH of a solution is the negative of the logarithm to the base 10 of its hydrogen ion concentration.

$$pH = -\log [H^+]$$

If the $[H^+]$ is written as a power of 10, then the corresponding pH value is the index of this exponential term without its negative sign; e.g. a $[H^+]$ of $10^{-3.72}$ mol dm^{-3} is equivalent to a pH value of 3.72.

The pH of a neutral solution (where $[H^+] = 10^{-7}$) is 7. In fact the usual pH range for most experiments of interest is from 0 (corresponding to a 1 mol dm^{-3} solution of H^+ ions) to 14 (corresponding to a 1 mol dm^{-3} solution of OH^- ions).

Although pH values less than 0 (negative values), and greater than 14, are theoretically possible, the range of pH values from 0 to 14 covers all hydrogen ion concentrations found in dilute aqueous solutions (and biological media).† In pure water at 298 K, $[H^+] = 10^{-7}$ mol dm^{-3} and the pH is 7 (often called 'neutral pH').

EXAMPLE 6.1 *Calculate (a) the pH of a solution whose hydrogen ion concentration is 2.3×10^{-9} mol dm^{-3}; (b) the hydrogen ion concentration in a solution of pH 4.31.*

Solution

(a) Since $[H^+] = 2.3 \times 10^{-9}$ mol dm^{-3} and $pH = -\log [H^+]$

$$pH = -\log (2.3 \times 10^{-9}) = -(\bar{9}.36) = -(-9 + 0.36)$$
$$= -(-8.64)$$
$$pH = 8.64$$

(*Note*: If in doubt concerning the use of logarithms, consult p.7. There you will find that the logarithm of the product of two terms is equal to the sum of the logarithms of each term taken separately, i.e. log ab = log a + log b.

Thus, $\log (2.3 \times 10^{-9}) = (\log 2.3 + \log 10^{-9}) = (0.36 + \bar{9}) = \bar{9}.36$.)

(b) Since pH = 4.31 and $pH = -\log [H^+]$, then $\log [H^+] = -4.31$
$= \bar{5}.69$

$$\log [H^+] = \bar{5}.69 \text{ and } [H^+] = \text{antilog } \bar{5}.69 = 4.9 \times 10^{-5}$$

∴ in a solution of pH 4.31 the $[H^+] = 4.9 \times 10^{-5}$ mol dm^{-3}

(*Note*: To convert a logarithm whose value is – 4.31 into the conventional form having a positive mantissa (p. 5), proceed as follows:

$$-4.31 = (-5 + (1 - 0.31)) = (-5 + 0.69) = \bar{5}.69)$$

---

† Always remember when using the pH notation, that (a) the pH of a solution decreases as its $[H^+]$ increases, and vice versa, and (b) a tenfold change in $[H^+]$ is represented by a pH difference of 1 unit.

In the same way that hydrogen ion concentrations are converted into pH values, so hydroxyl ion concentrations may also be represented as pOH values, where pOH = – log [OH^-]. Thus, in aqueous solutions when [H^+][OH^-] = 10^{-14}, log [H^+] + log [OH^-] = log 10^{-14} = –14. Reversing the signs throughout, – log [H^+] – log [OH^-] = 14

so that, pH + pOH = 14

Relationship between [H^+] [OH^-] and pH is given in Table 6.1

ACIDS AND BASES

Solution of an acid in water raises the hydrogen ion concentration above 10^{-7} mol dm^{-3} and correspondingly lowers the pH. Solution of a base in water raise the hydroxyl ion concentration above 10^{-7} mol dm^{-3}, and correspondingly decreases the hydrogen ion concentration (i.e. pH rises). This behaviour suggests that acids and bases may be defined as compounds which dissociate in water to yield H^+ and OH^- ions respectively:

$$\text{Acid: } HA \rightleftharpoons H^+ + A^-$$

$$\text{Base: } BOH \rightleftharpoons OH^- + B^+$$

A compound which can yield both H^+ and OH^- ions by dissociation, and which may therefore act both as an acid and a base, is termed an *amphoteric substance* or *ampholyte*.

This dissociation theory is simple, but it gives rise to several anomalies in its appraisal of bases. According to the theory, sodium hydroxide whose OH^- anion is liberated in aqueous solution, is obviously a base. Yet ammonia (NH_3), which cannot possibly liberate OH^- ions in this manner, also gives aqueous solutions that are strongly alkaline. The dissociation theory explains this by supposing that ammonia is hydrated in water to form a dissociable base, ammonium hydroxide (NH_4OH), and proposes that ammonia is an 'anhydro-base'. In fact, there is no good evidence for the existence of such a molecule as NH_4OH.

The most useful definition of acids and bases and of their interaction was proposed by Bronsted and by Lowry in 1923. In their view an acid is characterized by its tendency to lose H^+, whilst a base is characterized by its tendency to associate with H^+. In other words, *acids are proton donors and bases are proton acceptors*. Amphoteric compounds might, according to this theory, be termed *amphiprotic*

Table 6.1
Relationships among $[H^+]$, $[OH^-]$, and pH

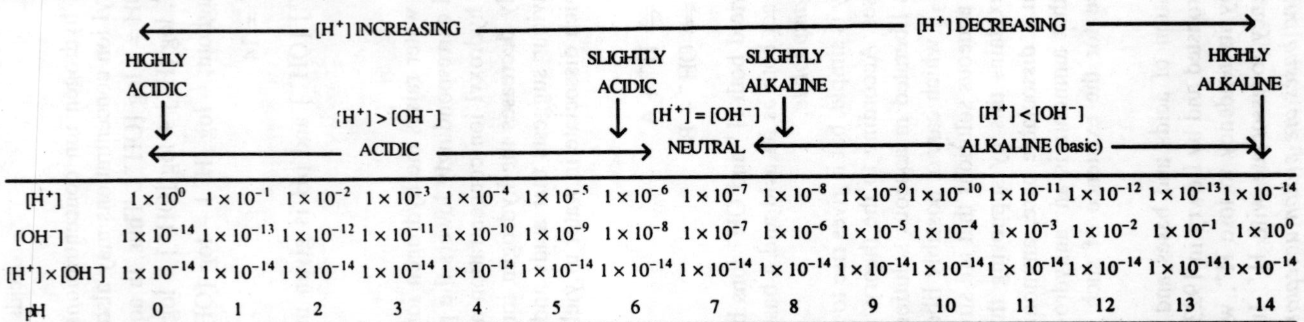

$[H^+]$	1×10^{0}	1×10^{-1}	1×10^{-2}	1×10^{-3}	1×10^{-4}	1×10^{-5}	1×10^{-6}	1×10^{-7}	1×10^{-8}	1×10^{-9}	1×10^{-10}	1×10^{-11}	1×10^{-12}	1×10^{-13}	1×10^{-14}
$[OH^-]$	1×10^{-14}	1×10^{-13}	1×10^{-12}	1×10^{-11}	1×10^{-10}	1×10^{-9}	1×10^{-8}	1×10^{-7}	1×10^{-6}	1×10^{-5}	1×10^{-4}	1×10^{-3}	1×10^{-2}	1×10^{-1}	1×10^{0}
$[H^+] \times [OH^-]$	1×10^{-14}	1×10^{-14}	1×10^{-14}	1×10^{-14}	1×10^{-14}	1×10^{-14}	1×10^{-14}	1×10^{-14}	1×10^{-14}	1×10^{-14}	1×10^{-14}	1×10^{-14}	1×10^{-14}	1×10^{-14}	1×10^{-14}
pH	0	1	2	3	4	5	6	7	8	9	10	11	12	13	14

compounds capable of acting both as proton donors and as proton acceptors.

Thus water is amphiprotic since it can both (a) behave as an acid by undergoing protolysis, $H_2O \rightleftharpoons H^+ + OH^-$, and (b) behave as a base by combining with protons to form hydronium ions, $H_2O + H^+ \rightleftharpoons H_3O^+$. This means that one molecule of water acting as an acid can donate a proton to another molecule of water acting as a base, $H_2O + H_2O \rightleftharpoons H_3O^+ + OH^-$. The behaviour of ammonia in aqueous solution is readily explained by this theory as being that of a base which accepts protons donated by the water,

$$NH_3 + H_2O \rightleftharpoons NH_4^+ + OH^-$$

It follows, from the Brönsted and Lowry definition of acid and base, that the product other than H^+ formed by the dissociation of an acid must be a base. For example, when the acid HA dissociates, $HA \rightleftharpoons H^+ + A^-$, its anion A^- is manifestly a base since it serves as a proton acceptor in the reverse reaction. The special relationship of base A^- to the parent acid HA is acknowledged by calling it the *conjugate base* of that acid, so that the acid HA and the base A^- comprise a *conjugate pair*. This means, for example, that the cyanide ion CN^- is the conjugate base of hydrocyanic acid HCN, and that the acetate ion CH_3COO^- is the conjugate base of acetic acid CH_3COOH.

$$\underset{\text{acid}_1}{HCN} + \underset{\text{bases}_2}{H_2O} \rightleftharpoons \underset{\text{conjugate acid}_2}{H_3O^+} + \underset{\text{conjugate base}_1}{CN^-}$$

Analogously, a base and its protonated derivative form a conjugate pair, so that if $B + H^+ \rightleftharpoons BH^+$, then BH^+ is the *conjugate acid* of base B. It follows that acetic acid is the conjugate acid of the acetate ion, and that the ammonium ion, NH_4^+, is the conjugate acid of ammonia, NH_3.

The Strength of Acids and Bases

One must be careful not to confuse the terms 'concentration' and 'strength' as applied to aqueous solutions of acids and bases, for these terms are not synonymous and cannot be used interchangeably. By *concentration* is meant the quantity of acid or base dissolved in a certain volume of water and usually expressed in mol dm^{-3}. The *strength* of an acid or base in an indication of the degree to which the compound demonstrates the properties of an acid or base relative to other compounds with similar properties. The acidic or base relative to other compounds with similar properties. The acidic or basic strength of a compound is therefore a measure of the effectiveness with which that compound behaves as an acid or base.

The Strength of an Acid

The strength of an acid is determined by the efficiency with which it acts as a proton donor. This in turn will be affected by the proton-accepting or proton-donating properties of the medium in which the acid is dissolved. Here we are concerned only with aqueous solutions and can assess the relative acidic strengths of compounds in terms of the extent to which each increases the hydrogen ion concentration when it is dissolved in water in known concentration and at a standard temperature. This can be recorded in any one of the following ways:

1. as the hydrogen ion concentration (or pH) of a dilute solution of stated concentration;
2. as the degree of dissociation of the acid, i.e. % of acid molecules dissociated at equilibrium;
3. as the dissociation constant, K_a, of the acid.

The last of these methods is undoubtedly the most convenient. According to the dissociation theory, the acid HA dissociates in water as follows,

$$HA \rightleftharpoons H^+ + A^-, \text{ whence at equilibrium, } \frac{[H^+][A^-]}{[HA]} = K_a$$

Being a ratio of equilibrium concentrations, K_a is independent of the total concentration of acid in any aqueous solution in which its strength is determined (so long as this is dilute and possesses a low ionic strength).

Although K_a is a *dissociation constant*, even according to the Brönsted and Lowry theory its value is a valid measure of the acid strength of a substance. In the Brönsted and Lowry view, a compound evidences its acidic nature in aqueous solution by donating protons to water (which here acts as a base). Thus,

$$HA + H_2O \rightleftharpoons H_3O^+ + A^-, \text{ and at equilibrium } \frac{[H_3O^+][A^-]}{[HA][H_2O]} = K_{eq}$$

But the water (as solvent) is present in such excess that its concentration is not measurably affected by the occurrence of this reaction, so that $[H_2O]$ is a constant, and at equilibrium, $[H_3O^+][A^-]/[HA] = K_{eq}$ $[H_2O]$ = a constant. Since the hydronium ion concentration in an aqueous solution equals its 'H^+ ion concentration',

$$\frac{[H_3O^+][A^-]}{[HA]} = \frac{[H^+][A^-]}{[HA]}$$

Thus, $$K_{eq}[H_2O] = K_a$$

Thus the constant $(K_{eq}[H_2O])$ is identical with K_a, the acid dissociation constant. Consequently acid strength is defined by the value of its dissociation constant Ka. Then, the stronger the acid the more fully does it dissociate in water, and the greater is the value of its acid dissociation constant K_a.

The pH of an Aqueous Solution of a Strong Acid

Strong acids give the appearance of being virtually completely dissociated in *dilute* aqueous solution. Consequently their values of K_a approach infinity, and they are all of approximately equal strength. The pH of a dilute aqueous solution of such an acid is very easily calculated, as shown in the following example.

EXAMPLE 6.2 *Calculate the pH of a 0.025 M solution of monobasic strong acid at 298 K.*

Solution

A 1 mol dm^{-3} solution of an ideal, monobasic strong acid will contain 1 mol dm^{-3} of H^+.

$\therefore$ a 0.025 mol dm^{-3} solution will contain 2.5×10^{-2} mol dm^{-3} of H^+

Since pH = – log $[H^+]$, the pH of the 0.025 mol dm^{-3} solution is – log 2.5×10^{-2}

$$\therefore \qquad pH = -(\bar{2}.4) = -(-1.6) = 1.6.$$

The pH of an Aqueous Solution of a Weak Acid

Amongst weak acids there exists a gradation of strengths evidenced by the wide spectrum covered by their values of K_a at a given temperature. All these values of K_a, like H^+ ion concentrations, involve negative powers of 10 and they are therefore frequently expressed as pK_a

values, where $pK_a = -\log K_a$. For example, the statement that 'monochloroacetic acid ($K_a = 1.4 \times 10^{-3}$) is a stronger acid than acetic acid ($K_a = 1.82 \times 10^{-5}$)' can be restated as, 'monochloroacetic acid ($pK_a = 2.86$) is a stronger acid than acetic acid ($pK_a = 4.74$)'. The stronger the acid, the larger is the value of its K_a but the smaller is its pK_a value (if no temperature is mentioned, the values of K_a are assumed to have been determined at 298 K, i.e. 25°C).

In the special case of a *very* weak acid, it is possible to calculate the pH of its dilutes solution from the values of K_a and of c (its concentration in mol dm^{-3}). This calculation assumes that the extent of dissociation is so small that the concentration of undissociated acid at equilibrium is negligibly less than the total concentration of acid present. If we consider the dissociation of the very weak acid HX in an aqueous solution in which it is present in a total concentration of c mol dm^{-3}, then,

$HX \rightleftharpoons H^+ + X^-$, and at equilibrium,

$$\frac{[H^+][X^-]}{[HX]} = K_a$$

or,

$$[H^+]\,[X^-] = K_a\,[HX]$$

But, $[X^-]$ equals $[H^+]$ at equilibrium, and if we now assume that [HX] approximately equals c,

$$[H^+][X^-] = K_a\,[HX] \quad \text{becomes} \quad [H^+]^2 = K_a c \;\; or \;\; [H^+] = \sqrt{K_a c}$$

Taking logarithms of both sides of this last expression, we obtain the new equation,

$$\log\,[H^+] = \frac{\log K_a}{2} + \frac{\log c}{2}$$

or,

$$-\log\,[H^+] = -\tfrac{1}{2}\log K_a - \tfrac{1}{2}\log c$$

∴

$$pH = \tfrac{1}{2}pK_a - \tfrac{1}{2}\log c$$

Example 6.3 *Calculate the pH of a 0.01 mol dm^{-3} solution of the very weak acid HY (K_a of HY is 3.2×10^{-7}).*

Solution

Assuming that [HY] at equilibrium is negligibly less than the total concentration of HY in the solution (0.01 mol dm^{-3}), then the approximate pH of this solution will be given by the equation,

$$\text{pH} = \tfrac{1}{2}\text{p}K_a - \tfrac{1}{2}\log c \quad \text{where} \begin{cases} \text{p}K_a = -\log 3.2\times 10^{-7} \\ c \quad = 0.01 \text{ mol dm}^{-3} \end{cases}$$

$$\therefore \qquad \text{pH} = \tfrac{1}{2}(-\log 3.2\times 10^{-7}) - \tfrac{1}{2}\log 0.01$$

$$= \tfrac{1}{2}(-\bar{7}.505) - \tfrac{1}{2}\bar{2}.0$$

$$= \tfrac{1}{2}(+6.495) - \tfrac{1}{2}(-2)$$

$$= 3.247 + 1$$

$$\therefore \qquad \text{pH} = 4.25$$

The pH of a dilute aqueous solution of a more extensively dissociated weak acid may equally simply be calculated from its degree of dissociation (α) in the aqueous solution in which it is present at a concentration of c mol dm^{-3}.

Example 6.4 *Calculate the pH of (a) a 0.1 mol dm^{-3} solution of hydrochloric acid which is 83% dissociated at 298 K and (b) a 0.1 mol dm^{-3} solution of acetic acid which is 1.35% dissociated at the same temperature.*

Were the acids fully dissociated, the [H$^+$] in a 0.1 mol dm^{-3} solution would be 0.1 mol dm^{-3}.

Solution

(a) Since the HCl is 83% dissociated, the [H$^+$] in its 0.1 mol dm^{-3} solution = 83/100 × 0.1 = 8.3 × 10^{-2} mol dm^{-3}.

$\therefore$ since pH = – log [H$^+$], its pH = – log (8.3 × 10^{-2})

$$= -(\bar{2}.92) = -(-2 + 0.92)$$

$$= -(-1.08)$$

$$= 1.08$$

(*Note*: If it was assumed that HCl was an ideal, strong acid and was 100% dissociated in its 0.1 mol dm^{-3} solution, the pH would be assumed to be 1.00, i.e. not very different from the actual pH of 1.08.)

(b) Since the acetic acid is 1.35% dissociated, the [H$^+$] in its 0.1 mol dm^{-3} solution = 1.35/100 × 0.1 = 1.35 × 10^{-3} mol dm^{-3}.

$\therefore$ its $\quad \text{pH} = -\log(1.35 \times 10^{-3}) = -(\bar{3}.13) = -(-3 + 0.13)$

$= -(-2.87)$

$= 2.87$

Evidently the degree of dissociation (α) of an acid in aqueous solution at a concentration c mol dm^{-3} will be related to its dissociation constant (K_a).

For example, if $HA \rightleftharpoons H^+ + A^-$, then at equilibrium, $[HA] = (1 - \alpha)c$,

$$[H^+] = [A^-] = \alpha c, \text{ whence } K_a = \frac{[H^+][A^-]}{[HA]} = \frac{(\alpha c)^2}{(1-\alpha)c} = \frac{\alpha^2 c}{1-\alpha}$$

This simple equation, relating the values of α and K_a, is one expression of ***Ostwald's Dilution Law***.

Strength of a Base

According to the dissociation theory, the strength of a base is determined by the extent to which it dissociates in aqueous solution to yield hydroxyl ions. If base B dissociates as follows,

$$B \rightleftharpoons B^+ + OH^-$$

then at equilibrium, $[B^+][OH^-]/[B] = K_b$, where the basic dissociation constant K_b measures the extent of dissociation and hence the basic strength of B.

In the Brönsted and Lowry view of the same base B, its basic strength is determined by the efficiency with which it accepts protons. In aqueous solution, the protons are donated by the water (here acting as an acid) so that,

$$B + H_2O \rightleftharpoons BH^+ + OH^-$$

The value of the equilibrium constant is therefore given by the equation

$$K_{eq} = \frac{[BH^+][OH^-]}{[B][H_2O]}$$

Once again, the concentration of water is so great that it can be assumed to remain constant. Therefore

$$\frac{[BH^+][OH^-]}{[B]} = K_{eq} \times [H_2O] = \text{Constant}$$

Since $[BH^+]$ in this equation is the same as $[B^+]$ in the equilibrium expression obtained by using the dissociation theory, the constant $K_{eq} \times [H_2O]$ derived from the Bronsted and Lowry theory is identical with K_b, the basic dissociation constant. This means that while the Brönsted and Lowry theory may provide the more satisfactory explanation of the true nature of a base, the strength of the base in aqueous solution is still defined by its value of K_b (even though the hydroxyl ions which appear in the solution need not have originated in its own molecules).

We have noted that the species formed by the association of a base with a proton is called its conjugate acid,

$$\underset{\text{(base)}}{B} + H^+ \rightleftharpoons \underset{\text{(conjugate acid)}}{BH^+}$$

When B is a strong base having a high affinity for protons, BH^+ will show little tendency to dissociate; that is to say, it will be a weak acid. Conversely, when B is a weak base it will form a strong conjugate acid. This inverse relationship between the strength of a base and the strength of its conjugate acid makes it possible, and valid, to define the strength of a base in terms of the K_a of its conjugate acid.

There are thus two methods of expressing the strength of a base in aqueous solution:

(i) as its basic dissociation constant K_b (or pK_b)

$$K_b = \frac{[BH^+][OH^-]}{[B]}$$

(ii) as the acid dissociation constant of its conjugate acid, K_a (or pK_a)

$$K_a = \frac{[B][H^+]}{[BH^+]}$$

The two dissociation constants are related in a simple manner, for K_a is inversely proportional to K_b and the proportionality constant is K_w, the ion product of water , i.e.:

$$K_a \times K_b = \frac{[B][H^+]}{[BH^+]} \times \frac{[BH^+][OH^-]}{[B]}$$

$$= [H^+][OH^-] = K_w$$

$$\therefore \qquad K_a \times K_b = K_w$$

or $\quad pK_a + pK_b = pK_w$

The stronger the base, the larger is its K_b and the smaller is the K_a of its conjugate acid. In other words, the stronger the base, the smaller is its pK_b value and the larger is the pK_a value of its conjugated acid.†

The pH of an Aqueous Solution of Strong Base (or Alkali)

A strong base can be defined *either* as a base which yields an infinitely weak, conjugate acid, *or* as a base with a K_b value which approaches infinity. Since $K_b = [BH^+][OH^-]/[B]$, this means in effect that a strong base, when dissolved in water, promotes the formation of an equivalent concentration of hydroxyl ions.

This raises the question of whether sodium hydroxide should be considered to be a base (as it is by the dissociation theory), or as a 'reservoir' of the true base OH^-. Such hydroxides of the alkali metals (e.g. sodium, potassium) are extremely soluble, and are fully ionized, both in the solid state and in water. Their metal cations (e.g. Na^+, K^+), are 'neutral' in the sense that they possess nil acidic or basic strength, and it is therefore legitimate to consider the basicity of these hydroxides to be due entirely to their content of OH^-, which is present independently of the presence of water. Since their solutions in water contain concentrations of hydroxyl ions equivalent to their molarity, even dilute solutions are very alkaline in pH, and in recognition of this fact, plus their somewhat special status, these hydroxides are called *alkalis*. Use of this term avoids the necessity of calling the Na^+ OH^- molecule a base, and thus satisfies those who would prefer to consider it as a salt possessing a 'neutral' cation and the strongly basic OH^- anion.

The pH of a dilute aqueous solution of alkali is easily calculated if the normality or molarity of the solution is known.

EXAMPLE 6.5 *Calculate the pH of a 0.56 mol dm^{-3} aqueous solution of potassium hydroxide (assuming ideal behaviour of the alkali).*

---

† K_a of a weak acid and K_b of a weak base are equilibrium constants and their values are temperature dependent. Thus, although you can assume that the values of K_a an K_b quoted throughout this Chapter refer to 298 K (i.e. 25°C), in practice you must employ the values possessed by these constants at the actual temperature of the solution under consideration.

Solution

A 0.56 mol dm^{-3} solution of K^+OH^- will contain 0.56 mol dm^{-3} of OH^-

$\therefore$ $[OH^-] = 5.6 \times 10^{-1}$ mol dm^{-3}

whence $pOH = -\log [OH^-] = -\log (5.6 \times 10^{-1})$

$= -(\bar{1}.75) = -(-1 + 0.75)$

$= -(-0.25)$

$= 0.25$

In any aqueous solution at 298 K, pH = 14 – pOH, $\therefore$ pH = 13.75.

The composition of aqueous solutions of ammonia

Having just mentioned alkalis, it is appropriate that we now consider why it is wrong to speak of 'ammonium hydroxide solutions' when we are really referring to aqueous solutions of ammonia. Ammonia (NH_3) is a highly soluble, weak base which is capable of accepting protons to form its conjugate acid (NH_4^+). When dissolved in water, the protons are donated by the water (here acting as a weak acid), i.e.

$$NH_3 + H_2O \rightleftharpoons NH_4^+ + OH^-$$

The NH_4^+ and OH^- ions that are produced when ammonia is dissolved in water are therefore formed as the direct result of the action of ammonia as a base, and the extent to which they are produced is a function of the K_b of ammonia, where $K_b = [NH_4^+][OH^-]/[NH_3]$. It is misleading to call this an 'ammonium hydroxide solution', for this suggests quite wrongly that the only unique components are the NH_4^+ and OH^- ions (thus overlooking its content of NH_3), and furthermore infers the separate existence of the compound ammonium hydroxide (in the absence of NH_3 and H_2O). It is therefore much better to refer to these solutions as 'aqueous solutions of ammonia' for attention is then drawn to the actual, weakly basic component that determines their ionic composition.

The pH of an Aqueous Solution of a Weak Base

We have seen that a weak base is characterized by its relatively small affinity for protons; this is evident in aqueous solutions from the small value of its K_b. If $B + H_2O \rightleftharpoons BH^+ + OH^-$, then at equilibrium in an aqueous solution $[BH^+][OH^-]/[B] = K_b$. Assuming that the solution contains a total concentration of B equal to c mol dm^{-3}, then, since the extent to which it accepts protons and is transformed into its strong, conjugate acid BH^+, is relatively small, [B] at equilibrium will be insignificantly less than c. The BH^+ and OH^- ions are produced in equal concentration; therefore at equilibrium in the dilute aqueous solution of base B, $[B] = c$, $[BH^+] = [OH^-]$, and therefore,

$$K_b = \frac{[BH^+][OH^-]}{[B]} = \frac{[OH^-]^2}{c}$$

But in any aqueous solution $[OH^-] = K_w/[H^+]$, and substituting this term for $[OH^-]$ in the above equation, we obtain the expression,

$$K_b = \frac{K_w^2}{[H^+]^2 \cdot c}, \text{ whence } [H^+]^2 = \frac{K_w^2}{K_b \cdot c} \text{ or } [H^+] = \frac{K_w}{\sqrt{K_b \cdot c}}$$

Taking logarithms,

$$\log [H^+] = \log K_w - \tfrac{1}{2} \log K_b - \tfrac{1}{2} \log c$$

Reversing the signs throughout,

$$-\log [H^+] = -\log K_w + \tfrac{1}{2} \log K_b + \tfrac{1}{2} \log c$$

so that, $$pH = pK_w - \tfrac{1}{2} pK_b + \tfrac{1}{2} \log c$$

EXAMPLE 6.6 *The pK_b values of ammonia and trimethylamine are, respectively, 4.74 and 4.21 at 298 K. Calculate:*

(1) the pH of 0.05 mol dm^{-3} aqueous solutions of ammonia and trimethylamine,

(2) the acid dissociation constants of ammonium and trimethylammonium ions.

Solution

(1) Ammonia is a weak base that in aqueous solution forms its conjugate acid NH_4^+, according to the equation: $NH_3 + H_2O \rightleftharpoons NH_4^+ + OH^-$. In a 0.05 mol dm^{-3} solution, the total concentration of 'ammonia' is sufficiently great, in comparison with its K_b value, for $[NH_3]$ at equilibrium to be assumed to be negligibly less than 0.05 mol dm^{-3}.

The pH of this solution is approximately given by the equation,

$$\text{pH} = \text{p}K_w - \tfrac{1}{2}\text{p}K_b + \tfrac{1}{2}\log c \quad \text{where} \begin{cases} \text{p}K_w = 14 \\ \text{p}K_b = 4.74 \\ c = 0.05 \text{ mol dm}^{-3} \end{cases}$$

$$\therefore \quad \text{pH} = 14 - \tfrac{1}{2}(4.74) + \tfrac{1}{2}\log 0.05$$

$$= 14 - 2.37 + \tfrac{1}{2}(\bar{2}.7) = 14 - 2.37 + \tfrac{1}{2}(-1.3)$$

$$= 14 - 2.37 - 0.65 = 14 - 3.02 = 10.98$$

By similar reasoning, the pH in a 0.05 mol dm^{-3} solution of trimethylamine is given by the equation,

$$\text{pH} = 14 - \tfrac{1}{2}(4.21) + \tfrac{1}{2}\log 0.05$$

$$= 14 - 2.11 - 0.65$$

$$= 11.24$$

(2) The pK_a of the conjugate acid of a base is related to its pK_b value according to the equation, pK_a = pK_w − pK_b;

$$\therefore \quad \text{p}K_a \text{ of } NH_4^+ = 14 - \text{p}K_b \text{ of } NH_3$$

$$= 14 - 4.74 = 9.26$$

$$\therefore \quad -\log K_a = 9.26, \text{ and } \log K_a = -9.26 = \overline{10}.74$$

$$\therefore \quad K_a \text{ of } NH_4^+ = \text{antilog } \overline{10}.74 = 5.5 \times 10^{-10}$$

Similarly, $\text{p}K_a$ of $(CH_3)_3NH^+ = 14 - 4.21 = 9.79$

$$\therefore \quad \log K_a = -9.79 = \overline{10}.21$$

$$\therefore \quad K_a \text{ of } (CH_3)_3NH^+ = \text{antilog } \overline{10}.21 = 1.62 \times 10^{-10}$$

Note that ammonia is a weaker base than trimethylamine, and NH_4^+ is a stronger acid than $(CH_3)_3NH^+$.

THE INTERACTION OF AN ACID WITH A BASE

Accepting the definition of acids and base that is proposed by the dissociation theory, the interaction of any acid with any base can be considered as the formation of H_2O by the association of H^+ ions liberated by the acid, with OH^- ions derived form the base,

$$HA \rightleftharpoons H^+ + A^-$$
$$Base \rightleftharpoons B^+ + OH^-$$
$$H^+ + OH^- \rightleftharpoons H_2O$$

Net reaction: $HA + Base \rightleftharpoons B^+ + A^- + H_2O$

The ions (B^+, A^-) produced by this interaction of acid and base, are the ions of the salt BA, and in elementary textbooks a salt is still often defined as 'the product, other than water, that is formed by the interaction of an acid with a base'. Similarly the term 'neutralization' is generally used to describe the interaction of acid and base to yield a salt.

The Brönsted and Lowry conception of the interaction between acids and bases offers a much broader view of the process of neutralization. In this view, neutralization is the process of proton transference from an acid to a base; this need not involve water, and need not result in the formation of a recognizable salt.

HA	+	B	⇌	BH^+	+	A^-
$(acid)_1$		$(base)_2$		(conjugate acid)$_2$		(conjugate base)$_1$

Therefore, even though we shall only be considering the interaction between acids and bases in aqueous solution, we would be better employed in identifying the actual reactant acid and base in any neutralization process (and hence the conjugate acid and base produced) than in attempting to identify a salt among the products.

Neutralization of a Strong Acid with a Strong Base

The manner in which the pH changes when an aqueous solution of strong acid (HCl) is titrated with a strong base (NaOH) provides an insight into the nature of the neutralization process. We have noted that NaOH can be assumed to act as an equivalent concentration of OH^- ions. Therefore, according to the proton transfer view of neutralization, the interaction of HCl and NaOH can be represented by the equation,

HCl	+	OH^-	⇌	H_2O	+	Cl^-
(acid)		(base)		(conjugate acid)		(conjugate base)

The products of neutralization are the exceedingly weak acid H_2O and the infinitely weak base Cl^-. Thus neutralization will be complete, for the reverse reaction will not occur to any significant extent, and the neutralization equation can be written,

$$HCl + OH^- \rightarrow H_2O + Cl^-$$

This means that:

1. When less than an equivalent quantity of NaOH has been added, the hydrogen ion concentration will be equal to the concentration of HCl that remains unneutralized, since HCl may be assumed to be completely dissociated in water, $HCl \rightarrow H^+ + Cl^-$.
2. At the equivalence point, the pH of the mixture will be 7 for it will contain, besides water, only Na^+ and Cl^- ions that are negligibly acidic or basic. (pH 7 is therefore the pH of an aqueous solution of sodium chloride.)
3. When more than an equivalent quantity of NaOH has been added, the residual hydrogen ion concentration is inversely proportional to the concentration of excess NaOH present (since this is a measure of the excess $[OH^-]$).

The relationship between added NaOH and the pH. which is generally called a *titration curve.*

The shape of this curve is explained by the logarithmic nature of the pH scale, which means that, compared with the quantity of alkali required to raise the pH from 1 to 2, it takes only one tenth this amount of alkali to raise the pH from 2 to 3, one hundredth the quantity to raise the pH from 3 to 4, and one hundred thousandth as much to change the pH by one unit from 6 to 7. Over 95% of the total quantity of alkali needed to raise the pH to 7 is consumed between pH 1 and 3. This zone, where relatively large additions of alkali produce comparatively small changes in pH, is called the *acid buffer zone*. Addition of extremely small quantities of alkali to a mixture containing almost equivalent concentrations of HCl and NaOH produces a very marked increase in pH. These mixtures of pH between 3 and 11, are insignificantly buffered compared with mixtures containing either a great excess of strong acid (pH < 3) or a great excess of alkali (pH > 11), which are said to possess *buffering ability* because of their relatively great resistance to pH change on the addition of a small quantity of alkali or strong acid.

EXAMPLE 6.7 *To 10 cm^3 of 0.1 mol dm^{-3} HCl was added 9.6 cm^3 of 0.1 mol dm^{-3} NaOH. Calculate the pH of the final solution.*

Solution

9.6 cm^3 of 0.1 mol dm^{-3} NaOH will completely neutralize an equivalent quantity of the strong acid, HCl.

$$HCl + (Na^+)OH^- \rightarrow H_2O + (Na^+)Cl^-$$

Thus if 10 cm^3 of 0.1 mol dm^{-3} HCl was initially present, a quantity equivalent to 0.4 cm^3 of 0.1 mol dm^{-3} HCl will remain unneutralized in the solution whose total volume is now 19.6 cm^3.

∴ concentration of HCl remaining unneutralized

$$= 0.4/19.6 \times 0.1 \text{ mol dm}^{-3}$$

$$= 2.04 \times 10^{-3} \text{ mol dm}^{-3}$$

$$[H^+] \text{ in } 1 \text{ mol dm}^{-3} \text{ HCl} = 1 \text{ mol dm}^{-3}$$

$$\therefore \quad [H^+] \text{ in } 2.04 \times 10^{-3} \text{ mol dm}^{-3} \text{ HCl} = 2.04 \times 10^{-3} \text{ mol dm}^{-3}$$

$$\therefore \text{pH} = -\log [H^+] = -\log (2.04 \times 10^{-3}) = -(\bar{3}.31) = -(-3 + 0.31)$$

$$= -(-2.69) = 2.69$$

Neutralization of a Weak Acid with a Strong Base

When an aqueous solution of acetic acid is titrated with sodium hydroxide, the neutralization process can be written as

$$\underset{\text{(acid)}}{HAc} + OH^- \rightleftharpoons H_2O + \underset{\text{(conjugate base)}}{Ac^-}$$

The product of neutralization, other than water, is acetate ion Ac^-, which is the appreciably basic conjugate base of the weak acid HAc, i.e.

$$Ac^- + H^+ \rightleftharpoons HAc$$

This means that the pH of a solution of acetic acid, to which less than an equivalent of NaOH has been added, will depend not only on the quantity of acetic acid that remains unneutralized, but also upon the extent to which it is dissociated, which will be determined by the $[Ac^-]$ in the solution. Thus the course of the pH change during the titration will reflect:

1. the removal, by neutralization, of acetic acid, with the concurrent production of Ac^-;

2. the progressively increasing repression of the dissociation of the residual acetic acid in the face of the mounting $[Ac^-]$ produced by (1).

The quantitative effect of these events upon the pH during neutralization can be calculated from the equation that defines the acid dissociation constant of the weak acid, HA; i.e.

$$K_a = \frac{[H^+][A^-]}{[HA]}$$

whence, $$[H^+] = K_a \times \frac{[HA]}{[A^-]}$$

and $$\log [H^+] = \log K_a + \log \frac{[HA]}{[A^-]}$$

or, $$-\log [H^+] = -\log K_a - \log \frac{[HA]}{[A^-]}$$

But, $$-\log \frac{[HA]}{[A^-]} = +\log \frac{[A^-]}{[HA]}$$

∴ $$-\log [H^+] = -\log K_a + \log \frac{[A^-]}{[HA]}$$

i.e. $$pH = pK_a + \log \frac{[A^-]}{[HA]}$$

This means that the pH in any aqueous solution which contains a significant concentration of weak acid HA in the presence of its conjugate base A^-, will depend (1) on the pK_a value of the acid and (2) on the ratio of what concentrations of the acid and its conjugate base it contains, as described by the equation,

$$pH = pK_a + \log \frac{[\text{conjugate base}]}{[\text{acid}]}$$

This most important relationship is known as the *Henderson–Hasselbalch Equation*. It means in practice that the weak acid-strong base titration curve will have the characteristic shape for the titration of acetic acid and NaOH.

It can be seen from this titration curve that the pH at the equivalence point is greater than 7. This is due to the basicity of the acetate ion, and the pH of the solution at equivalence can readily be calculated. Secondly, the zone of rapid pH change about the equivalence point is less extensive than that in the titration of a strong acid with NaOH However, the most striking feature of this titration curve is that on the acid side of equivalence it has a characteristic sigmoidal shape with an inflection at half equivalence. At this point, half the acid initially present will have been converted into its conjugate base, and half will be unneutralized. Thus at this half-equivalence point [conjugate base] = [acid], and the pH equals (pK_a + log 1) = pK_a (since log 1 equals 0). Therefore, by measuring the pH at the half-equivalence point when a weak acid is titrated with alkali, an experimental (apparent) value is obtained for the pK_a of the acid.

It further appears, from this titration curve, that for a considerable range on either side of half equivalence, the pH changes only slowly on addition of alkali (or of strong acid), though near the start, and again when approaching equivalence, the pH change is more rapid. In other words, mixtures in the middle range of the pre-equivalence curve possess considerable buffering ability, while those at both extremes are deficient in this respect. The sigmoidal shape of the titration curve is a consequence of the factors embodied in the Henderson-Hasselbalch equation. The change in pH brought about by the addition of a quantity of alkali will depend solely upon the change made by this addition in the value of the term log [conjugate base]/[acid], since pK_a is a constant in any titration. At the half-equivalence point, [conjugate base]/[acid] is 1, and the addition of, say, one-tenth of an equivalent of alkali to such a mixture would produce a relatively small change in the ratio, and consequently only a slight increase in the pH (the ratio would become 1.5, and log 1.5 = 0.176). On the other hand, addition of this amount of alkali to mixtures in which [conjugate base]/[acid] is initially 10 or 1/10 would produce a much larger change in the ratio and hence in the pH. In faet, the alteration in pH produced by a given small quantity of alkali (or of strong acid) is minimal at the half-equivalence point and increases the more the ratio [conjugate base]/[acid] differs from 1; hence the sigmoidal shape of the titration curve. So, too, we can conclude that:

(a) a mixture with a [conjugate base]/[acid] ratio of 1, in which the pH is equal to the pK_a of the acid component, constitutes the most efficient buffer mixture,

(b) the buffering ability of the mixture diminishes the more its pH differs from the pK_a of the weak acid component.

In practice, it is considered that a mixture of a weak acid and its conjugate base makes a satisfactory buffer over the pH range of pH = (pK_a – 1) to pH = (pK_a + 1). Examination of the titration curve enables us to make rather more accurate predictions concerning the relative buffering abilities of mixtures containing equal total concentrations of the same weak acid and conjugate base, namely that:

1. The mixture whose pH equals the pK_a of the weak acid, in which the ratio [conjugate base]/[acid] is 1, is the optimal buffer mixture in the sense that it equally minimizes pH change on addition of equivalent, small quantities of either alkali or strong acid.
2. A mixture whose pH equals (pK_a – 1), in which the ratio [conjugate base]/[acid] is 0.1 is an effective buffer 'against' alkali, but is much less effective 'against' strong acid.
3. The mixture at the other extreme of the buffer range, whose pH equals (pK_a + 1) and in which the ratio [conjugate base]/[acid] is 10, is an effective buffer 'against' strong acid, but is much less effective as a buffer 'against' alkali.

However, the actual buffer capacity of a solution will be determined not only by the initial ratio of [conjugate base]/[acid], but also by the actual magnitudes of these concentrations. Note also that the same 'pre-equivalence' titration curve would be obtained if a solution of sodium acetate (equivalence point mixture) were titrated with strong acid, though now the pH would decrease, moving 'back down the curve'.

We are now in a position to calculate the pH of a mixture prepared by partial neutralization of a weak acid with alkali. It should, however, be pointed out that values calculated by use of the Henderson-Hasselbalch equation are always approximate, and are extremely unreliable if the equation is applied to solutions in which the ratio [conjugate base]/[acid] is very large or small, or in which the strength of the acid component is too great or too small. In practice this means that it is generally unwise to apply the Henderson-Hasselbalch equation to aqueous solutions of pH less than 4 or greater than 10.

EXAMPLE 6.8 *To 35 cm^3 of 0.02 mol dm^{-3} acetic acid were added 10 cm^3 of 0.05 mol dm^{-3} sodium hydroxide. What is the pH of the resulting solution? (pK_a of acetic acid = 4.74).*

Solution

The neutralization of acetic acid (HAc) by sodium hydroxide can be represented as:

$$\underset{\text{(acid)}}{\text{HAc}} + (\text{Na}^+)\text{OH}^- \rightleftharpoons \text{H}_2\text{O} + \underset{\text{(conjugate base)}}{(\text{Na}^+)\,\text{Ac}^-}$$

Thus a quantity of NaOH will neutralize an equivalent quantity of HAc to form Ac^-. In the initial [HAc] = *a* mol dm^{-3}, and [NaOH] = *x* mol dm^{-3} was added with no volume change, then,

Initially	*Finally*
$[Ac^-]$ = negligible	$[Ac^-] = x$
$[HAc] = a$	$[HAc] = (a - x)$

The pH of the final solution is given by the Henderson-Hasselbalch equation,

$$\text{pH} = \text{p}K_a + \log \frac{[\text{conjugate base}]}{[\text{acid}]}$$

$$= 4.74 + \log \frac{[\text{Ac}^-]}{[\text{HAc}]} = 4.74 + \log \frac{x}{(a-x)}$$

The problem set by the present example is a little more complicated in that the addition of alkali in solution brought about a significant change in the volume of the solution. Two methods of calculation are possible:

(a) Calculating actual final concentrations of the conjugate base and acid.

(b) Recognizing that since what is used in the Henderson-Hasselbalch equation is the *ratio* of concentrations of conjugate base and acid, this will be identical with the ratio of their amounts in the final solution, whatever its volume might be.

We can compare the merits of these methods by applying both to the present example.

Method (a)

Initially: Volume = 35 cm^3

$[Ac^-]$ = negligible

$[HAc] = 0.02$ mol dm^{-3}

Finally: Volume = 45 cm^3

$$[Ac^-] = [NaOH] = \frac{10}{45} \times 0.05 = 1.11 \times 10^{-2} \text{ mol dm}^{-3}$$

$$[HAc] = [HAc \text{ if original solution was diluted to } 45 \text{ cm}^3] - [Ac^-]$$

$$= \left(\frac{35}{45} \times 0.02\right) - (1.11 \times 10^{-2}) \text{ mol dm}^{-3}$$

$$= (1.55 \times 10^{-2} - 1.11 \times 10^{-2}) \text{ mol dm}^{-3}$$

$$= 0.44 \times 10^{-2} \text{ mol dm}^{-3}$$

The pH of the final solution is given by the equation

$$\text{pH} = 4.74 + \log \frac{[Ac^-]}{[HAc]} = 4.74 + \log \frac{1.11 \times 10^{-2}}{0.44 \times 10^{-2}}$$

$$= 4.74 + \log 2.52$$

$$= 4.74 + 0.4$$

$$= 5.14$$

Method (b)

Let us express all quantities as their equivalent in cm^3 of a 0.02 mol dm^{-3} solution.

Initially	*Finally*
Ac^- = negligible	$Ac^- \equiv$ NaOH added = 10 cm^3 of 0.05 mol dm^{-3}
	$\equiv$ 25 cm^3 of 0.02 mol dm^{-3}
	= 25
HA = 35	HAc = Initial HAc − Ac^- produced
	= 35 − 25 = 10

Thus in the final solution

$[Ac^-] \equiv$ 25 cm^3 of 0.02 mol dm^{-3} solution diluted to 45 cm^3

[HAc] ≡ 10 cm^3 of 0.02 mol dm^{-3} solution diluted to 45 cm^3

$$\therefore \qquad \frac{[Ac^-]}{[HAc]} = \frac{25}{10} = 2.5$$

Since in the final solution, $pH = 4.74 + \log \frac{[Ac^-]}{[HAc]}$,

$$pH = 4.74 + \log 2.5$$
$$= 5.14$$

Method (b), when carefully used, is simpler than method (a), at least when it is applied to elementary problems.

Neutralization of a Weak Base with a Strong Acid

The titration curve that is obtained when an aqueous solution of ammonia is titrated with HCl. The pH at equivalence is now considerably less than 7 and the whole curve appears to be almost the mirror image of the weak acid-alkali titration curve.

The neutralization reaction occurring during the titration can be represented by the general equation,

$$\underset{\text{(base)}}{B} + HCl \rightleftharpoons \underset{\text{(conjugate acid)}}{BH^+} + Cl^-$$

(in this instance, B is NH_3). The conjugate acid BH^+, formed by the neutralization of B, is a moderately strong acid and will appreciably dissociate to re-form B. Equilibrium at any stage in the titration is established with B and BH^+ present in concentrations that conform to the equation $\frac{[BH^+]}{[B][H^+]} = \frac{1}{K_a}$, where K_a is the acid dissociation constant of BH^+. Thus, so long as both B and BH^+ are present in significantly large concentrations, the $[H^+]$ in the solution will be given by the equation, $[H^+] = K_a \times \frac{[BH_+]}{[B]}$,

whence, $$pH = pK_a + \log \frac{[B]}{[BH^+]}$$

This means that the Henderson-Hasselbalch equation is also applicable to solutions containing a weak base and its conjugate acid. To emphasize one's primary concern with the weak base in such a mixture, the equation may be written in the form,

$$pH = pK_a + \log \frac{[\text{Base}]}{[\text{conjugate acid}]}$$

Note that the pH of the mixture at the half-equivalence point in the titration of a weak base with a strong acid will equal the apparent pK_a value of *the conjugate acid of the weak base*. The pK_a of NH_4^+ is approximately 9.3.

Such mixtures of a weak base with its conjugate acid are just as effective pH buffers as the weak acid-conjugate base mixtures formed during the titration of a weak acid with alkali. For identical reasons substantial buffering ability will be confined to mixtures whose pH's are within approximately 1 unit of the pK_a of the acid component. Since the pK_a values of the conjugate acids of weak bases are usually greater than 7, this means that these weak base-conjugate acid mixtures are useful buffers alkaline pH's.

Neutralization of a Weak Acid with a Weak Base

In the titration of a weak acid with a weak base in aqueous solution, the pH at the equivalence point may be acid, alkaline or neutral depending on the relative strengths of the acid and base being titrated. Furthermore, it is a feature of this type of titration that the equivalence point may be difficult to determine since the rate of change in pH as it is approached is much less striking than in any titration that involves either a strong acid or a strong base.

Mixtures of a weak acid with a weak base will buffer either at acid or at alkaline pH's, depending on which component is present in excess. This can be illustrated with mixtures of acetic acid and ammonia in water. When acetic acid is present in such excess that the pH is about 3.7 to 5.7, the mixture buffers due to its content of the components of the acetic acid-acetate conjugate pair in suitable proportion, when

$$pH = 4.74 + \log \frac{[Ac^-]}{[HAc]}$$

(The $[Ac^-]$ in this mixture is provided by the neutralization of HAc with NH_3;

$$HAc + NH_3 \rightleftharpoons NH_4^+ + Ac^-)$$

When equivalent amounts of HAc and NH_3 are mixed together in water, the resulting ammonium acetate solution has a pH of 7. Addition of excess ammonia to produce a mixture of pH 8.3 to 10.3 (approx.) also confers pH buffering ability on the solution which now contains the components of the $NH_4^+ - NH_3$ conjugate pair in suitable proportion $\left(\text{since pH} = 9.26 + \log \frac{[NH_3]}{[NH_4^+]}\right)$

This means that although an ammonium acetate solution is not itself a good buffer, it can be converted into two quite different buffer mixtures: (a) by the addition of a very small quantity of a strong acid, or of a somewhat larger amount of HAc, (b) by the addition of a very small quantity of alkali, or of a somewhat larger amount of NH_3.

BUFFER MIXTURES AND THEIR BUFFER CAPACITY

From the appearance of the various titration curves, we have concluded that aqueous solutions of various types demonstrate buffering ability. Even solutions of strong acids or of strong bases exhibit such behaviour at their respective extremes of the normal pH range, viz. < 3 and > 11. Yet the solutions that make the best buffers in the pH range 4 to 10 are those that contain a weak acid with its conjugate base, or a weak base with its conjugate acid. Amongst the properties of such solutions that we have already mentioned, the following are particularly noteworthy (N.B. *Here 'acid' and 'base' refer to the members of the conjugate pair present in the mixture whether this is weak acid-conjugate base or weak base-conjugate acid*):

(a) that mixture with a [base]/[acid] ratio of 1, is optimally buffered against both strong acid and strong base, and its pH equals the pK_a of the acid component:

(b) mixtures with [base]/[acid] ratios between 0.1 and 10 are significantly buffering and their pH will fall within 1 unit of the pK_a value of their acid component;

(c) the pH of any mixture of this type can be calculated by applying the Henderson-Hasselbalch equation, $pH = pK_a + \log \frac{[Base]}{[acid]}$, where pK_a is the pK_a value of its acid component.

The *buffer capacity* of a solution is an indication of its effectiveness in minimizing the pH change that results from the addition of a standard quantity of strong acid or strong base. A quantitative measure of this is given by the *buffer value*, a unit introduced by Van Slyke. If the addition of 1 mol of monoacidic strong base (e.g. OH^-) to 1 dm^3 of a solution causes its pH to increase by 1 pH unit, then that solution is said to have a buffer value of 1 unit, so that buffer value = β = db/dpH, where db is a quantity of monoacidic strong base in mol added to 1 dm^3 of buffer solution, and dpH is the resultant increase in pH. A quantity of monobasic strong acid would, in its effect, be equivalent to the addition of an equal but negative quantity of monoacidic strong base, i.e. – db, and would causes a decrease in pH equal to – dpH. Therefore, whether the buffer capacity is tested against strong acid or strong base, the buffer values, though they may be numerically different, will always be positive.

The actual choice of a buffer to stabilize the pH in a reaction mixture is always restricted by the requirement that its components should not interfere with the reaction in any other way. From an assortment of harmless buffer mixtures, it would be wise to choose that mixture whose acid component possessed at the working temperature a pK_a value very close to the desired pH. The actual composition of the required buffer could be approximately calculated using the Henderson-Hasselbalch equation, and the concentration of its components chosen to yield a mixture of sufficiently large buffer capacity to be able to maintain a constant pH during the course of the experiment.

To prepare a buffer mixture of this type, instead of partially neutralizing a weak acid with alkali, it is frequently more convenient to mix together a quantity of the weak acid in solution, with a quantity of its salt whose anion is its conjugate base and whose cation is 'neutral'. Thus an acetic acid-acetate buffer can be prepared by mixing together acetic acid and sodium acetate solutions. Similarly a weak base can be mixed with a solution of its salt whose cation is its conjugate acid and whose anion is 'neutral'. In this way, an ammonia-ammonium ion buffer can be made by mixing solutions of ammonia and ammonium chloride in calculated proportions. The salts that are used to prepare these buffer mixtures, e.g. salts of weak acids with alkalis (e.g sodium acetate), or of weak bases with strong acids (e.g. ammonium chloride), can be considered to act as sources of an equimolar concentration of 'conjugate acid' or 'conjugate base'. Because of this, and because of the widespread use of such buffer mixtures, you will frequently find the Henderson-Hasselbalch equation written in the following forms:

(a) when applied to mixtures of a weak acid and its alkali metal salt, e.g. acetic acid-sodium acetate,

$$\mathrm{pH} = \mathrm{p}K_a + \log \frac{[\mathrm{salt}]}{[\mathrm{acid}]}$$

(b) when applied to mixtures of a weak base and its salt with a strong acid, e.g. ammonia-ammonium chloride

$$\mathrm{pH} = \mathrm{p}K_a + \log \frac{[\mathrm{base}]}{[\mathrm{salt}]}$$

Although Henderson in 1908 derived the equation in this form, since the Brönsted and Lowry theory of 1923 directed attention to the primary determinants of the pH of the mixture and suggested that the equation might be written as $\mathrm{pH} = \mathrm{p}K_a + \log \frac{[\text{conjugate base}]}{[\text{Brönsted acid}]}$, the use of a [salt] term in this equation has become redundant and, indeed, rather confusing.

EXAMPLE 6.9 *Calculate the change in pH caused by dissolving 1.025 g of anhydrous sodium acetate in 100 cm^3 of 0.25 mol dm^{-3} acetic acid (M. Wt. of anhydrous sodium acetate = 82; assume the apparent pK_a of acetic acid to be 4.74*)

Solution

The initial concentration of acetic acid = c = 0.25 mol dm^{-3}. Since acetic acid is a reasonably weak acid, the pH of its aqueous solution will be given (approximately) by the equation, $\mathrm{pH} = \frac{1}{2}\mathrm{p}K_a - \frac{1}{2}\log c$

$$\therefore \quad \mathrm{pH} = \tfrac{1}{2}(4.74) - \tfrac{1}{2}\log 0.25$$

Now, $\quad \log 0.25 = \bar{1}.4 = (-1 + 0.6) = -0.6$

$$\therefore \quad \mathrm{pH} = \tfrac{1}{2}(4.74) - \tfrac{1}{2}(-0.6) = 2.37 + 0.3 = 2.67$$

Sodium acetate is completely ionized, and 1.0.25 g dissolved in 100 cm^3 yields a solution containing 1.025/82 × 100/1000 mol dm^{-3} of acetate^- ions = 0.125 mol dm^{-3}.

∴ in the final solution,

$$[Ac^-] = 0.125 \text{ mol dm}^{-3}$$
$$[HAc] = [\text{initial HAc}] = 0.25 \text{ mol dm}^{-3}$$

According to the Henderson-Hasselbalch equation,

$$pH = pK_a + \log \frac{[\text{conjugate base}]}{[\text{acid}]}$$

$$= 4.74 + \log \frac{[Ac^-]}{[HAc]} = 4.74 + \log \frac{0.125}{0.25}$$

$$= 4.74 + \log 0.5$$

Now, $\log 0.5 = \bar{1}.7 \text{ (approx.)} = (-1 + 0.7) = -0.3$

∴ $pH = 4.74 - 0.3 = 4.44$

∴ increase in pH due to the addition of the salt = (4.44 – 2.67) = 1.77

Although the Henderson-Hasselbalch equation is of considerable use in preparing buffer mixtures, it is always advisable to measure the accurate pH of the final mixture *at the temperature at which it is to be used.* Fine adjustment of its pH can then be made by the addition of small quantities of strong acid or alkali.

THE DISSOCIATION OF POLYPROTIC WEAK ACIDS

So far, we have considered only monoprotic, weak acids whose molecules 'contain' only one proton that is available for the neutralization of a strong base. In biological media we frequently encounter polyprotic, weak acids (e.g. carbonic, phosphoric, citric acids), whose complete neutralization requires the addition of two or more equivalents of sodium hydroxide. The titration curve obtained when the triprotic, weak acid, phosphoric acid, is titrated with alkali.

Three distinct 'steps' are observable, each accomplished by the addition of one equivalent of alkali, and each evidencing a buffer zone centered on its 'half-equivalence' point. The first dissociation of phosphoric acid (pK_a about 2) must yield an amphiprotic conjugate base. This in turn dissociates as an acid (pK_a about 6.8) to form yet another amphiprotic conjugate base. This may in its turn dissociate (pK_a about 12) to form its conjugate base. Ascribing formulae and apparent pK_a values to these three dissociation 'steps', we obtain the following sequence for the dissociation of phosphoric acid (H_3PO_4).

1st dissociation $H_3PO_4 \rightleftharpoons H^+ + H_2PO_4^-$ .... $pK_a = 1.96$

2nd dissociation $H_2PO_4^- \rightleftharpoons H^+ + HPO_4^{2-}$ .... $pK_a = 6.80$

3rd dissociation $HPO_4^{2-} \rightleftharpoons H^+ + PO_4^{3-}$ .... $pK_a = 12$ (approx.)

Therefore, when we talk of the 2nd dissociation phosphoric acid (or of its 2nd acid dissociation constant), we are specifically referring to the dissociation of $H_2PO_4^-$ to yield its conjugate base HPO_4^{2-}. The multiple dissociation of the polyprotic, weak acid is thus seen to be a sequence of monoprotic dissociations yielding products of diminishing acid strength. The components of these reactions can be classified as follows: H_3PO_4 is a polyprotic (polybasic) acid; $H_2PO_4^-$ and HPO_4^{2-} are amphiprotic; and PO_4^{3-} is a polyacidic base.

The titration curve illustrates another important feature of the polyprotic, weak acid; namely that if its apparent pK_a values are sufficiently different, then the dissociation of any one acid intermediate is completed before its amphiprotic conjugate base is substantially dissociated (otherwise the 'steps' in the curve would tend to merge). At any one pH therefore, only the members of a single conjugate pair will be present in significant concentrations. At each of the 'equivalence points' one intermediate will greatly predominate over all others. At the first equivalence point when only the amphiprotic $H_2PO_4^-$ species is present, the pH will be 'half way' between the previous and subsequent 'half-equivalence' points; the pH will equal $\frac{pK_{a_1} + pK_{a_2}}{2}$ $= \frac{1.96 + 6.80}{2} = 4.38$. Similarly, at the second equivalence point, when the solution contains only HPO_4^{2-}, its pH will be $\frac{pK_{a_2} + pK_{a_3}}{2}$ $= \frac{6.8 + 12}{2} = 9.4$.

In a solution whose pH equals the pK_a value of one of the contributory weak acid intermediates, that weak acid and its conjugate base will be present in equal concentrations. Therefore, since

$$H_3PO_4 \underset{pK_{a_1}}{\rightleftharpoons} H^+ + H_2PO_4^- \underset{pK_{a_2}}{\rightleftharpoons} H^+ + HPO_4^{2-} \underset{pK_{a_3}}{\rightleftharpoons} H^+ + PO_4^{3-},$$

a solution whose pH is equal to pK_{a_1} will contain equal concentrations of H_3PO_4 and $H_2PO_4^-$, while a solution of pH equal to pK_{a_2} will contain equal concentrations of $H_2PO_4^-$ and HPO_4^{2-}, etc.

EXAMPLE 6.10 *To 100 cm^3 of 0.1 mol dm^{-3} phosphate buffer pH 7.1 were added:*

(a) 1 cm^3 of 1.0 mol dm^{-3} NaOH

(b) 5 cm^3 of 1.0 mol dm^{-3} NaOH

(c) 5 cm^3 of 1.0 mol dm^{-3} HCl

Calculate the new pH in each case. (Apparent pK_a values of phosphoric acid are 1.96, 6.8 and 12)

Solution

(*Note*: It is often helpful, in any calculation concerning a polyprotic weak acid, to draw a probable titration curve of the acid with alkali, positioning the 'steps' and equivalence points in relation to its apparent pK_a values (which will be that pH values at the medial points of the buffer zones). The ionic species that predominate at the 'equivalence' points should be indicated. This diagram now aids one in deciding which species are present in the solution at any given pH, and also helps one to identify the pK_a value to be employed in calculating the effect on the pH of the addition of strong acid, alkali or salts.)

In phosphate buffer of pH 7.1, the predominant conjugate pair will be $H_2PO_4^-$ and HPO_4^{2-}

$$\underset{\text{(acid)}}{H_2PO_4^-} \underset{pK_{a_2} = 6.8}{\rightleftharpoons} H^+ + \underset{\text{(conjugate base)}}{HPO_4^{2-}}$$

According to the Henderson-Hasselbalch equation, the pH of this solution is related to pK_{a_2} by the equation, $\text{pH} = pK_{a_2} + \log \dfrac{[HPO_4^{2-}]}{[H_2PO_4^-]}$. Therefore, since pH = 7.1, and $pK_{a_2} = 6.8$,

$$\log \frac{[HPO_4^{2-}]}{[H_2PO_4^-]} = 0.3, \quad \text{and} \quad \frac{[HPO_4^{2-}]}{[H_2PO_4^-]} = \text{antilog } 0.3 = 2.0$$

Thus the initial 100 cm^3 of 0.1 mol dm^{-3} phosphate buffer pH 7.1 is, in effect, a mixture of 66.7 cm^3 of 0.1 mol dm^{-3} HPO_4^{2-} with 33.3 cm^3 of 0.1 mol dm^{-3} $H_2PO_4^-$ (from $x + y = 100$ and $x/y = 2.0$).

Now,

1. Strong acid (HCl) would neutralize the conjugate base HPO_4^{2-} according to the equation,

$$HPO_4^{2-} + HCl \rightarrow H_2PO_4^- + Cl^-$$

The complete neutralization of 66.7 cm^3 of a 0.1 mol dm^{-3} solution of HPO_4^{2-} would consume 6.67 cm^3 of 1.0 mol dm^{-3} HCl.

2. Alkali (NaOH) would neutralize the acid component of this buffer mixture ($H_2PO_4^-$) according to the equation,

$$H_2PO_4^- + (Na^+)OH^- \rightarrow H_2O + (Na^+)HPO_4^-$$

The complete neutralization of 33.3 cm^3 of a 0.1 mol dm^{-3} solution of $H_2PO_4^-$ would consume 3.33 cm^3 of 1.0 mol dm^{-3} NaOH.

(a) *Addition of 1cm³ of 1 mol dm⁻³* NaOH

Let us express all quantities in terms of cm^3 of a solution containing 0.1 mol dm^{-3}.

∴ NaOH added ≡ 10 cm^3 of 0.1 mol dm^{-3} NaOH which will convert an equivalent quantity of $H_2PO_4^-$ into HPO_4^{2-}.

Initially	*Finally*
Volume = 100 cm^3	Volume = 101 cm^3
HPO_4^{2-} = 66.7	$HPO_4^{2-} \equiv (66.7 + 10) = 76.7$
$H_2PO_4^-$ = 33.3	$H_2PO_4^- \equiv (33.3 - 10) = 23.3$

The pH in the final solution is given by the equation

$$pH = 6.8 + \log \frac{[HPO_4^{2-}]}{[H_2PO_4^-]}$$

where,

$HPO_4^{2-}] \equiv 76.7$ cm^3 of a 0.1 mol dm^{-3} solution made up to 101 cm^3

$H_2PO_4^-] \equiv 23.3$ cm^3 of a 0.1 mol dm^{-3} solution made up to 101 cm^3

$$\therefore \quad \log \frac{[HPO_4^{2-}]}{[H_2PO_4^-]} = \log \frac{76.7}{23.3} = \log 3.3 = 0.52$$

$$\therefore \quad pH = 6.8 + 0.52 = 7.32$$

b) *Addition of 5 cm³ of 1 mol dm⁻³* NaOH

As noted above, addition to the buffer mixture of 3.33 cm^3 of 1 mol dm^{-3} NaOH is sufficient to convert it into the equivalent of 100 cm^3 of a 0.1 mol dm^{-3} solution of HPO_4^{2-}. Thus, when 5 cm^3 of 1 mol dm^{-3} NaOH is added, 1.67 cm^3 of 1 mol dm^{-3} NaOH remains to be neutralized by the HPO_4^{2-} acting as an acid when

$$\underset{\text{(acid)}}{HPO_4^{2-}} + (Na^+)OH^- \rightleftharpoons \underset{\text{(conjugate base)}}{PO_4^{3-}} + Na^+ + H_2O$$

The pH of the resulting solution will therefore be determined by the 3rd dissociation constant of phosphoric acid,

$$HPO_4^{2-} \underset{pK_{a_3} = 12}{\rightleftharpoons} H^+ + PO_4^{3-}$$

and in a solution containing the conjugate pair HPO_4^{2-} and PO_4^{3-},

$$pH = 12.0 + \log \frac{[PO_4^{3-}]}{[HPO_4^{2-}]}$$

Let us express all quantities in terms of cm^3 of a solution containing 0.1 mol dm^{-3}.

Then, NaOH added ≡ 50, of which 33.3 are consumed in converting all the $H_2PO_4^-$ initially present, into HPO_4^{2-}. The remaining 16.7 of NaOH will convert an equivalent quantity of HPO_4^{2-} into PO_4^{3-}.

Initially	*After addition of 33.3 NaOH*	*After further addition of 16.7 NaOH*
PO_4^{3-} = negligible	PO_4^{3-} = negligible	PO_4^{3-} = 16.7
HPO_4^{2-} = 66.7	HPO_4^{2-} = 100	HPO_4^{2-} = (100 – 16.7) = 83.3
$H_2PO_4^-$ = 33.3	$H_2PO_4^-$ = negligible	$H_2PO_4^-$ = negligible

∴ in the final solution formed by the addition of 5 cm^3 of 1 mol dm^{-3} NaOH,

$[PO_4^{3-}]$ ≡ 16.7 cm^3 of a 0.1 mol dm^{-3} solution made up to 105 cm^3

$[HPO_4^{2-}]$ ≡ 83.3 cm^3 of a 0.1 mol dm^{-3} solution made up to 105 cm^3

$$\therefore \quad \log \frac{[PO_4^{3-}]}{[HPO_4^{2-}]} = \log \frac{16.7}{83.3} = \log 0.2 = \bar{1}.3 = -0.7$$

∴ in this solution, $pH = 12.0 + (-0.7) = 11.3$

(c) *Addition of 5 cm^3 of 1 mol dm^{-3}* HCl

As noted above, it would take 6.67 cm^3 of 1 mol^{-3} HCl to neutralize completely the HPO_4^{2-} (conjugate base) present in 100 cm^3 of 0.1 mol dm^{-3} phosphate buffer pH 7.1. Addition of only 5 cm^3 of 1 mol

dm^{-3} HCl would produce a solution in which the conjugate pair is still $H_2PO_4^-$ and HPO_4^{2-}, where

$$H_2PO_4^- \underset{pK_{a2} = 6.8}{\rightleftharpoons} HPO_4^{2-} + H^+, \text{ and the pH} = 6.8 + \log \frac{[HPO_4^{2-}]}{[H_2PO_4^-]}.$$

Let us express all quantities in terms of cm^3 of a solution containing 0.1 mol dm^{-3}.

Then HCl added ≡ 50, which is entirely consumed in converting an equivalent quantity of HPO_4^{2-} into $H_2PO_4^-$ according to the equation,

$$\underset{\text{(base)}}{HPO_4^{2-}} + HCl \rightarrow \underset{\text{(conjugate acid)}}{H_2PO_4^-} + Cl^-$$

Initially	*Finally*
$HPO_4^{2-} = 66.7$	$HPO_4^{2-} = (66.7 - 50) = 16.7$
$H_2PO_4^- = 33.3$	$H_2PO_4^- = (33.3 + 50) = 83.3$

∴ in the final solution formed by the addition of 5 cm of 1 mol dm^{-3} HCl,

$[HPO_4^{2-}]$ ≡ 16.7 cm^3 of a 0.1 mol dm^{-3} solution made up to 105 cm^3

$[H_2PO_4^-]$ ≡ 83.3 cm^3 of a 0.1 mol dm^{-3} solution made up to 105 cm^3

$$\therefore \qquad \log \frac{[HPO_4^{2-}]}{[H_2PO_4^-]} = \log \frac{16.7}{83.3} \doteq -0.7$$

∴ in this solution, $\quad pH = 6.8 + (-0.7) = 6.1$

pH INDICATORS

A number of substances are available, known as 'acid-base indicators' or 'pH indicators', whose colour in solutions more acid than a characteristic pH differs markedly from their colour in solutions more alkaline than this. Very conveniently, this colour change is not abrupt but is gradual over a certain range of pH (usually of about 2 pH units), which is called the 'colour change interval' of the indicator. The pH range over which an indicator responds in this fashion is characteristic, evidently being determined by its chemical constitution.

In general, pH indicators are very weak organic acids or bases, and their behaviour is explained by the fact that the Brönsted acid form of the indicator is strikingly different in colour from its conjugate base, i.e.

$$\underset{\text{(colour 1)}}{HIn} \rightleftharpoons H^+ + \underset{\text{(colour 2)}}{In^-}$$

The Henderson-Hasselbalch equation is applicable to their aqueous solutions, and if these were titrated with strong acid or alkali as required, the pH at the half-equivalence point would be the pK_a value of the indicator (often termed the pK_{In}). At the half-equivalence point (known as the 'change-over point' in the case of an indicator), the concentrations of the indicator weak acid and of its conjugate base, or of the indicator weak base and of its conjugate acid, will be equal. Therefore the colour adopted by the indicator at its change-over point is that of an equal mixture of the acid and alkaline colours of that indicator. These colours are so intense that a pH indicator may be used in such low concentration that it confers negligible buffer capacity on the solution to which it is added. It should be possible to assay spectrophotometrically the concentrations of HIn and In- present in a solution whose pH falls within the colour change interval of the indicator. The pH of this solution could then be calculated from the equation, $pH = pK_{a\,(In)} + \log [In^-]/[HIn]$. Much more often, pH indicators are used to denote visually the approximate pH of a solution.

The above explanation applies to those indicators that are monoprotic, weak acids or monoacidic weak bases. In fact, some indicators are polyprotic with more than one pK_a value, others exist in tautomeric forms of differing colour, and still others adopt colours that depend in part on the ionic strength of the solution in which they are employed. Such complications need not concern us here.

Indicators are frequently used in volumetric analyses when the concentration of an acid solution is to be determined by titration against a standard concentration of base (or vice versa). To obtain an accurate 'end point' in such a titration, an indicator should be used whose pK_a is very nearly equal to the pH at the equivalence point. The more marked the pH change as equivalence is approached, the sharper will be the visual colour change of the indicator. Furthermore, a greater latitude is allowable in the choice of indicators for a titration in which the pH rapidly changes as the end point is approached. Thus, in the strong acid-alkali titration, although an indicator of $pK_a = 7$ would theoretically be best, others whose pK_a values fall within the range 5 to 9 would be acceptable. In the weak acid-alkali titration, the pH change about the equivalence point is not so extensive, and the choice of indicators is restricted to those whose pK_a values fall within a smaller range–perhaps within 1 unit on either side of the pH value at the true end point. The anticipated pH at the end point of an acid-base titration can be calculated as described below.

THE pH's OF DILUTE, AQUEOUS SOLUTIONS OF SALTS

Many salts form acid or alkaline aqueous solutions. The dissociation theory explains this as being due to hydrolysis of the salt by interaction with one or other of the ions of water, which then creates an imbalance in the residual concentrations of H^+ and OH^- ions. The extent of this hydrolysis is expressed as a hydrolysis constant whose value is correlated with the strengths of the parent acid and base by whose interaction the salt was formed.

The Brönsted and Lowry concept offers a much clearer explanation, attributing the effect of the salt on the pH of its aqueous medium to the relative acid and basic strengths of its component ions. This view is greatly to be preferred, if only because it avoids the needlessly confusing use for the term 'hydrolysis' to describe what is, after all, a neutralization process.

1. *A salt both of whose ions are 'neutral'*, e.g. sodium chloride

This type of salt is created by the interaction of a strong acid with a strong base (generally OH^- added as an alkali). Its ions are negligibly acidic and basic and yield an aqueous solution of neutral pH (7.0).

2. *A salt, one of whose ions is basic, the other 'neutral'*, e.g. sodium acetate

This type of salt is formed by the neutralization of a weak acid by a strong base (or alkali). One of its ions will be the significantly strong conjugate base of the parent weak acid, the other will be negligibly acidic or basic. The pH of its aqueous solution will be alkaline due to proton-acceptance from water by the basic ion; e.g. in the case of an aqueous solution of sodium acetate,

$$(Na^+)Ac^- + H_2O \rightleftharpoons HAc + OH^-$$

Thus the pH of a dilute aqueous solution of this type of salt will be given by the equation applicable to the solution of any weak base i.e.

$$pH = pK_w - \frac{1}{2}pK_b + \frac{1}{2}\log c$$

where K_b is the basic dissociation constant of the salt's basic ion, c is its concentration in mol dm^{-3}, and K_w is the ion product constant of water.

EXAMPLE 6.11 *What, approximately, is the pH of a 0.02 mol dm^{-3} aqueous solution of sodium propionate at 298 K? (pK_a of propionic acid = 4.85.)*

Solution

Of the component ions of sodium propionate, Na^+ is 'neutral' and propionate$^-$ being the conjugate base of the relatively weak acid, propionic acid, is significantly basic. Thus the pH of an aqueous solution of sodium propionate will be given by the equation,

$$pH = pK_w - \frac{1}{2} pK_b + \frac{1}{2} \log c$$

where K_b is the basic dissociation constant of the propionate$^-$ ion present in c mol dm^{-3}.

The pK_b of propionate$^-$ is related to the pK_a of its conjugate acid (propionic acid) according to the equation,

$$pK_b = pK_w - pK_a$$

$$\therefore \quad pK_b \text{ of propionate}^- \text{ anion} = (14 - 4.85) = 9.15$$

Since 1 mol of sodium propionate dissolved in 1 dm^3 of water yields a solution containing 1 mol dm^{-3} of propionate$^-$ anions, the concentration of propionate$^-$ in a 0.02 mol dm^{-3} solution of sodium propionate = 0.02 mol dm^{-3} = c.

$\therefore$ pH of 0.02 mol dm^{-3} sodium propionate at 298 K

$$= pK_w - \frac{1}{2}(9.15) + \frac{1}{2} \log 0.02$$

Now, $\log 0.02 = \bar{2}.3 \text{ (approx.)} = (-2 + 0.3) = -1.7$

$$\therefore \quad pH = 14 - \frac{1}{2}(9.15) + \frac{1}{2}(-1.7)$$

$$= 14 - 4.58 - 0.85$$

$$= 8.57$$

---

3. ***A salt, one of whose ions is acidic and the other 'neutral'***, **e.g. ammonium chloride**

This type of salt is formed by the interaction of a weak base with a strong acid. One of its ions will be the significantly strong conjugate acid of the parent weak base, the other will be the infinitely weak conjugate base of the parent strong acid. Aqueous solutions of these salts are acid, their actual pH depending on the concentration of salt present, as well as on the pK_a value of the acid ion, i.e. $pH = \frac{1}{2} pK_a - \frac{1}{2} \log c$, where pK_a is the pK_a of the acid ion and c is its concentration in mol dm^{-3}.

EXAMPLE 6.12 *Calculate the pH of a 0.05 mol dm^{-3} solution of 'tris' hydrochloride at 298 K. ('Tris' is the weak base, tris (hydroxymethyl)-aminomethane, $(CH_2OH)_3.C.NH_2$, pK_b = 5.92 at 298 K.)*

Solution

The salt 'tris' hydrochloride contains the two ions, $(CH_2OH)_3.C.NH_3^+$ and Cl^-. The former is the significantly strong conjugate acid of the weak base 'tris', the latter is the infinitely weak conjugate base of the strong acid HCl and is therefore 'neutral'. Thus the pH in an aqueous solution of 'tris' hydrochloride of concentration c mol dm^{-3}, will be given by the equation,

$$pH = \frac{1}{2} pK_a - \frac{1}{2} \log c$$

where K_a is the acid dissociation constant of the acid cation $(CH_2OH)_3.C.NH_3^+$.

The pK_b of 'tris' and the pK_a of its conjugate acid cation are related according to the equation, $pK_a = pK_w - pK_b$. Therefore the pK_a of the acid cation = (14 – 5.92) = 8.08. In a 0.05 mol dm^{-3} solution of 'tris' hydrochloride the [cation] = 0.05 mol dm^{-3}.

$$\therefore \quad \frac{1}{2} \log c = \frac{1}{2} \log 0.05 = \frac{1}{2} (\bar{2}.7) = \frac{1}{2} (-1.3) = -0.65$$

∴ pH of the 0.05 mol dm^{-3} solution of 'tris' hydrochloride

$$= \frac{1}{2} (8.08) - (-0.65) = 4.04 + 0.65 = 4.69$$

---

4. *A salt, one of whose ions is acidic and the other basic*, e.g. ammonium acetate

This type of salt is formed by the interaction of a weak acid with a weak base. Its solution in water may be acid, neutral or alkaline in pH, depending on the relative strengths of its acid and basic ions. In the example mentioned, i.e. ammonium acetate, the acid strength of the NH_4^+ ion ($pK_a = 9.26$) is exactly matched by the basic strength of the acetate$^-$ ion ($pK_b = 9.26$), and the mutual neutralization of these ions means that a dilute aqueous solution of ammonium acetate, whatever its actual concentration may be, will possess a pH of 7.

5. *A salt, one of whose ions is amphiprotic, the other 'neutral'*, e.g. sodium bicarbonate, sodium dihydrogen phosphate, or a diamine monohydrochloride.

This type of salt is formed when one 'acidic group' of a weak, diprotic acid is neutralized with alkali, or when one 'basic group' of a diacidic base is neutralized with strong acid. The pH of its aqueous solution will depend on the relative acid and basic strengths of its amphiprotic ion. These may be assessed as follows:

Basic strength: as the acid dissociation constant of the conjugate acid formed from the amphiprotic ion acting as a base $= K_{a_1}$

Acid strength: as the acid dissociation constant of the amphiprotic ion $= K_{a_2}$

It can be shown that when the values of K_{a_1} and K_{a_2} are small in comparison with the concentration of the amphiprotic ion, the pH of its dilute aqueous solution will be given by the equation

$$pH = \frac{pK_{a_1} + pK_{a_2}}{2}$$

EXAMPLE 6.13 *What is the pH of a 0.03 mol dm^{-3} solution of sodium bicarbonate at 298 K? (pK_{a_1} and pK_{a_2} of carbonic acid = 6.37 and 10.25.)*

Solution

A sodium bicarbonate solution contains 'neutral' Na^+ ions and amphiprotic HCO_3^- ions.

As an acid, HCO_3^- dissociates as follows,

$$HCO_3^- \rightleftharpoons H^+ + CO_3^{2-} \; ... \; K_a \text{ of } HCO_3^- = K_{a_2} \text{ of } H_2CO_3$$

As a base, HCO_3^- associates with protons as follows,

$$HCO_3^- + H^+ \rightleftharpoons H_2CO_3 \; ... \; K_a \text{ of conjugate acid} = K_{a_1} \text{ of } H_2CO_3$$

Since pK_{a_1} of carbonic acid = 6.37, K_{a_1} must be between 10^{-6} and 10^{-7} and is therefore negligibly small in comparison with the concentration of HCO_3^- ions in the solution (0.03 mol dm^{-3}). Thus the pH of the sodium bicarbonate solution will be given approximately by the equation,

$$pH = \frac{pK_{a_1} + pK_{a_2}}{2}$$

$$\therefore \quad pH = \frac{6.37 + 10.25}{2} = \frac{16.62}{2} = 8.31$$

We will consider 'multidissociable' compound and their amphiprotic ions in more detail when we discuss the acidic and basic properties of amino acids and proteins in Chapter 6.

We are now in a position to calculate the pH at the equivalence point in the titration of an acid with a base, and hence to select indicators which will demonstrate the correct end point.

EXAMPLE 6.14 *You are provided with the following pH indicators: methyl orange (pK_a = 3.7), methyl red (pK_a = 5.1), bromothymol blue (pK_a = 7.0), and α-naphthylphthalein (pK_a = 8.4). Which of these would you select to demonstrate the end point in the following titrations, in which the reagents are used in 0.2 mol dm^{-3} concentration?*

(a) hydrochloric acid with sodium hydroxide

(b) formic acid with sodium hydroxide

(c) pyridine with hydrochloric acid

(Assume pK_a of formic acid = 3.75; pK_b of pyridine = 8.85.)

Solution

(a) *HCl with NaOH*

At the end point of the titration, the solution contains the 'neutral' ions Na^+ and Cl^- and neutralization would be complete,

$$HCl + (Na^+)OH^- \rightarrow H_2O + (Na^+)Cl^-$$

The pH at the equivalence point would be 7, as in a solution of sodium chloride, and, of the indicators listed, bromothymol blue (pK_a = 7) should be chosen. (Yet, since the pH change is very great as the equivalence point is approached, methyl red (pK_a = 5.1) or α-naphthylphthalein (pK_a = 8.4) might also be used.)

(b) *Formic acid (HCOOH) with NaOH*

At the end point of this titration, the solution contains the 'neutral' ion Na^+ and the appreciably basic ion $HCOO^-$ (i.e. the conjugate base of the weak acid HCOOH), for

$$HCOOH + (Na^+)OH^- \rightleftharpoons H_2O + Na^+ + HCOO^-$$

Thus the pH of the solution at the end point will be a function of the pK_b of the $HCOO^-$ ion and of its concentration, i.e.

$$pH = pK_w - \tfrac{1}{2}pK_b + \tfrac{1}{2}\log c$$

Now

$$pK_b \text{ of } HCOO^- = pK_w - pK_a \text{ of } HCOOH = 14 - 3.75 = 10.25$$

In the final solution produced by mixing the 0.2 mol dm^{-3} formic acid with an equal volume of 0.2 mol dm^{-3} NaOH (necessary to attain the equivalence point), the $[HCOO^-] = 0.1$ mol dm^{-3} = [HCOOH] had this been merely diluted to twice its original volume.

$$\therefore \quad pH = 14 - \tfrac{1}{2}(10.25) + \tfrac{1}{2}\log 0.1$$

$$\tfrac{1}{2}\log 0.1 = \tfrac{1}{2}(-1) = -0.5$$

$$\therefore \quad pH = 14 - 5.13 - 0.5 = 8.35$$

Therefore, α-naphthylphthalein ($pK_a = 8.4$) would best indicate this end point.

(c) *Pyridine with HCl*

The neutralization process can be written as,

$$C_5H_5N + HCl \rightleftharpoons C_5H_5N^+H + Cl^-$$

(weak base) (conjugate acid) = pyridinium ion

The solution at the end point in this titration would contain the 'neutral' Cl^- ion and the quite strongly acidic pyridinium ion,

pK_a of the pyridinium ion = $pK_w - pK_b$ of its conjugate base, pyridine.

Thus the pH of the solution at the end point will be a function of the

pK_a of the pyridinium ion and its concentration; i.e.

$$pH = \frac{1}{2} pK_a - \frac{1}{2} \log c$$

where $pK_a = 14 - 8.85 = 5.15$, and c = [pyridinium$^+$] = 0.1 mol dm^{-3} (see [$HCOO^-$] above).

$$\therefore \quad pH = \frac{1}{2}(5.15) - \frac{1}{2} \log 0.1 = 2.58 - (-0.5)$$
$$= 2.58 + 0.5 = 3.08$$

Thus methyl orange (pK_a = 3.7) would, of those indicators given, be the best able to demonstrate this end point.

The equations which relate the pH of a solution of salt to the pK_a and pK_b values of its component ions, and to their concentration, have been recorded chiefly for reference purposes.

THE pH-DEPENDENT IONIZATION OF AMINO ACIDS

All amino acids are amphiprotic compounds and the majority of those present in protein hydrolysates conform to the general formula R.CH.(NH_2)COOH. The amino group is known to be weakly basic, associating with protons to form a positively charged ion, which in aliphatic amines has a pK_a of about 10.6 ($R—NH_2 + H^+ \rightleftharpoons R—NH_3^+$). The carboxylic group, on the other hand , is weakly acidic, and when present as the chain terminating group in a fatty acid it dissociates, with a pK_a of about 4.5, to yield a negatively charged ion ($R—COOH \rightleftharpoons H^+ + RCOO^-$). In amino acids, the attachment of the amino and carboxylic groups to the same (α) carbon atom enhances their acidity, the pK_a of the α-carboxyl group now being about 1.7 to 2.4, whilst the pK_a of the α-NH_3^+ group is about 9 to 10.5. Thus all amino acids will be ionized in aqueous solution in a manner determined by the prevailing pH.

Those amino acids whose side chain (R) bear no dissociable groups may exist in three ionic forms as shown:

$$\underset{\text{cation}}{\begin{array}{c} R \\ | \\ CHNH_3^+ \\ | \\ COOH \end{array}} \overset{K_{a_1}}{\rightleftharpoons} \underset{\text{zwitterion}}{\begin{array}{c} R \\ | \\ CHNH_3^+ \\ | \\ COO^- \end{array}} \overset{K_{a_2}}{\rightleftharpoons} \underset{\text{anion}}{\begin{array}{c} R \\ | \\ CHNH_2 \\ | \\ COO^- \end{array}}$$

increasing pH →

At low pH's the amino acid will exist as a cation in which only the α-amino group is ionized. As this solution is made more alkaline, so will the α-carboxyl group dissociate and become ionized. The product is the dipolar form of the amino acid which bears no net electrical charge. This is called the *zwitterion* (i.e. hybrid ion). Raising the pH of the solution still further, causes the α-NH_3^+ group to dissociate and thereby lose its charge. The product which predominates at very alkaline pH's is therefore the anion in which only the α-carboxyl group is ionized. At the pH equal to the pK_a value of the carboxyl group (pK_{a_1} of this amino acid), cation and zwitterion will coexist in equal concentration. Similarly, at the pH equal to the pK_a of the $-NH_3^+$ group (pK_{a_2} of this amino acid), zwitterion and anion will be present in equal concentration. In its crystalline state, or in solution in pure water, the amino acid will be predominantly in its zwitterionic form.

Take for example, the simplest amino acid, glycine ($H_2N.CH_2.COOH$). In solution in pure water it exists as the zwitterion, $^+H_3N.CH_2.COO^-$, and will behave as a base when titrated with hydrochloric acid.

The neutralization process can be represented by the equation,

$$^+H_3N.CH_2.COO^- + HCl \rightleftharpoons {}^+H_3N.CH_2.COOH + Cl^-$$

and the titration curve is that of a weak base; the pH at half equivalence is the apparent pK_a value of the carboxylic group. When titrated with alkali, the zwitterion of glycine behaves as a weak acid, the neutralization equation being,

$$^+H_3N.CH_2.COO^- + (Na^+)OH^- \rightleftharpoons H_2N.CH_2.COO^- + Na^+ + H_2O$$

Now the pH at the half-equivalence point equal the apparent pK_a value of the α-NH_3^+ group. Thus the acid group of the zwitterion of glycine is its α-NH_3^+ group, whilst its basic group is its α-COO^- group. The pH of a dilute aqueous solution of this zwitterion, as of any other amphiprotic substance , is determined by its basic and acidic strengths, and is given by the equation, $pH = \dfrac{pK_{a_1} + pK_{a_2}}{2}$, where K_{a_1} and K_{a_2} are the acid dissociation constants of its basic and acidic groups respectively. For glycine, pK_{a_1} equal 2.4, and pK_{a_2} is 9.8, so that the pH of a dilute aqueous solution of glycine is $(2.4 + 9.8)/2 = 6.1$. In this solution the zwitterion predominates, and pH 6.1 is the *isoionic point* of glycine where the number of negative charges on the molecule produced by protolysis equals the number of positive charges acquired by proton gain. It is also the *isoelectric point* of glycine, for at pH 6.1 its molecules carry no net charge and are electrophoretically immobile.

The ionization of glycine can therefore be summarized as follows,

$$\underset{\text{cation}}{\begin{matrix} NH_3^+ \\ | \\ CH_2 \\ | \\ COOH \end{matrix}} \underset{}{\overset{pK_{a_1} = 2.4}{\rightleftharpoons}} \underset{\text{zwitterion}}{\begin{matrix} NH_3^+ \\ | \\ CH_2 \\ | \\ COO^- \end{matrix}} \overset{pK_{a_2} = 9.8}{\rightleftharpoons} \underset{\text{anion}}{\begin{matrix} NH_2 \\ | \\ CH_2 \\ | \\ COO^- \end{matrix}}$$

$\Leftarrow$ Acid ——— pH 6.1 ——— Alkali $\Rightarrow$

(isoionic point)

(isoelectric point)

EXAMPLE 6.15 *1.068 g of a crystalline α-amino acid with pK_{a_1} and pK_{a_2} values of 2.4 and 9.7 gave a solution of pH 10.4 when dissolved in 100 cm^3 of 0.1 mol dm^{-3} sodium hydroxide. Calculate the molecular weight of the amino acid. (Atomic weights: C = 12, N = 14, O = 16, and H = 1.)*

Solution

Let the amino acid be represented by the formula $R.CH(NH_2)COOH$. Then it will dissociate in aqueous solution as follows:

$$\underset{\text{cation}}{\begin{matrix} R \\ | \\ CHNH_3^+ \\ | \\ COOH \end{matrix}} \overset{pK_{a_1} = 2.4}{\rightleftharpoons} \underset{\text{zwitterion}}{\begin{matrix} R \\ | \\ CHNH_3^+ \\ | \\ COO^- \end{matrix}} \overset{pK_{a_2} = 9.7}{\rightleftharpoons} \underset{\text{anion}}{\begin{matrix} R \\ | \\ CHNH_2 \\ | \\ COO^- \end{matrix}}$$

Alkali $\Rightarrow$

The crystalline amino acid will be in its zwitterionic form, and when added to alkali it will behave as a weak acid, dissociating according to the equation:

$$\underset{\substack{\text{zwitterion} \\ \text{(acid)}}}{\begin{matrix} R \\ | \\ CHNH_3^+ \\ | \\ COO^- \end{matrix}} \overset{pK_{a_2} = 9.7}{\rightleftharpoons} \underset{\substack{\text{anion} \\ \text{(conjugate base)}}}{\begin{matrix} R \\ | \\ CHNH_2 \\ | \\ COO^- \end{matrix}} + H^+$$

The quantity of alkali used was evidently insufficient to neutralize the zwitterion completely, since the pH of the final solution is within 1 pH unit of the pK_a value for this dissociation. Therefore the pH of the final solution defines its composition, for, by the Henderson-Hasselbalch equation,

$$\text{pH} = \text{p}K_a + \log \frac{[\text{conjugate base}]}{[\text{acid}]}$$

so that, $$10.4 = 9.7 + \log \frac{[\text{anion}]}{[\text{zwitterion}]}$$

whence, $\log \dfrac{[\text{anion}]}{[\text{zwitterion}]} = 0.7$, and $\dfrac{[\text{anion}]}{[\text{zwitterion}]} = \text{antilog } 0.7 = 5$

Since $\text{zwitterion}^{\pm} + (\text{Na}^+)\text{OH}^- \rightarrow \text{anion}^- + \text{Na}^+ + H_2O$
the [anion] in the final solution

$$\equiv [\text{Na}^+ \text{OH}^-] \text{ 'consumed'} = 0.1 \text{ mol dm}^{-3}$$

$\therefore$ $[\text{anion}] = 0.1 \text{ mol dm}^{-3}$

$\therefore$ [zwitterion] being equal to $\frac{1}{5}$[anion] $= 0.02 \text{ mol dm}^{-3}$

$\therefore$ [total amino acid] = [anion] + [zwitterion] $= 0.12 \text{ mol dm}^{-3}$

$\therefore$ [total amino acid] $= 0.12 \text{ mol dm}^{-3}$

Thus 100 cm^3 of 0.12 mol dm^{-3} amino acid solution contains 1.068 g of amino acid

$\therefore$ 1 dm^3 of 1 mol dm^{-3} solution contains

$$\left(1.068 \times \frac{1000}{100} \times \frac{1}{0.12}\right) \text{g} = 89 \text{ g}$$

$\therefore$ molecular weight of the amino acid = 89

The Ionization of Amino Acids which carry Supplementary Dissociable Groups

The naturally-occurring amino acids have distinctive side chains (R–), some of which carry additional basic or acidic groups that give the molecule an overall basic or acidic character; e.g.[†]

The presence of supplementary, dissociable groups complicates the ionic behaviour of an amino acid. This is evident in the titration curves that are obtained when glutamic acid is titrated with (a) HCl and (b) NaOH.

†

$(CH_3)_2CH$	$COOH.(CH_2)_2$	$NH_2.(CH_2)_4$
$CHNH_2$	$CHNH_2$	$CHNH_2$
$COOH$	$COOH$	$COOH$
valine	**glutamic acid**	**lysine**
('neutral')	**(acidic')**	**('basic')**

The aqueous solution of glutamic acid again contains the zwitterionic form of the amino acid but its pH is 3.1, which makes it considerably more acid than the solution of the 'neutral' amino acid, glycine. Comparison of the titration curves of glutamic acid with those of glycine. Comparison of the titration curves of glutamic acid with those of glycine, shows that tow equivalents of alkali are required to neutralize glutamic acid. The new dissociation step apparent in the titration curve of glutamic acid with NaOH demonstrates the existence of a buffer zone centered on the half-equivalence point of pH 4.1. This is the apparent pK_a of the γ-carboxyl group so that glutamic acid ionizes according to the following dissociation sequence,

$$
\begin{matrix} COOH \\ | \\ (CH_2)_2 \\ | \\ CHNH_3^+ \\ | \\ COOH \end{matrix}
\quad \underset{}{\overset{pK_{a_1} = 2.1}{\rightleftharpoons}} \quad
\begin{matrix} COOH \\ | \\ (CH_2)_2 \\ | \\ CHNH_3^+ \\ | \\ COO^- \end{matrix}
\quad \overset{pK_{a_2} = 4.1}{\rightleftharpoons} \quad
\begin{matrix} COO^- \\ | \\ (CH_2)_2 \\ | \\ CHNH_3^+ \\ | \\ COO^- \end{matrix}
\quad \overset{pK_{a_3} = 9.5}{\rightleftharpoons} \quad
\begin{matrix} COO^- \\ | \\ (CH_2)_2 \\ | \\ CHNH_2 \\ | \\ COO^- \end{matrix}
$$

zwitterion

← Acid—pH 3.1 ———— Alkali →

When crystalline glutamic acid is dissolved in water, a proportion of its zwitterions will dissociate, and the pH will fall due to the ionization of their γ-carboxyl groups. Equilibrium is achieved when the pH has fallen to pH 3.1, with the bulk of the amino acid still present in its zwitterionic form. When glutamic acid is dissolved in buffers of pH 2.2 to 4.0, its zwitterionic form will again predominate, though in buffers of pH about pH 7 the amino acid will be present almost entirely as the net negatively charged species, $^-OOC.(CH_2)_2.CH(NH_3^+).COO^-$.

The 'basic' amino acid, lysine, behaves quite differently when it is titrated with HCl and NaOH.

Lysine is a diamino-monocarboxylic acid which yields an alkaline solution in pure water (isoionic point of lysine is pH 10). Two equivalents of HCl are consumed in the titration of lysine which commences at this isoionic point, and the values of pK_a (apparent) supplied by this titration (9.2 and 2.2) are the values to be expected of its α-NH_3 group and α-carboxyl group respectively. On the other hand, one equivalent of NaOH is consumed in the titration of lysine from its isoionic point, the pH at half equivalence being 10.8. It follows that 10.8 must be the apparent pK_{a_3} of lysine, characteristic of its ε-NH_3^+ group. It further

NH_3^+		NH_3^+		NH_3^+		NH_2
$(CH_2)_4$	pK_{a_1} = 2.2 $\rightleftharpoons$	$(CH_2)_4$	pK_{a_2} = 9.2 $\rightleftharpoons$	$(CH_2)_4$	pK_{a_3} = 10.8 $\rightleftharpoons$	$(CH_2)_4$
$CHNH_3^+$		$CHNH_3^+$		$CHNH_2$		$CHNH_2$
COOH		COO^-		COO^-		COO^-
				zwitterion		

⟵ Acid ——— pH 10 — Alkali ⟶

follows that it will be this ε-NH_3^+ group that will be ionized in the zwitterion of lysine, for the amino acid must dissociate.

In buffer at pH 7, at which pH glycine carries no net charge and glutamic acid is negatively charged overall, the 'basic' amino acid, lysine, will carry a net positive charge since predominant in its solution will be the cation $^+H_3N.(CH_2)_4.CH(NH_3^+).COO^-$.

Note that the isoionic point is always determined by the pK_a values that define the basic and acidic strengths of the *zwitterion*. Thus, for glutamic acid the isoionic point is $\dfrac{pK_{a_1} + pK_{a_2}}{2}$, but for lysine the isoionic point is $\dfrac{pK_{a_2} + pK_{a_3}}{2}$. Yet, although we calculate an isoionic *point* in this way, the fact is, that when the contributory pK_a values are very different, there exists a considerable range of intermediate pH values in which the molecule will be virtually entirely in its zwitterionic form.

EXAMPLE 6.16 *Calculate the isoionic points of the following amino acids*:

(*a*) *aspartic acid* (*p*K_{a_1} = *1.99*; *p*K_{a_2} = *3.90*; *p*K_{a_3} = *9.90*);

(*b*) *arginine* (pK_{a_1} = *1.82;* pK_{a_2} = *8.99;* pK_{a_3} = *12.48);*
(*c*) *histidine* (pK_{a_1} = *1.80;* pK_{a_2} = *6.04;* pK_{a_3} = *9.33);*
(*d*) *tyrosine* (pK_{a_1} = *2.20;* pK_{a_2} = *9.11;* pK_{a_3} = *10.13).*

Solution

(*Note*: It is always advisable to obtain the formula of the compound and then to write out equations for the sequential dissociations that it undergoes, using your knowledge of the probable pK_a values of its dissociable groups to place these in the correct order of diminishing acid strength. Then the zwitterion, and the pK_a values that define its acidic and basic strengths, are readily identified.)

(a) *Aspartic acid*. Formula: $HOOC.CH_2.CH(NH_2).COOH$

This dicarboxylic, monoamino acid will dissociate as follows:

$$
\begin{array}{c}
\text{COOH} \\ | \\ \text{CH}_2 \\ | \\ \text{CHNH}_3^+ \\ | \\ \text{COOH}
\end{array}
\underset{}{\overset{pK_{a_1} = 1.99}{\rightleftharpoons}}
\begin{array}{c}
\text{COOH} \\ | \\ \text{CH}_2 \\ | \\ \text{CHNH}_3^+ \\ | \\ \text{COO}^- \\ \text{zwitterion}
\end{array}
\overset{pK_{a_2} = 1.9}{\rightleftharpoons}
\begin{array}{c}
\text{COO}^- \\ | \\ \text{CH}_2 \\ | \\ \text{CHNH}_3^+ \\ | \\ \text{COO}^-
\end{array}
\overset{pK_{a_3} = 9.9}{\rightleftharpoons}
\begin{array}{c}
\text{COO}^- \\ | \\ \text{CH}_2 \\ | \\ \text{CHNH}_2 \\ | \\ \text{COO}^-
\end{array}
$$

$$\therefore \quad \text{isoionic point} = \frac{pK_{a_1} + pK_{a_2}}{2} = \frac{1.99 + 3.9}{2} = 2.95$$

(b) *Arginine*. Formula: $H_2N.\underset{\underset{\text{NH}}{||}}{C}.NH.(CH_2)_3.CH(NH_2).COOH$

Because of its guanidyl group, this is a 'basic ' amino acid that will dissociate as follows:

$$
\begin{array}{c}
\text{NH}_2 \\ | \\ \text{C}{=}\text{NH}_2^+ \\ | \\ \text{NH} \\ | \\ (\text{CH}_2)_3 \\ | \\ \text{CHNH}_3^+ \\ | \\ \text{COOH}
\end{array}
\overset{pK_{a_1} = 1.82}{\rightleftharpoons}
\begin{array}{c}
\text{NH}_2 \\ | \\ \text{C}{=}\text{NH}_2^+ \\ | \\ \text{NH} \\ | \\ (\text{CH}_2)_3 \\ | \\ \text{CHNH}_3^+ \\ | \\ \text{COO}^-
\end{array}
\overset{pK_{a_2} = 8.99}{\rightleftharpoons}
\begin{array}{c}
\text{NH}_2 \\ | \\ \text{C}{=}\text{NH}_2^+ \\ | \\ \text{NH} \\ | \\ (\text{CH}_2)_3 \\ | \\ \text{CHNH}_2 \\ | \\ \text{COO}^- \\ \text{zwitterion}
\end{array}
\overset{pK_{a_3} = 12.48}{\rightleftharpoons}
\begin{array}{c}
\text{NH}_2 \\ | \\ \text{C}{=}\text{NH} \\ | \\ \text{NH} \\ | \\ (\text{CH}_2)_3 \\ | \\ \text{CHNH}_2 \\ | \\ \text{COO}^-
\end{array}
$$

$$\therefore \qquad \text{isoionic point} = \frac{\mathrm{p}K_{a_2} + \mathrm{p}K_{a_3}}{2} = \frac{8.99 + 12.48}{2} = 10.74$$

pH-Dependent Properties of Amino Acids

The properties of an amino acid which are most likely to be affected by changes in pH are those that are possessed in different measure by its variously ionized forms, since the pH determines which ionic forms are present, and in what proportions.

As one example of this, the solubility of an amino acid is often very evidently affected by the pH in a manner that suggests that the zwitterion is its least soluble form.† This is obviously true of cystine and tyrosine, for these amino acids are sparingly soluble in the pH range in which their zwitterions predominate (pH 4 to pH 9), and are much more soluble in both more acid or more alkaline media in which they exist as cations or anions.

Other properties which may be possessed by the various ionic forms of an amino acid in differing degrees, include optical rotation, ultraviolet absorbance at a given wavelength, metal chelating ability, even biological activity, so that these properties also can be expected to be affected by changes in environmental pH.

One compensation for the behavioural complications that stem from the way in which the pH determines the predominant ionic state of an amino acid is the practical use to which this effect can be put in separating and identifying amino acids by electrophoresis and ion-exchange chromatography.

Electrophoresis of Amino Acids

This technique exploits the migration of ions in an electrical field. When a direct current is passed between two electrodes in an ion-containing medium, the negative ions (anions) move towards the positive electrode (anode), while the positive ions (cations) move towards the negative electrode (cathode). The rate of migration of an ion will depend, among other things, on the size of the charge that it carries, and the magnitude of the current employed. In the version of the technique that is known as high voltage, paper electrophoresis, a mixture

† Note, however, that since the charges on the amino acid molecule may be neutralized by becoming electrostatically bound to alien anions or cations, the ionic strength of the medium can also affect the solubility of an amino acid.

is applied in a small spot or transverse band in the centre of a long strip of filter paper which is then saturated with a buffer of suitable pH. The ends of the paper strip are placed in electrical contact with electrodes across which is applied a voltage (about 100 V/cm) sufficient to cause the passage of a current of about 40 milliamperes through the buffered medium. A compound that is homogeneously charged at the pH used will migrate at a characteristic rate towards the electrode of opposite charge, and will have travelled a characteristic distance relative to other compounds when the current is switched off. After the paper (electrophoretogram) has been dried, the new sites now occupied by the charged components of the starting mixture can be revealed by spectrophotometric or staining techniques.

Paper electrophoresis is particularly suitable for the separation of ionized compounds of relatively small molecular weight such as amino acids. The method can also be used to study the way in which the net charge on the molecules varies with the pH of the medium, which may in turn indicate the number of dissociable groups in the molecule and their approximate pK_a values. The isoelectric point of a compound, at a known ionic strength, is particularly easy to determine by finding (quite empirically) at what pH it does not move from the starting line, i.e. the pH at which it is electrophoretically immobile. However, the most common use of paper electrophoresis is as a means of fractionating mixtures for analytical purposes. Here the pH of the electrophoresis medium must be most carefully chosen. For example, to separate glycine, glutamic acid and lysine by paper electrophoresis, one should use a buffer of about pH 6, for at this pH:

(a) glycine is in its zwitterionic form and will be immobile;
(b) glutamic acid will be negatively charged and will move towards the positive electrode (anode);
(c) lysine will be positively charged and will move towards the negative electrode (cathode).

A very clean separation of the differently charged components will therefore be obtained. Yet resolution of a mixture of similarly charged compounds can also be accomplished by this technique, since the rate of migration of an ion in a constant electrical field is determined not only by its net charge but also by its size—which in an aqueous medium generally means the size of the hydrated ion. Indeed, mixtures of amino acids are frequently separated into their components by paper electrophoresis at pH 1.8 to 2, at which pH all amino acids carry a net positive charge and all migrate towards the negative electrode. The rate of migration of any one amino acid at these low pH's will be determined by the size of its hydrated cation, and the degree to which it is charged (since values of pK_{a_1} differ, though all are in the range of 1.7 to 2.4).

Ion-exchange Chromatography of Amino Acids

This procedure makes use of ion-exchange materials that consist of a water-insoluble polymeric matrix whose surface carries one type of acidic or basic group. The pH-controlled dissociation of these groups determines the number of *fixed charges* on the ion exchange material which are available to hold (by electrostatic attraction) an equivalent number of oppositely charged *mobile ions* supplied by the medium.

Ion-exchange materials are classified according to the charge of their *mobile ions*.

(a) *Cation exchangers*

These ionize to reveal fixed negative charges that 'bind' mobile cations. They are also called 'acidic' ion-exchange materials since their fixed negative charges are produced by the protolysis of acidic groups; they are in fact catalogued according to the strength of these acidic groups.

1. Strongly Acidic Cation Exchangers For example, a resin lattice of the polystyrene type, bearing sulphonic acid groups (Dowex 50)

$$\vdash SO_3H \rightleftharpoons \vdash SO_3^- + H^+$$

These strongly acidic groups will be ionized at all except very low pH's .

2. 'Intermediate' and Weakly Acidic Cation Exchangers For example, a polyacrylic resin lattice bearing carboxylic acid groups, or a polysaccharide lattice bearing carboxymethyl groups (CM-Cellulose).

$$\vdash O{-}CH_2{-}COOH \rightleftharpoons \vdash O{-}CH_2{-}COO^- + H^+$$

(b) *Anion exchangers*

These bear positive fixed charges that bind mobile anions. They are also known as 'basic' ion-exchange materials since their fixed positive charges are generally the product of association of H^+ ions with fixed basic groups.

1. **Strongly Basic Anion Exchangers** For example, a polystyrene lattice bearing quaternary ammonium groups (Dowex 50). Such materials are fully ionized at all save very alkaline pH's. An 'intermediate' strength material is given by a polysaccharide lattice bearing diethylamino groups (DEAE-cellulose).

$$\int\!\!-O-C_2H_4-N\begin{matrix} C_2H_5 \\ C_2H_5 \end{matrix} + H^+ \rightleftharpoons \int\!\!-O-C_2H_4-NH^+\begin{matrix} C_2H_5 \\ C_2H_5 \end{matrix}$$

2. **Weakly Basic Anion Exchangers** For example, a polysaccharide lattice bearing *p*-aminobenzyl groups.

$$\int\!\!-O-CH_2-C_6H_4-NH_2 + H^+ \rightleftharpoons \int\!\!-O-CH_2-C_6H_4-NH_3^+$$

The usefulness of these materials can be illustrated by considering a cationic exchanger which, at a pH where it is completely ionized, is fully saturated with a mobile cation X^+; i.e.

$$\underset{\text{Acidic cation exchanger}}{\int\!\!-CH} \xrightarrow[\text{pH}]{\text{appropriate}} \underset{\text{Ionized cation exchanger displaying fixed – ve charge}}{\int\!\!-C^-} + \underset{\text{Mobile cation}}{X^+} \longrightarrow \underset{\text{Cation exchanger saturated with mobile cation } X^+}{\int\!\!-C^- \text{---} X^+}$$

Cation exchange takes place when this ion-exchange material, with its bound cation, is added to a solution containing a different cation Y^+ at a pH at which the exchange material remains fully ionized. Some X^+ will be displaced by Y^+ until equilibrium is established,

$$\int\!\!-C^-\text{---}X^+ + Y^+ \rightleftharpoons \int\!\!-C^-\text{---}Y^+ + X^+$$

The composition of the equilibrium mixture will depend on several factors, including the temperature, the initial concentrations of $\int\!\!-C^-$

--- X^+ and of Y^+, and the relative affinity of the ion-exchange material for X^+ and Y^+. It is because the exchange of cations proceeds to an equilibrium, and not to completion, that column treatment with ion-exchange materials is generally more effective than batch treatment. If instead of proceeding as in the batch treatment just described, the solution of Y^+ is allowed to percolate slowly through a column of ⌠— C^---- X^+, then progressive removal of Y^+ ensures that successive layers of fresh ⌠—C^---- X^+ are exposed to diminishing concentrations of Y^+, until Y^+ is virtually completely removed from solution. By the correct choice of ion-exchange material and its initial mobile ion, and with suitably controlled operating conditions of pH, ionic strength, temperature and flow rate, Y^+ can be 'bound' within a relatively narrow band on the column material and will remain bound while the column is washed with a buffer of low ionic strength, and of the same pH as the applied solution of Y^+. To recover Y^+, it can be displaced and eluted by passing a suitable buffer solution through it can be displaced and eluted by passing a suitable buffer solution through the column. This eluting buffer should differ from the loading buffer by being:

(a) of lower pH—sufficiently low to cause the fixed negative charges to be lost by association with H^+ ions;

or (b) of identical pH but increased ionic strength;

or (c) of identical pH and ionic strength, but containing a cation for which the ion-exchange material demonstrates a much greater affinity than it does for Y^+.

When the net charge carried by the mobile ions is also pH-determined, as it is when these are amino acids, then the operating pH has to be even more carefully chosen. In fact, the ion-exchange material most frequently employed for the column chromatography of amino acids is a strongly acidic, cation-exchange resin whose fixed charges are sulphonate ions (i.e. Dowex 50). This material is completely ionized at all save very acid pH's and is generally used in its Na^+ form (i.e. saturated with Na^+ mobile cations). Since all amino acids will be positively charged in a solution buffered at pH 2 to pH 3, when such a solution is loaded onto a Dowex 50 (Na^+) column, the amino acid cations will exchange with Na^+ and will be bound to the resin. The amino acids may now be individually eluted from the column by applying buffer solutions of gradually increasing pH. As the pH increases, so will each amino acid progressively lose its net positive charge at a rate determined by the strengths of its distinctive acidic groups. As it loses its net positive charge, so too the amino acid becomes progressively less firmly bound to the negatively charged resin and will be washed down (and eventually out of) the column. Amongst the first amino acids to be eluted in this way from a strongly acidic resin, would be the most acidic amino acids, i.e. aspartic and glutamic acids (possibly at pH 3.5 to 4, depending on the ionic strength, etc.). The basic amino acids will retain a net positive charge in media of much higher pH, so that lysine and arginine would be the last to emerge in an eluate of perhaps pH 10 to 11.

Although one can in general predict the behaviour of an ion-exchange material and its mobile ions at various pH's and ionic strengths, complete separation of all the components of a protein hydrolysate on a single column of ion-exchange material, relies greatly on somewhat empirical observations; e.g. whether it is preferable at a certain stage in the elution to raise the pH, or to increase the ionic strength, or both, or to do neither but instead raise the temperature. Yet skilled use of ion-exchange methods requires above all else an understanding of the pH-dependence of the fixed and mobile charges that are involved.

EXAMPLE 6.17 *Suggest how you might separate the components of the following pairs of amino acids by ion exchange chromatography:*

(*a*) *aspartic acid and glycine;*

(*b*) *aspartic acid and glutamic acid;*

(*c*) *lysine and arginine.*

Solution

(a) *Aspartic acid and glycine*

To separate these amino acids, we can exploit the fact that in buffer of pH 6, glycine will bear no net charge ($^+H_3N.CH_2.COO^-$) whilst aspartic acid will be negatively charged ($^-OOC.CH_2.CH.(NH_3^+).COO^-$). We might use an anion-exchange material which is completely ionized (positive fixed charges) at pH 6. A suitable resin would be the strongly basic anion exchanger Dowex 1 in its chloride form.

A column of Dowex 1 (Cl^-) is equilibrated with a suitable buffer of low ionic strength and pH 6, and the amino acid mixture in solution in this buffer is loaded onto the column. The aspartate anions will be bound to the Dowex 1 by anion exchange, i.e.

$$\int\!\!-N^+(CH_3)_3\text{---}Cl^- + asp^- \rightleftharpoons \int\!\!-N^+(CH_3)_3\text{---}\,asp^- + Cl^-$$

The glycine, of nil net charge, will not be bound, and will appear in the effluent and in the first 'washings' obtained when the loading buffer is used to wash to column. The aspartate can then be eluted with a buffer of greater ionic strength, and /or lower pH (when the ionization of its β-carboxyl group is repressed by association with H^+).

(b) *Aspartic and glutamic acids*

Both of these amino acids are negatively charged at pH 6 and will be bound at this pH to Dowex 1 anion-exchange resin as described in (a). However, the γ-carboxyl group of glutamate (pK_a 4.1) is a little less strongly acidic than the β-carboxyl group of aspartate (pK_a 3.9), and glutamate will lose its net charge in advance of aspartate as the pH of the eluting buffer is gradually lowered. Together with the somewhat larger size of the hydrated glutamate ion, this means that glutamate can be completely eluted from the column in advance of aspartate if a long enough column of the ion-exchange material is used, together with a slow flow rate of eluting buffer, and a 'gentle' gradient of decreasing pH.

(c) *Lysine and arginine*

To separate these amino acids, we can exploit the fact that the ε-NH_3^+ group of lysine (pK_a 10.8) is more strongly acidic than the guanidino = NH_2^+ group of arginine (pK_a 12.5), and that therefore the isoionic point of lysine (pH 10) is lower than that of arginine (pH 10.7) In buffer at pH 7, both amino acids will be positively charged, and will be bound by a cation exchanger which is fully ionized (– ve fixed charges) at this pH. A suitable material would be the strongly acidic, cation-exchange resin Dowex 50, which could be used in its Na^+ form.

Thus, a column of Dowex 50 (Na^+) is equilibrated with a low ionic strength buffer of pH 7 and the amino acid mixture loaded in solution in this same buffer. Both amino acids will be bound by cation exchange

$$\begin{matrix} -SO_3^- \ldots Na^+ \\ -SO_3^- \ldots Na^+ \end{matrix} + \begin{matrix} lys^+ \\ arg^+ \end{matrix} \rightleftharpoons \begin{matrix} -SO_3^- \ldots lys^+ \\ -SO_3^- \ldots arg^+ \end{matrix} + 2Na^+$$

After washing the column with the loading buffer, the amino acids will be differentially eluted when the pH and ionic strength of the eluting buffers are gradually increased. As the pH rises, lysine will lose its net positive charge sufficiently ahead of arginine to ensure that with care, and the use of an adequately long column, the lysine could be fully eluted before any arginine appeared in the eluant.

THE pH-DEPENDENT IONIZATION OF PROTEINS

The multidissociable polypeptide chains of large molecular weight that make up a protein molecule are composed of uniquely ordered sequences of amino acids linked together by peptide bonds. As these bonds are formed by the condensation of the α-amino group of one amino acid with the α-carboxyl group of its neighbour, the majority of these groups will no longer be ionizable in the completed polypeptide.

Only the terminal, free α-carboxyl group at one end of the polypeptide chain, and the terminal, α-carboxyl group at the other end, remain as potential sources of the ionized α-NH_3^+ and α-COO^- groups that were so characteristic of the component amino acids. The very large number of dissociable groups that are actually possessed by a polypeptide must mainly be those carried by the protruding side chains.

The final three-dimensional protein molecule results from the very specific coiling and aggregation of its polypeptide chains, and it can be very large, with a molecular weight of from ten thousand to several millions. Although other types of bond are involved in maintaining the stability of the final structure, an important contribution is made by electrostatic bonding between ionized groups of opposite charge. As the number of these is determined by the pH of the medium, the stability of all proteins is likely to be affected by pH. Furthermore, it is conceivable that in the three-dimensional protein molecule the polypeptide chains may be folded in such a manner that amino acids with polar side chains are predominant in one part of the molecule, while those with non-polar side chains are congregated in another area. This is the case in the molecule of myoglobin, wherein, those amino acids with ionizable, or other polar side chains (e.g. glutamate, lysine, serine) are virtually entirely excluded from its centre. This means that a hydrophobic micro-environment is created within the molecule by the central congregation of those amino acids with non-ionizable, non-polar side chains (e.g. valine, tryptophan). The surface of the molecule will, on the other hand, be hydrophilic since it is here that the ionizable and polar side chains are located. As we find out more about the three-dimensional structure of protein molecules and its relation to their special properties and functions, it is likely that the importance of the congregation of polar and of non-polar side chains to create hydrophilic and hydrophobic areas will be emphasized. Indeed it might be one determinant of the manner in which the constituent polypeptide chains must fold to give the stable, biologically functional, three-dimensional structure. It has been possible to show that the quantitatively most important sources of the charges borne by proteins around neutral pH are the following groups, which ionize within the pH range indicated:

(a) Acidic groups that dissociate to create *negatively* charged sites on the protein molecule

β-carboxyl of aspartic acid	pH 4 to 4.8
γ-carboxyl of glutamic acid	pH 4 to 4.8

(b) Basic groups that associate with protons to create *positively* charged sites on the protein molecule

imidazolyl of histidine	pH 6.4 to 7.4
ε-amino of lysine	pH 9.5 to 10.5
guanidyl of arginine	pH > 12

A quantitatively smaller contribution is made by the chain-terminating α-carboxyl groups (pH 3.3 to 3.8) and α-amino groups (pH 7.5 to 8). The amide groups of glutamine and asparagine are extremely weakly basic and will not be ionized at usual pH's, whilst the weakly acidic sulphydryl group of cysteine and the hydroxyl group of tyrosine (both pH 9 to 10), will only ionize in moderately alkaline solution.

The Isoionic and Isoelectric Points of a Protein

The *isoionic point* of a protein can be defined as that pH at which the protein molecule bears equal numbers of negative charges formed by the dissociation of its acidic groups and of positive charges resulting from the association of its basic groups with protons. In other words it is the pH at which the protein molecule carries no net charge because of the complete self-neutralization of its 'fixed' positive charges by its own 'fixed' negative charges.

The *isoelectric point* is also a pH at which the protein molecule bears no net charge, for it is that pH at which the protein is electrophoretically immobile. Why is it then that this isoelectric point is generally not identical with the isoionic point of the same protein? The answer lies in the fact that, to determine the electrophoretic mobility of a protein, it is dissolved in a buffered medium containing small molecular weight cations and anions. The large multi-ionized protein molecule tends to bind these buffer ions after the manner of an ion-exchange material, and the equal balance of charges on its molecule at the isoelectric point could in part be due to there being more 'mobile' anions than cations bound at this pH (or vice versa); this could mask an actual imbalance of fixed charges on the molecule. Thus the isoelectric point of a protein is not constant, but is likely to depend on the nature and concentration of the ions present in the medium in which its electrophoretic mobility is tested. In turn, this means that the isoionic point of a protein can only be experimentally determined by measuring the pH of a solution of the protein in salt-free water, which in practice requires that the protein solution be exhaustively dialysed against 'de-ionized' water. But even at its isoionic point, thanks to extensive hybridization, a protein solution will not be homogeneous. For the relatively small molecule of insulin (51 component amino acids, less than a quarter of which carry ionizable side chains) it has been calculated that there are over 8000 possible structures of nil net charge that can coexist at the isoionic point.

Practical Importance of the Isoelectric Point of a Protein

As proteins are generally handled in buffered media, we are usually interested in their isoelectric points (pI). Fairly reliable values have been obtained for many proteins in media of comparatively low ionic strength and it has been discovered that the pI values of most proteins are generally less than 7, though they cover a very wide range, from 1.1 for pepsin to 12 for some protamines. The pI value of a protein is an indication of its net acidic or basic character which is also reflected in the net charge that it bears in solution at pH 7. At this pH, positive charges are contributed by the side chains of arginine and lysine and by approximately half of its histidine residues (since the apparent pK_a of the imidazole ring of histidine in protein is close to 7). Also at pH 7, negative charges are contributed by the side chain carboxyl groups of aspartate and glutamate, while although the chain-terminating α-amino and α- carboxyl groups will be fully ionized their charges will usually 'balance out'. Thus an 'acidic' protein with a low pI will be negatively charged at pH 7, whilst a 'basic' protein with a high pI will be positively charged at neutral pH. In fact, knowing the amino acid composition of a protein molecule, you can make an intelligent guess at its probable isoelectric point (though it is much easier to determine the pI experimentally). Calf thymus histone possesses 11 lysine, 15 arginine and 2 histidine residues in its molecule but only 1 aspartate and 1 glutamate residue. Thus it will be positively charged at pH 7 with each molecule bearing 27 positive charges and only 2 negative charges derived from these ionized side chains. One can predict from this that the histone is a 'basic' protein with a high pI value somewhere near the pK_{a_3} of arginine, i.e. 10 to 12.

The interaction between the basic proteins of a cell's nucleus and its nucleic acid (DNA) to form nucleoprotein complexes is well known, and might prove to be of importance in 'masking' certain portions of the DNA of higher organisms so as to control their genetic expression. It is, however, less widely recognized that proteins with greatly different pI values can also interact. Thus insulin (pI about 5) can bind with a protamine (pI about 12) to create a protamine insulinate that would bear a much smaller net charge at pH 7 than either of its components. Indeed the high pI value of protamine makes it very useful as a precipitant at near neutral pH's of acidic compounds of high molecular weight, such as nucleic acids, and proteins of low pI.

In general, the isoelectric point of a protein is also the pH at which it is least soluble and most easily precipitated. This is probably due to mutual attraction between oppositely charged groups of neighbouring molecules, for the number of opposing charges will be greatest at the isoelectric point. Factors other than the pH also affect a protein's solubility, e.g. the hydrophilic character of its ionized groups, the ionic strength of the medium. Even so, it is frequently possible to precipitate, and so purify, a protein by suitably adjusting the ionic strength of its solution at a pH somewhat removed from the isoelectric point, and then gradually altering the pH until it is identical with the pI, causing precipitation of the required protein but not of others with different pI values.

We can summarize such ionic behaviour of a protein by stating that, if it is in solution at a pH higher than its isoelectric point in that medium, it will carry a net negative charge, while at pH's lower than its pI value it will be positively charged; i.e.

$$\text{Protein}^{\oplus} \rightleftharpoons \text{Protein}^{\oplus\!\!\!\!\ominus} \rightleftharpoons \text{Protein}^{\ominus}$$

$$\xleftarrow{\text{Acid}} \text{pI} \xrightarrow{\text{Alkaline}}$$

The ability to vary the charge carried by a protein molecule by changing the pH of its solution can be exploited to purify and characterize proteins by electrophoresis and ion-exchange chromatography.

Electrophoresis of Proteins

The same principles are involved in the electrophoresis of proteins as in the electrophoretic separation of smaller molecules. By careful adjustment of the pH, ionic strength, and temperature of the electrophoresis medium, it has proved possible to separate proteins that differ only slightly in isoionic points and molecular size. The electrophoresis can be performed in rigid media in which the buffer solution is supported by a polymer matrix, e.g. starch and polyacrylamide gels, or cellulose acetate strips. Using these media, it is usual to complete the electrophoresis and then identify the separated bands either by staining, or by sectioning the medium, eluting the sections with buffer, and assaying each eluate for its content of protein or characteristic biological activity. However, it is also possible to follow the migration of 'bands' of protein during electrophoresis in unsupported buffered media, by making use of the refractivity of their boundaries or their characteristic ultraviolet absorbancy. In this form of electrophoresis, which is usually performed in a vertical vessel, among the factors other than the size of the current and net charge on the molecules that will affect their rate of migration will be the size and shape of the hydrated molecules and the viscosity of the medium. Tiselius (1937) demonstrated with this 'moving boundary' method that it was possible to derive a value for the molecular weight of a protein from its observed electrophoretic mobility if some idealized molecular shape was assumed. Yet, so many 'imponderables' are involved in the calculation that this has been superseded by more reliable methods for the determination of protein molecular weights (e.g. from their sedimentation velocities during ultracentrifugation).

Isoelectric Focusing of Proteins

Whereas in conventional electrophoresis the protein molecules retain their initial charge during their passage through a medium of constant pH, in the technique known as *isoelectric focusing* they progressively lose their charge as they migrate along a stable pH gradient, eventually coming to rest (and being concentrated) in that pH zone in which they have nil net charge (i.e. when the pH equals their isoelectric point). In practice, the stable pH gradient is produced and maintained by the electrolysis of an aqueous solution of suitably low molecular weight ampholytes (called carrier ampholytes), the pH within the vessel increase progressively from anode to cathode. If a protein is then introduced at a pH lower than its pI, its molecules will carry a net positive charge, and as they migrate towards the cathode will move through zones of increasing pH. The positive charge on the molecules will in consequence progressively diminish until eventually (when pH = pI) the molecules will bear nil net charge and will cease their migration. By this means, separation of the components of a mixture of proteins can be achieved, since each will accumulate at that pH which is its isoelectric point. To counteract convective remixing of the separated components a suitable 'density gradient' (e.g. of glycerol or polyglycan) can be concurrently employed to aid stabilization of the focused bands of protein. In this way isoelectric focusing can be exploited as a useful analytical and preparative separation technique which has the additional advantage of allowing the pI values of the separated proteins to be quite accurately determined.

Ion Exchange Chromatography of Proteins

The most useful ion-exchange materials for protein chromatography have proved to be weak cationic and 'intermediate strength' anionic exchangers whose fixed charges are carried on a polysaccharide lattice of cellulose or dextran. Suitably charged proteins are relatively weakly bound to these materials and can be eluted by employing buffers of slightly altered pH, or increased ionic strength. Again, no new principle is involved, though to maintain the native structure and biological activities of the proteins the chromatography is usually performed at low temperatures (0° – 4 °C) and no use is made of highly acid or alkaline buffers. The technique can be illustrated by considering the use of the anionic-exchange material DEAE-cellulose, and the weak cationic exchanger CM-cellulose.

(a) *DEAE-cellulose*—an intermediate strength anion exchanger

This material is very suitable for the chromatography of proteins with pI values less than 7. The DEAE-cellulose is pre-equilibrated with a buffer of comparatively low ionic strength and of pH about 8, whilst the protein is applied in solution in the same buffer. At this pH, DEAE-cellulose is appreciably ionized and bears positive fixed charges, while the protein, being 'on the alkaline side' of its pI, will bear a net negative charge and will be bound by anion exchange. Lowering the pH will decrease the net negative charge on the protein and it may be eluted by the use of the buffers of decreasing pH. Alternatively, elution may be accomplished at the loading pH using buffers of increased ionic strength, or buffers containing a competing di- or trivalent anion, i.e.

$$\underbrace{\int\!\!-\text{DEAE}^{\oplus} + \text{Pr}^{\ominus}}_{\text{pH 8}} \xrightarrow{\text{`Bound'}} \int\!\!-\text{DEAE}^{\oplus}\text{---}\text{Pr}^{\ominus}$$

$$\xrightarrow[\text{(a) decreasing pH} \\ \text{(b) increasing ionic strength} \\ \text{(c) competing anions}]{\text{Eluted by}} \text{Protein}$$

(b) *CM-cellulose*—a weak cation exchanger

This material is generally most suitable for the chromatography of proteins with pI values greater than 7. The CM-cellulose is pre-equilibrated with a buffer of comparatively low ionic strength at a pH of about 5, and the protein is loaded onto the column in solution in the same buffer. At this pH the CM-cellulose bears negative fixed charges whilst the protein, being 'on the acid side' of its pI, will carry a net positive charge and will be bound by cation exchange. Increasing the pH will decrease the net positive charge on the protein and it may be eluted using buffers of increasing pH. Alternatively, it might be eluted at the loading pH by employing buffers of gradually increased ionic strength, or buffers to which a competing di- or trivalent cation has been added, i.e.

$$\underbrace{\int\!\!-\text{CM}^{\oplus} + \text{Pr}^{\oplus}}_{\text{pH 5}} \xrightarrow{\text{`Bound'}} \int\!\!-\text{CM}^{\ominus}\text{---}\text{Pr}^{\oplus}$$

$$\xrightarrow[\text{(a) increasing pH} \\ \text{(b) increasing ionic strength} \\ \text{(c) competing anions}]{\text{Eluted by}} \text{Protein}$$

Proteins with pI values close to 7 that are stable at moderately acid or alkaline pH's could be chromatographed on either anionic or cationic exchangers. More often, the choice of ion-exchange material is then dictated by the instability of the protein either (a) in mildly acid solution, when DEAE-cellulose would be used, or (b) in mildly alkaline solution, when CM-cellulose would be employed.

pH and the Biological Activities of Proteins

Since electrostatic bonds contribute to the maintenance of the 'native' three-dimensional structure of every protein molecule, it is not surprising that a change in pH which alters the proportions of positive and negative charges on the molecule can lead to unfolding and disorientation of the component polypeptide chains. Minor conformational changes caused by small pH changes may not markedly affect the physical properties of the protein, and restitution of a favourable pH may be followed by re-establishment of the usual protein structure. More drastic pH changes may cause more profound and irreversible structural changes. If the pH were sufficiently lowered, the negatively charged COO^- groups would be neutralized, leaving positively charged —NH_3^+ groups free to exercise electrostatic repulsion, thus aggravating distortion and unfolding of the molecule. Were the pH to be sufficiently raised, the COO^- groups, in the absence now of nullifying—NH_3^+ groups, might exert this same disruptive tendency. Thus a drastic change in pH can so alter the conformation of a protein molecule that it undergoes irreversible precipitation and coagulation (i.e. denaturation). Yet the effects of pH changes on proteins cannot be attributed solely to alterations in the number and distribution of their intramolecular electrostatic (ionic) bonds, for at least three other types of bond are of major importance in the maintenance of the secondary and tertiary structures of proteins, viz: apolar (hydrophobic) bonds, hydrogen bonds between peptide groups, and hydrogen bonds between side-chain groups. Since the pH also affects the extent of hydrogen bonding, some of the pH-induced denaturation of proteins can be ascribed to the 'weakening' of these structure-stabilizing hydrogen bonds.

A few proteins are unusual in maintaining their structure at extremely high or low pH's and it is quite often found that they also demonstrate optimal biological activity at these pH's; for example, the optimal pH for the enzymic activity of pepsin is about pH 1 to 2. Biological activity is generally an attribute of one specific conformation of the protein molecule so that pH changes that drastically affect the physical characteristics of a protein are likely to have at least an equivalently profound effect on its biological activity. Frequently, the biological activity of a protein is even more sensitive to pH changes than would be expected from physical measurements of the distortion of the molecule that these cause. Indeed, this is one of the many pieces of evidence which together suggest that the biological activity of a protein may primarily be the function of only part of its molecule, which can be called its biologically active site. Should a slight change in pH affect the ionization of a single dissociable group at this active site, then it could greatly change the biological activity without significantly altering the overall structure of the protein molecule. Examination of the effect of pH changes on the biological activity of a protein may therefore provide evidence of its possession of an active site, and might even suggest what dissociable groups this contains.

Proteins as pH Buffers

All proteins, irrespective of their specialist function, must contribute to the general buffer capacity of the cellular contents by virtue of their high content of weakly acidic and basic groups. Haemoglobin provides a spectacular example of a protein of specialist function being called upon to undertake the role of an efficient pH buffer in an unusual manner.

The occurrence of O_2-consuming and CO_2-releasing cellular respiration in tissues of the body far distant from the lungs, requires that O_2 be transported in the arterial blood supplied to these tissues and that CO_2 be carried from the tissues to the lungs in the venous blood. In man, arterial O_2 transport is accomplished by the combination of haemoglobin with O_2 at the lungs to form oxyhaemoglobin. Arriving at the respiring tissue, the oxyhaemoglobin delivers up its O_2 and reverts to haemoglobin. At first sight the problem of CO_2 transport appears less formidable, since CO_2 is much more soluble in aqueous media than is O_2 and erythrocytes are known to contain an enzyme (carbonic anhydrase) which promotes the rapid reaction of CO_2 with water to form carbonic acid. At the pH of blood (about pH 7.4), carbonic acid (pK_{a_1} 6.1), will be 96% dissociated into H^+ plus bicarbonate ions, i.e.

$$CO_2(g) \leftrightarrow CO_2 + H_2O \rightleftharpoons H_2CO_3 \rightleftharpoons H^+ + HCO_3^-$$

Respiring tissue: $\longrightarrow$

$\longleftarrow$ Lungs

Thus the carriage of considerable amounts of CO_2 in venous blood would tend to decrease its pH. The scale of the problem becomes evident from the calculation that a quantity of CO_2 equivalent to between 20 and 40 dm^3 of 1 mol dm^{-3} monobasic acid is excreted via the lungs of man in the course of 1 day. The observation that the pH of the CO_2-carrying venous blood is only marginally lower than that of the CO_2-depleted arterial blood argues the presence in the blood of concentrations of bases (supplied by 'blood buffers') sufficient to associate with all the H^+ ions which are formed in equivalent concentration to the HCO_3^- ions.

Although a portion of the buffer base required for CO_2 transport at pH 7.4 is supplied by plasma phosphates and plasma proteins (as HPO_4^{2-} and Pr^-), more than three-quarters is provided by haemoglobin.

We have noted that a portion of the haemoglobin that arrives at the respiring tissue is in its oxygenated form ($HHbO_2$) and that it there yields up its O_2 and so re-forms haemoglobin (HHb). To represent these complex, multidissociated protein molecules as the model, monoprotic weak acids $HHbO_2$ and HHb is a gross simplification, but one that serves to emphasize the fact that at the pH of the blood they will be dissociated to different extents, since it is known that oxyhaemoglobin is a stronger acid than haemoglobin. This is reflected in their isoelectric points (pI oxyhaemoglobin = 6.7; pI haemoglobin = 6.81) but is best expressed in terms of the apparent pK_a values of the representative model compounds,

$$HHbO_2 \rightleftharpoons H^+ + HbO_2^-, \qquad pK_a = 6.62$$

$$HHb \rightleftharpoons H^+ + Hb^-, \qquad pK_a = 8.18$$

From these values we can calculate that, at the normal pH of blood (pH 7.4), only 14% of oxyhaemoglobin will be in its undissociated state ($HHbO_2$), but 85% of haemoglobin will be present in this condition (HHb). Therefore, when at pH 7.4 oxyhaemoglobin loses its O_2 and becomes converted into haemoglobin, a quantity of H^+ ions must be 'taken up'.

$$\underset{\text{(predominantly } HbO_2^-\text{)}}{\text{Oxyhaemoglobin}} \xrightarrow[\text{At pH 7.4}]{(H^+)} \underset{\text{(predominantly HHb)}}{O_2 + \text{Haemoglobin}}$$

While the O_2^- consuming tissue by producing CO_2 causes H^+ ions to be liberated, by simultaneously obtaining its oxygen from oxyhaemoglobin it forms sufficient base (Hb^-) to associate with the majority of these H^+ ions. This briefly is the basis of the phenomenon known as *isohydric exchange*. You will notice that this differs from normal pH buffering which relies on the buffering capacity of a single conjugate pair (of one pK_a value). Instead, in isohydric exchange, one buffer ($HHbO_2$—HbO_2^-) is converted, at the site of H^+ 'loading', into another buffer of higher pK_a (HHb—Hb^-).

Example 6.18 ***What amount of CO_2 can be transported in blood as carbonic acid and bicarbonate ions without pH change at pH 7.4, as a consequence of the deoxygenation of 1 mole of oxyhaemoglobin? (Apparent pK_a values: oxyhaemoglobin = 6.62; haemoglobin = 8.18; carbonic acid = 6.1.)***

Solution

The 'solution' of CO_2 in blood at pH 7.4 proceeds by the following reaction sequence:

$$CO_2 + H_2O \rightarrow H_2CO_3 \rightarrow H^+ + HCO_3^-$$

Since H^+ and HCO_3^- ions are formed in equivalent amounts, the quantity of CO_2 that can be carried at pH 7.4 as a consequence of the deoxygenation of 1 mol of oxyhaemoglobin will be determined by the quantity of H^+ ions that will be 'taken up' by this process. This will equal the difference between the quantities of H^+ 'bound' in 1 mole of oxyhaemoglobin and in 1 mole of haemoglobin at this pH. Throughout this calculation we will assume that haemoglobin and its oxygenated derivative can be represented by the model compounds HHb and $HHbO_2$ respectively.

(a) ***H^+ 'bound' in 1 mole of oxyhaemoglobin at pH 7.4***

Assuming that $$\underset{\text{(acid)}}{HHbO_2} \rightleftharpoons H^+ + \underset{\text{(conjugate base)}}{HbO_2^-} \ \ldots\ pK_a = 6.62$$

then since, according to the Henderson–Hasselbalch equation,

$$pH = pK_a + \log \frac{[\text{conjugate base}]}{[\text{acid}]}$$ **, for oxyhaemoglobin at pH 7.4,**

$$7.4 = 6.62 + \log \frac{[HbO_2^-]}{[HHbO_2]}$$

i.e. in any given volume, $$\log \frac{HbO_2^-}{HHbO_2} = 0.78,$$

or $$\frac{HbO_2^-}{HHbO_2} = \text{antilog } 0.78$$

$$= 6.03$$

If 1 mol of oxyhaemoglobin contained x mol of $HHbO_2$, the HbO_2^- content = $(1 - x)$

but $$\frac{(1-x)}{x} = 6.03 \quad \text{or} \quad 1 - x = 6.03x$$

$$\therefore \quad 7.03x = 1 \quad \text{and} \quad x = \frac{1}{7.03} = 0.142$$

∴ 1 mol of oxyhaemoglobin at pH 7.4 contains 0.142 mol of $HHbO_2$; or, to state it somewhat differently, 1 mol of oxyhaemoglobin at pH 7.4 effectively has 'bound' to it 0.142 mol of H^+ ions (which have associated with HbO_2^- to produce $HHbO_2$).

(b) *H^+ 'bound' by 1 mole of haemoglobin at pH 7.4*

Assuming: $$\underset{\text{(acid)}}{HHb} \rightleftharpoons H^+ + \underset{\text{(conjugate base)}}{Hb^-} \ldots pK_a = 8.18$$

Then at pH 7.4, $$7.4 = 8.18 + \log\frac{[Hb^-]}{[HHb]}$$

and in any given volume,

$$\log\frac{Hb^-}{HHb} = -0.78 = \bar{1}.22, \text{ and } \frac{Hb^-}{HHb} = \text{antilog } \bar{1}.22 = 0.17$$

If 1 mol of haemoglobin contains x mol of HHb, then it contains $(1 - x)$ mol of Hb^- and,

$$\frac{(1-x)}{x} = 0.17 \text{ whence } 1.17\,x = 1 \text{ and } x = \frac{1}{1.17} = 0.85$$

∴ 1 mol of haemoglobin at pH 7.4 contains 0.85 mol of HHb , i.e. it effectively has 'bound' to it 0.850 mol of H^+ (associated with Hb^- to form HHb).

Thus the quantity of H^+ ions that can be taken up at pH 7.4 as a consequence of the loss of O_2 from 1 mol of oxyhaemoglobin = (Quantity of H^+ bound by 1 mol of haemoglobin) minus (Quantity of H^+ bound by 1 mol of oxyhaemoglobin) = (0.850 – 0.142) mol = 0.708 mol.

By extrapolation, this means that 0.708 mol of bicarbonate ions can be carried by blood at pH 7.4 without any change of pH as a consequence of the H^+ uptake which accompanies the deoxygenation of 1 mol of oxygenated haemoglobin. In turn, this quantity of bicarbonate represents the uptake of 0.708 mol of CO_2.

(c) *CO_2 'carried' as H_2CO_3*

At pH 7.4 bicarbonate ions will be in equilibrium with H_2CO_3,

$$\underset{\text{(acid)}}{H_2CO_3} \rightleftharpoons H^+ + \underset{\substack{\text{(conjugate}\\\text{base)}}}{HCO_3^-} \;\ldots\; pK_a = 6.1$$

Thus an extra load of CO_2 will be carried as undissociated H_2CO_3 as a consequence of O_2 loss from 1 mol of oxyhaemoglobin. Its quantity can be calculated by applying the Henderson-Hasselbalch equation to the dissociation of H_2CO_3, when at pH 7.4,

$$7.4 = 6.1 + \log \frac{[HCO_3^-]}{[H_2CO_3]}$$

and in a given volume,

$$\log \frac{HCO_3^-}{H_2CO_3} = 7.4 - 61 = 1.3, \text{ and } \frac{HCO_3^-}{H_2CO_3} = \text{antilog } 1.3 = 20$$

Thus at pH 7.4, in a volume of blood containing 0.708 mol of HCO_3^-, there must also be present 0.708/20 = 0.035 mol of H_2CO_3 derived from 0.035 mol CO_2. Therefore, total amount of CO_2 carried at pH 7.4 without pH change as a consequence of the liberation of 1 mol of O_2 from oxyhaemoglobin,

$$\begin{aligned} &= (CO_2 \text{ as } HCO_3^-) + (CO_2 \text{ as } H_2CO_3) \\ &= 0.703 + 0.035 \text{ mol} \\ &= 0.743 \text{ mol} \end{aligned}$$

pH AND METABOLIC REACTIONS INVOLVING PROTONS

Chemical reactions affected by pH change are chiefly of two types:

(a) reactions catalysed by H^+ ions

(b) reactions in which H^+ ions are reactants.

In metabolic systems, non-enzymic acid catalysis by H^+ ions is perhaps not as important as in non-biological systems. However, metabolic reactions of type (b) are very common and the possible effect of pH change on these is best illustrated by a specific example.

Yeast in its fermentation of glucose reduces acetaldehyde to yield ethyl alcohol. The reaction is catalysed by alcohol dehydrogenase and utilizes reduced pyridine nucleotide as hydrogen donor:

$$CH_3CHO + \text{reduced NAD} \rightleftharpoons CH_3CH_2OH + \text{oxidized NAD}$$

At equilibrium, by the Law of Mass Action,

$$\text{Equilibrium constant} = \frac{[CH_3CH_2OH]\,[NAD_{ox}]}{[CH_3CHO]\,[NAD_{red}]}$$

But the equilibrium constant calculated from these equilibrium concentrations (K_c') varies with the pH, being 10^4 at pH 7, but 10^2 at pH 9. This is explained by the participation of H^+ ions in this reaction, which should in fact be written as,

$$CH_3CHO + NADH + H^+ \rightleftharpoons CH_3CH_2OH + NAD^+$$

whence

$$\text{real } K_c = \frac{[CH_3CH_2OH]\,[NAD^+]}{[CH_3CHO]\,[NADH]\,[H^+]}$$

Since the concentration of hydrogen ions at equilibrium contributes to the value of K_c, the environmental pH will affect the extent to which acetaldehyde will have been converted into alcohol at equilibrium. The more acid the environment, the more complete will be the reduction of acetaldehyde to alcohol; the more alkaline the medium, the more will oxidation of ethanol be favoured. We will see later that this effect of pH on the value of the *apparent* equilibrium constant of a reaction that involves protons is mirrored in certain other of its characteristics, e.g. the modified standard free energy change, $\Delta G°'$, or electrode potential, E′, and that the extent of ionization of acidic and basic reactants (also affected by pH) can determine whether a reaction will proceed spontaneously.

In the living cell the situation is complicated by the establishment of an overall dynamic equilibrium in which any one reaction is merely one member of an integrated sequence of metabolic events. In such a situation it is highly unlikely that the steady state concentrations attained by the substrates and products of any individuals reaction will be identical with those concentrations achieved by the same reaction proceeding in isolation. Nevertheless, it should be obvious that a change in pH must disturb the overall dynamic equilibrium by virtue of its effect on those reactions that yield or utilize protons, even if the outcome of this disturbance is not entirely predictable in the metabolic situation.

MAINTAINING THE pH IN BLOOD

The pH of most intracellular fluids varies between 6.8 and 7.8. Among the large number of buffer systems needed to maintain the proper pH are HCO_3^-–H_2CO_3 and HPO_4^{2-}–$H_2PO_4^-$. Around pH 7.3 the predominant buffers are bicarbonate and carbonic acid HPO_4^{2-} and $H_2PO_4^-$. They react with acids and bases as follows:

$$HA + HCO_3^- \rightleftharpoons A^- + H_2CO_3$$
$$B + H_2CO_3 \rightleftharpoons BH^+ + HCO_3^-$$

and

$$HA + HPO_4^{2-} \rightleftharpoons A^- + H_2PO_4^-$$
$$B + H_2PO_4^- \rightleftharpoons BH^+ + HPO_4^{2-}$$

In an adult weighing 70 kg, it is estimated that about 0.1 mol of H^+ ions and 12 mol of CO_2 are produced everyday as a result of metabolism. The body has two mechanisms for handling the acid produced by metabolism to prevent a lowering of pH: buffering and excreting H^+. We shall concentrate first on the buffering action of blood. Blood consists essentially of two components: blood plasma, a complex solution containing many biochemically important compounds (carbohydrates, amino acids, proteins, enzymes, hormones, vitamins, and inorganic ions), and erythrocytes or red blood cells. The blood of the average adult contains about 5 million erythrocytes per milliliter of blood. Hemoglobin, the "respiratory" protein, is present inside the erythrocyte. There are about 2×10^8 hemoglobin molecules per erythrocyte. Blood plasma is maintained at pH 7.4 largely by the HCO_3^-/H_2CO_3 and $HPO_4^{2-}/H_2PO_4^-$ buffers, and by various plasma proteins. Proteins are polyamino acids which can themselves act as buffers. A particularly important plasma protein, called albumin, contains 16 histidine residues per molecule. The imidazole group of the histidine molecule in proteins can combine with and buffer H^+. Equation

$$pH = pK_a + \log \frac{[\text{conjugate base}]}{[\text{acid}]}$$

The *Henderson-Hasselbalch equation* can be used to estimate the relative concentrations of H_2CO_3 and HCO_3^- as follows. At pH 7.4 we write

$$7.4 = 6.1 + \log \frac{[HCO_3^-]}{[H_2CO_3]}$$

or $[HCO_3^-]/[H_2CO_3] = 20$. In the normal person, the concentration of CO_2 is about 1.2×10^{-3} M and HCO_3^- is about 0.024 M, giving a ratio of 24/1.2 or 20, as we calculated above. In the erythrocytes the buffers are HCO_3^-/H_2CO_3 and hemoglobins. (Hemoglobin molecules also possess histidine residues.) There the pH is about 7.25. Again from the above Eq. the $[HCO_3^-]/[H_2CO_3]$ is found to be about 14. The membrane of the red blood cell is more permeable to anions such as CHO_3^-, OH^-, and Cl^- than to K^+ and Na^+ cations.

Oxyhemoglobin, formed by the combination of oxygen with hemoglobin in the lungs, is carried in the arterial blood to the capillary beds, where the oxygen is unloaded to the tissues via myoglobin. Both hemoglobin and oxyhemoglobin are weak acids, although the latter is considerably stronger than the former:

$$HHb \rightleftharpoons H^+ + Hb^- \qquad pK_a = 8.2$$

$$HHbO_2 \rightleftharpoons H^+ + HbO_2^- \qquad pK_a = 6.95$$

where HHb and $HHbO_2$ represent "monoprotic" hemoglobin and oxyhemoglobin, respectively. This means that at pH = 7.25, about 65% of $HHbO_2$ is in the dissociated form, while only 10% of HHb is dissociated. The release of oxygen by $HHbO_2$ is strongly influenced by the presence of carbon dioxide. In metabolizing tissues, the partial pressure of CO_2, P_{CO_2}, is higher in the interstitial fluid (that is, fluid within the tissue space) than in the plasma. Thus it diffuses into the blood vessel and then into the erythrocytes. Here, most of the CO_2 is converted to H_2CO_3 by the enzyme *carbonic anhydrase*:

$$CO_2 + H_2O \rightleftharpoons H_2CO_3$$

The presence of H_2CO_3 lowers the pH, which has a direct effect on the release of oxygen. Oxygen may be released from either $HHbO_2$ or HbO_2^- as shown:

$$HHbO_2 \rightleftharpoons HHb + O_2$$

$$HHbO_2 \rightleftharpoons H^+ + HbO_2^-$$

$$HbO_2^- \rightleftharpoons Hb^- + O_2$$

Since $HHbO_2$ releases oxygen more readily than HbO_2^-, a lowering in pH increases the concentration of $HHbO_2$ and promotes the first step. The conjugate base Hb^- of the weaker acid HHb has a greater tendency to react with H_2CO_3 as follows:

$$Hb^- + H_2CO_3 \rightleftharpoons HHb + HCO_3^-$$

The bicarbonate ion formed passes through the membrane and is carried away in the plasma. This is the major mechanism for the elimination of CO_2. When the venous blood circulates back to the lungs, where P_{CO_2} is low and P_{O_2} is high, hemoglobin recombines with oxygen to form oxyhemoglobin:

$$HHb + O_2 \rightleftharpoons HHbO_2$$

The bicarbonate ions in blood plasma now diffuse into the erythrocyte to raise the pH, and we have

$$HHbO_2 + HCO_3^- \rightleftharpoons HbO_2^- + H_2CO_3$$

The H_2CO_3 is then converted to CO_2, catalyzed by carbonic anhydrase:

$$H_2CO_3 \rightleftharpoons CO_2 + H_2O$$

Because of the lower P_{CO_2} in the lungs, the CO_2 formed diffuses out of the erythrocyte and is then exhaled into the atmosphere.

What causes the bicarbonate ions formed in red blood cells to preferentially diffuse into plasma? The concentration of Hb^- and HbO_2^- is quite high in the erythrocytes; consequently there is an unequal distribution of the diffusible anions in the erythrocytes and in the plasma. Now, according to the Donnan equilibrium, the concentrations of the HCO_3^-, Cl^-, and OH^- ions are greater in the plasma than in the erythrocyte (assuming the proteins to be in the anion form). Further, for any given salt MX we can write

$$(\mu_{MX})^C = (\mu_{MX})^P$$

where C and P denote red blood cell and blood plasma. Following the same procedure as that used for the Donnan equilibrium, we arrive at the result

$$[M^+]_C[X^-]_C = [M^+]_P[X^-]_P$$

or

$$\frac{[M^+]_P}{[M^+]_C} = \frac{[X^-]_C}{[X^-]_P}$$

Thus for a given cation and varying the anions we can show that

$$\frac{[HCO_3^-]_C}{[HCO_3^-]_p} = \frac{[Cl^-]_C}{[Cl^-]_p} = \frac{[OH^-]_C}{[OH^-]_p}$$

At the capillary beds, the CO_2 diffuses into the red blood cell and is converted into H_2CO_3. The carbonic acid reacts with Hb^- and HbO^-_2 to form bicarbonate ion, causing the ratio $[HCO_3^-]_C/[HCO_3^-]_p$ to increase. For proper balance of ionic concentrations, the bicarbonate ions diffuse into the plasma, while the chloride and hydroxide ions diffuse into the cell, maintaining electrical neutrality until the above equalities are restored. There is also a corresponding decrease in pH in the erythrocyte caused by the departure of HCO_3^- ions, but this is balanced by the flow of OH^- ions in the reverse direction. Since

$$[H^+]_C[OH^-]_C = [H^+]_p[OH^-]_p$$

or

$$\frac{[OH^-]_C}{[OH^-]_p} = \frac{[H^+]_p}{[H^+]_C}$$

we see that the difference in pH is always maintained in the cell and in the plasma.

The phenomenon described above is sometimes called the bicarbonate chloride shift. In the lungs, the process is exactly reversed. There the bicarbonate ion reacts with oxygenated hemoglobin, causing the ratio $[HCO_3^-]_C/[HCO_3^-]_p$ to decrease. The HCO_3^- ions diffuse from the plasma into the erythrocyte, while the Cl^- and OH^- ions diffuse in the opposite direction, until all the ratios of concentrations are again equal.

The foregoing discussion outlines the efficient and fascinating ways our bodies make use of the buffering action. But buffering alone is not enough. The excretion of constantly forming H^+ ions also plays an important role in maintaining a normal blood pH. The kidneys can excrete H^+ ions and return HCO_3^- ions back to the blood, which can then combine with more H^+ ions. Normally, the pH of urine lies between 4.8 and 7.5. But from blood that has a pH of 7.4, the kidneys can produce a urine with a pH as low as 4.5! Two other mechanisms through which the kidneys can excrete H^+ ions are worth noting. The first is the excretion of anions of weak acids, particularly phosphoric acid. At pH 7.4, roughly one-third of the phosphate is present as $H_2PO^-_4$ and two-thirds as HPO^-_4. In an acid urine, however, most of the phosphate will be excreted as $H_2PO^-_4$. The second mechanism involves the formation of NH^+_4 ions. Amino acids are degraded by the kidneys to form ammonia, which combines with H^+ ions as follows:

$$NH_3 + H^+ \rightarrow NH^+_4$$

The ammonium ions are then excreted in the urine.

PROBLEMS

1. The heat of neutralization of a strong acid by a strong base is −56.8 kJ mol^{-1}. Calculate the ionic product (K_w) of water at 310 K, given that it is 0.61×10^{-14} at 291 K. What is neutral pH at 310 K? (R = 8.31 J K^{-1} mol^{-1}.)

2. Calculate the pH of the following solutions, given $K_w = 10^{-14}$.
   (a) 0.05 mol dm^{-3} HCl
   (b) 0.1 mol dm^{-3} acetic acid ($K_a = 1.75 \times 10^{-5}$ (mol dm^{-3}))
   (c) 0.1 mol dm^{-3} aniline ($K_b = 3.82 \times 10^{-10}$ (mol dm^{-3}))
   (d) A mixture of 0.1 mol dm^{-3} acetic acid and 0.001 mol dm^{-3} HCl
   (e) 10^{-8} mol dm^{-3} HCl.

3. What is the activity coefficient of the hydrogen ion (γ_{H^+}) in a 0.01 mol dm^{-3} solution of HCl if the PH is 2.08?

4. What do you understand by the term 'isoelectric point'? What is the isoelectric point of:
   (a) Glyclglycine (pK_{a_1} = 3.06, pK_{a_2} = 8.13)
   (b) Aspartylglycine (pK_{a_1} = 2.1, pK_{a_2} = 4.53, pK_{a_3} = 9.07)
   (c) Lysylglycine (pK_{a_1} = 2.1, pK_{a_2} = 9.0, pK_{a_3} = 10.6)
   (d) Cysteine (pK_{a_1} = 1.71, pK_{a_2} (SH) = 8.33, pK_{a_3} (NH^+_3) = 10.78).

5. Derive the relationship between pH, pK and the concentration of acidic and basic forms of a compound, by talking logarithms of the appropriate equilibrium constant expression.

   (The relationship is often called the Henderson–Hasselbalch equation.)

   Use this equation to determine the volumes of 0.1 mol dm^{-3} NaH_2PO_4 and 0.1 mol dm^{-3} Na_2HPO_4 solutions which are required to produce 100 cm^3 of a buffer solutions of PH 7.0.

   For $H_2PO^-_4 \rightleftharpoons H^+ + HPO^{2-}_4$, p$K_a$ = 7.2.

What is the ionic strength of the two solutions and also of the final buffer?

6. What is meant by the term buffer solution? The ionization constant of acetic acid is 1.75×10^{-5} (mol dm^{-3}) at 298 K. If a solution contains 0.16 moles of acetic acid per dm^{-3}, how many moles of sodium acetate must be added to give a solution of pH 4.2. Discuss the effects on this solution which result from changes in (1) temperature and (2) the ionic strength.

7. By considering the neutralization of a solution of a weak acid by added hydroxide ion, it can be shown that the buffer value of the solution, i.e.

$$\frac{\text{d (equivalents of base added)}}{\text{d(pH)}}$$

is given by

$$\beta = 2.3\, C\alpha\, (1 - \alpha)$$

where β = buffer value

C = concentration of weak acid

α = degree of dissociation.

Use this equation to evaluate the pH values at which solutions of phosphate and tris act as most effective buffers.

For $H_2PO_4^- \rightleftharpoons H^+ + HPO_4^{2-}$; $K_a = 6.3 \times 10^{-8}$ (mol dm^{-3}) (298 K)

for tris $H^+ \rightleftharpoons H^+$ + tris; $K_a = 8.3 \times 10^{-9}$ (mol dm^{-3}) (298 K).

What is the ratio of buffer values of the following solutions to that of 0.1 mol dm^{-3} phosphate buffer at pH 7.0?

(*a*) 0.1 mol dm^{-3} phosphate buffer at pH 8.0

(*b*) 0.1 mol dm^{-3} phosphate buffer at pH 8.5

(*c*) 0.01 mol dm^{-3} phosphate buffer at pH 7.0.

Comment on your answers.

8. ATP hydrolysis proceeds according to the following equation (at pH 8.0).

$$ATP^{4-} + H_2O \rightarrow ADP^{3-} + HPO_4^{2-} + H^+.$$

A 1 mmol dm^{-3} solution of ATP was hydrolyzed enzymatically in a 0.1 mol dm^{-3} tris buffer (pH 8.0). Calculate the new pH at the end of the reaction. What would have been the final pH if a 0.01 mol dm^{-3} tris buffer (pH 8.0) was employed and also if no buffer was used? (For tris, pK_a = 8.1.)

9. The pH of a sample of arterial blood is 7.50. 20 cm^3 of this sample liberated 12.2 cm^3 of CO_2 at 298 K (1 atm.) on acidification (after correction for dissolved CO_2). If the pK_a for the reaction

$$CO_2 + H_2O \rightarrow H^+ + HCO_3^-$$

is 6.1, determine the concentrations of CO_2 and HCO_3^- in the blood. (1 mol CO_2 occupies 24.45 dm^3 at 298 K.)

If the partial pressure of CO_2 (in atm) is related to the concentration of dissolved CO_2 (in mol dm^{-3}) by the equation

$$[CO_2]_{dissolved} = 0.031 \times P_{CO_2}$$

what is the pressure of CO_2 above the blood?

10. The following data were obtained for a titration of 0.1 mol dm^{-3} acetic acid with sodium hydroxide.

Equivalents NaOH added	0	0.10	0.20	0.40	0.60	0.80	0.90	0.95	1.00	1.05	1.10	1.30
pH	2.60	3.55	4.00	4.50	4.85	5.25	5.60	6.25	8.80	11.00	11.25	11.65

Explain which of the following indicators might be suitable for this titration:

Congo red (pK_{In} = 4.0); Methyl red (pK_{In} = 5.2);
Bromocresol purple (pK_{In} = 6.0); Neutral red (pK_{In} = 7.4);
Phenolphthalein (pK_{In} = 9.1); Thymolphthalein (pK_{In} = 9.9).

14. The mobility (m) of a peptide in electrophoresis (relative to aspartic acid) is believed to follow the relationship

$$m = KeM^{-\frac{2}{3}}$$

where K is a constant, e = charge (irrespective of sign) and M = molecular weight. At pH 6.5 the following results were obtained:

Peptide	Molecular weight	Mobility
Asp-Leu	246.3	0.65
Leu-Gly-Arg	344.5	0.53
Ileu-Ala-Ser-Lys-Phe	565	0.40
Asp-Gly-Asp	305.3	0.98
Gly-Arg-Lys	359.5	0.90
Asp-Gly-Leu-Asp	418.5	0.80

Plot $\log_{10}$(mobility) against $\log_{10}$(molecular weight) for the peptides of different charge classes.

Of the amino acids above, Asp carries a charge of −1 at pH 6.5. Arg and Lys a charge of +1 and the others (Leu, Gly, Ileu, Ala, Ser, Phe, Asparagine) carry no net charge.

Two peptides after acid hydrolysis were each found to contain only Asp and Leu in equimolar proportions. One peptide has a mobility of 0.75, and the other a mobility of 0.45. Use the above plot to deduce the probable formulae of the peptides.

(During acid hydrolysis the amide group of asparagine (Asp-NH_2) is converted to an acid group, yielding aspartic acid.)

12. If $\Delta H°$ for the uptake of a proton by tris-(hydroxymethyl) methylamine (tris) is – 46.0 kJ mol^{-1}, what is the ratio of K_a at 298 K to that at 273 K? $\Delta H°$ for the uptake of a proton by the phosphate dianion is – 4.2 kJ mol^{-1}. Calculate the change in K in this case and comment on the relative merits of these two buffers. (R = 8.31 J K^{-1} mol^{-1}.)

Solution to problems

1. K_w(310 K) = 2.57×10^{-14}

This is a straightforward application of the van't Hoff isochore, noting that $\Delta H°$ for the equilibrium $H_2O \rightleftharpoons H^+ + OH^-$ is +56.8 kJ mol^{-1}. Neutral pH at 310 K is 6.79 (i.e. when $H^+ = OH^-$)

2. (*i*) pH = 1.3

(*ii*) pH = 2.88 ($[H^+]$ = [acetate$^-$] = 1.31×10^{-3} mol dm^{-3})

(*iii*) pH = 8.79 ($[OH^-]$ = $[C_6H_5NH_3]$ = 6.18×10^{-6} mol dm^{-3})

(*iv*) pH = 2.72

We derive this result by noting that HCl is completely dissociated and thus contributes 10^{-3} mol dm^{-3} H^+ to the solution. This partially suppresses the ionization of the acetic acid. If the concentration of acetic acid dissociated is x mol dm^{-3}, then [acetate$^-$] = x mol dm^{-3}, [acetic acid] = $(0.1 - x)$ mol dm^{-3} and $[H^+] = (10^{-3} + x)$ mol dm^{-3}. Inserting these values in the dissociation constant expression, we obtain x and hence the total $[H^+]$.

(*v*) pH = 6.98.

This result is derived in a similar manner to part (iv) above. The HCl contributes 10^{-8} mol dm^{-3} H^+. However the ionization of the water (to give H^+ and OH^-) is suppressed. We add the H^+ contributions from the HCl and the H_2O to obtain the final pH.

3. $\gamma_{H^+} = 0.83$ ($\alpha_{H^+} = 8.31 \times 10^{-3}$, $[H^+] = 10^{-2}$).

4. The pH at which the overall charge is zero.

(*a*) pI = 5.59

(*b*) pI = 3.32

(*c*) pI = 9.8

(*d*) pI = 5.02

Note that in each case the pI is the average of the two pK_as of the *zwitterion* form.

5. $\text{pH} = \text{p}K + \log \frac{[\text{base}]}{[\text{acid}]}$

   From this the ratio $[H_2PO_4^-]/[HPO_4^{2-}] = 1.585$ and thus 61.3 cm^3 of the NaH_2PO_4 and 38.7 cm^3 of the Na_2HPO_4 solution are required.
   For 0.1 mol dm^{-3} NaH_2PO_4, $I = 0.1$.
   For 0.1 mol dm^{-3} Na_2HPO_4, $I = 0.3$.
   For the final buffer, $I = 0.177$.

6. A buffer solution is a solution of a weak acid and its conjugate base which can resist changes in pH on addition of H^+ or OH^- ions.

   The concentration of sodium acetate to be added is 0.044 mol dm^{-3}.

   (*i*) The pK_a of acetic acid will change with temperature (in accordance with the van't Hoff isochore).

   (*ii*) Increasing the ionic strength will increase the dissociation of the acetic acid (i.e. lower the pK_a). This arises from the decrease in activity coefficients at high ionic strength.

7. The solution acts as the most effective buffer when pH = pK_a (at this pH, $\alpha = 0.5$ and the expression $\alpha(1 - \alpha)$ is a maximum).

   For $H_2PO_4^-/HPO_4^{2-}$, pH = 7.2 = pK_a.

   For tris H^+/tris, pH = 8.08 = pK_a.

   (*a*) 0.499.
   (*b*) 0.192.
   (*c*) 0.1.

   The results in the parts (*a*) and (*b*) are derived by evaluating α (and hence $\alpha(1 - \alpha)$) at the various pH values, using the Henderson–Hasselbalch equation. In part (*c*), α is the same, but C (the concentration) of buffer has decreased.

   The results clearly show that a buffer should only be used within small range (normally ± 1 unit) of pH around the pK value, and should be as concentrated as possible.

8. 0.1 mol dm^{-3} buffer, new pH = 7.98.
   0.01 mol dm^{-3} buffer, new pH = 7.82.

   The Henderson–Hasselbalch equation is used to derive the concentrations of tris H^+ and tris at pH 8.0. Addition of 1 mmol dm^{-3} H^+ from the enzyme reaction will increase [tris H^+] by 1 mmol dm^{-3} and decrease [tris] by this amount. The new pH can then be calculated.

If there were no buffer present, we might expect the final $[H^+]$ to be 1 mmol dm^{-3} (i.e. pH = 3). However of course the ADP and P_i present will act as buffers and we could calculate the new pH approximately by assuming that these were equivalent buffers with a pK_a of 7.2 (see Question 5). If the total $[P_i] = 2$ mmol dm^{-3} the final pH after the reaction would be 6.96.

9. $[HCO_3^-] = 24.00$ mmol dm^{-3}

   $[CO_2] = 0.95$ mmol dm^{-3}

   On acidification, the total CO_2 (i.e. $CO_2 + HCO_3^-$) is measured. This is 24.95 mmol dm^{-3}. The ratio $[HCO_3^-]/[CO_2]$ is given by the Henderson–Hasselbalch equation.

$$[P_{CO_2}] = 0.0306 \text{ atm.}$$

10. Plotting the titration data we see the pH is changing most rapidly in the pH range 7–10 (i.e. at the equivalence point). The most suitable indicators would thus be *neutral red, phenolphthalein,* or *thymolphthalein.*

11. The plots of log (mobility) vs. log (molecular weight) for the two charge classes (i.e. singly charged and doubly charged) are both linear with a slope of $-\frac{2}{3}$, in accordance with the equation.

    The peptides might be expected to be of general formula $(\text{Asp-Leu})_n$. These would have a charge of n and a molecular weight of approximately $240n$. From the plots, we see that the peptide of mobility 0.75 (i.e. log $m = -0.125$) would have a molecular weight 480 if its charge were 2 and a molecular weight of 190 if its charge were 1. Clearly the peptide is **(Asp-Leu)$_2$**.

    The peptide of mobility 0.45 (i.e. log $m = -0.35$) is more of a problem. A peptide of this mobility with a charge of 2 would have a very high molecular weight (approximately 1500), and with a charge of 1 would have a molecular weight of 480. Bearing in mind that asparagine (with no net charge) is converted to aspartic acid (with a charge of 1) on acid hydrolysis, it could be proposed that the probable formula of the peptide is **Asp-Asn-Leu$_2$** (Asn ≡ asparagine).

    This problem illustrates some of the difficulties involved in deciding whether a certain amino acid residue in a sequence is either aspartic acid or asparagine. A similar problem occurs with glutamic acid and its amide, glutamine.

12. Using the van't Hoff isochore, we find that

    For tris

$$\frac{K_a\,(273)}{K_a\,(298)} = 0.182$$

This means that the change in pK_a on going from 298 K to 273 K is +0.74 units.

For the phosphate dianion

$$\frac{K_a\,(273)}{K_a\,(298)} = 0.865,$$

i.e. the change in pK_a is +**0.067** units on going from 298 K to 273 K.

This shows that the pH of tris buffers is far more sensitive to changes in temperature than is the pH of phosphate buffers, and this should be borne in mind when using buffers at a variety of temperatures.

7

Chemical Kinetics

The Pathway of a Reaction

The study of thermodynamics enables us to predict which processes or reactions might occur spontaneously. However, it cannot tell us about the rate at which such processes occur, and it is this aspect which is encompassed in the subject of *kinetics*. For example, the hydrolysis of ATP to give ADP and phosphate is known to be highly favourable thermodynamically ($\Delta G^{\circ\prime} = -30.5$ kJ mol^{-1} at pH 7.0 and 298 K), yet a solution of ATP at pH 7.0 is fairly stable.

The solid line indicates the energy content as the reaction proceeds. The term *'reaction coordinate'* is plotted as the *x*-axis, and can be thought of as a 'measure of the extent of reaction', although its precise meaning is difficult to define. For instance, in the case of ATP hydrolysis, it could be taken to refer to the distance between the P and O atoms forming the bond which is being broken, indicated by the zig-zag line.

```
                O       O       O
                ‖       ‖ ⌇     ‖
Ad — O — P — O — P ⌇ O — P — O
                |       | ⌇     |
                O       O       O
```

Breaking such a bond would require the input of a considerable amount of energy and it is essentially this 'energy barrier' which is responsible for the stability of ATP. The term *transition state* refers to the maximum in the energy curve, and if there is a minimum this is termed an *intermediate*. The difference between the energy of the reactants and that of the highest energy transition state is known as the *activation energy* of the reaction. A catalyst would lower this activation energy and hence speed up the reaction, though it would not cause any change in the energy levels of the reactants and products.†

In the case of ATP hydrolysis, the catalyst might be an enzyme (ATPase).

We are interested in the properties of the transition state and yet, by its very nature, this is an unstable state. If we can isolate an intermediate (and this depends on how marked the dip in the energy curve is) this may give us some clue as to the reaction pathway, but the study of transition states clearly requires an indirect approach. By contrast thermodynamics is a more exact subject because it deals only with the properties of the reactants and products independent of the pathway by which they are interconverted.

Kinetics is a subject in which the rate of a reaction is studied as certain parameters are varied; such as the concentration of reactants, temperature, pressure, pH, etc. The data accumulated is then analysed so as to give certain *rate laws*, and hence to yield some information about the *mechanism* of the reaction. The postulated mechanism (usually consisting of the elementary steps involved in the reaction, with some idea of their rates and energy changes) can then be used as a basis for predicting other features of the reaction (such as the dependence of its rate on ionic strength), which are open to experimental test.

We shall now examine certain aspects of the kinetics of reactions.

The Order and Molecularity of a Reaction

The order is defined as the power to which the concentration of a reactant is raised in the *rate law*. Thus in the reaction below where x moles of A combine with y moles of B

$$x\mathrm{A} + y\mathrm{B} \rightarrow \text{products}$$

the expression for the rate of formation of product (P), $\mathrm{d}[\mathrm{P}]/\mathrm{d}t$, might be

$$\frac{\mathrm{d}[\mathrm{P}]}{\mathrm{d}t} = k[\mathrm{A}]^a[\mathrm{B}]^b,$$

where [ ] represents concentration, and k is a constant known as the *rate constant*.

The reaction would then be ath order in A, bth order in B, with an overall order $(a + b)$.

† This means that the catalyst does not change the position of equilibrium of the reaction, only the rate of the reaction.

The *order* of a reaction is an experimental quantity and can have integral values (0, 1, 2, . . .) or non-integral values. For instance the decomposition of gaseous acetaldehyde to give CO and CH_4 is of approximately three halves order in acetaldehyde, and many enzyme-catalysed reactions are of a nonintegral order in substrate in certain ranges of substrate concentration.

Order should not be confused with *molecularity*. The molecularity of a reaction is the minimum number of species involved in the slowest (or rate-determining) step of the reaction. Reference to the energy profile diagram shows that this is the same as saying that the molecularity is the minimum number of species involved in the transition state. We see that molecularity depends on the proposed mechanism of the reaction and thus cannot be deduced from a simple experiment (as can *order*). It is also clear that the molecularity of a reaction must be integral, whereas the order is often non-integral.

Various types of rate law can now be considered.

TYPES OF RATE PROCESSES

Zeroth-order Processes

In this case the rate of the reaction does not depend on the concentration of the reactant A.

$$-\frac{d[A]}{dt} = k.$$

On integration $[A]_0 - [A]_t = kt$, where $[A]_0$ is the initial value of [A], and $[A]_t$ is the value at time t. Thus the concentration of A falls in a linear fashion with time.

Another example of an apparent or pseudo zeroth-order process involves the reaction between glucose and haemoglobin which proceeds at a constant rate in the early stages since the proportion of reagents consumed is very small. The measurement of glycosylated Hb is fairly routine in hospitals now to estimate blood sugar levels during diabetic treatments.

Zeroth-order processes are not very common. Some examples are found in reactions of gases on metal surface, and in enzyme-catalysed reactions at high concentrations of substrate. In these types of processes we have a catalyst which is saturated with the reactant, so that further increases in the concentration of the bulk reactant do not change the reaction rate.

First-order Processes

$$A \rightarrow \text{products}$$

The rate law for this is:

$$-\frac{d[A]}{dt} = k[A]$$

By integration we obtain $\ln[A] = -kt + c$
where c is the constant of integration.

If $[A]_0$ is the initial concentration of A (i.e. when $t = 0$) then

$$c = \ln[A]_0$$

$$\therefore \quad \boxed{\ln\frac{[A]}{[A_0]} = -kt\,.} \tag{7.1}$$

A plot of ln [A] vs. t gives a straight line of slope $-k$.

We have considered the case where we are measuring the disappearance of the reactant [A]. However we often study the formation of product [P]. In this case the rate law becomes modified as follows.

In the process $A \rightarrow P$ let $[P]_t$ be the concentration of P at time t and $[P]_\infty$ be the final concentration of P. If the reaction is considered irreversible and stoichiometric, $([P]_\infty - [P]_t)$ is the concentration of [A] remaining at time t.

Then

$$\frac{d[P]}{dt} = -\frac{d[A]}{dt}$$

$$= k[A]_t = k([P]_\infty - [P]_t)$$

$$\therefore \quad \frac{d[P]}{([P]_\infty - [P]_t)} = k\,dt.$$

$\therefore$ by integration

$$\ln([P]_\infty - [P]_t) = -kt + c$$

when $t = 0$, $[P] = 0$ since we start with pure A.

$$\therefore \quad c = \ln[P]_\infty$$

$$\therefore \quad \ln\left(\frac{[P]_\infty}{[P]_\infty - [P]_t}\right) = kt$$

i.e. a plot of

$$\ln\left(\frac{[P]_\infty}{[P]_\infty - [P]_t}\right) \text{ vs. } t$$

gives a straight line of slope k.

The end point of the reaction can be measured directly, or calculated from the time course of the reaction.

A very useful quantity is the *half-time* for the reaction. This is the time taken for [A] to fall to half its initial value. Let this time be denoted by $t_{\frac{1}{2}}$. Then from the eqn. (7.1)

$$\ln 2 = k t_{\frac{1}{2}}$$

$$\therefore \qquad t_{\frac{1}{2}} = \frac{\ln 2}{k}$$

$$\boxed{t_{\frac{1}{2}} = \frac{0.693}{k}}$$

Thus for first-order reactions the half-time is independent of the concentration of A. The time taken for the concentration of A to fall from $[A]_0$ to $0.5[A]_0$ is as long as the time taken for it to fall from $0.5[A]_0$ to $0.25[A]_0$, and so on.

EXAMPLE 7.1 *Radioactivity decays in a first-order manner. The following data were obtained for the decay of the isotope ^{24}Na.*

Time (h)	*0*	*4*	*8*	*12*	*16*	*20*	*24*
Activity (disintegrations per min)	*478*	*395*	*329*	*272*	*226*	*187*	*155*

Calculate the rate constant for the decay and half-time (half-life) of the isotope.

Solution

A plot of ln(activity) against time is linear.

The slope of this graph ($-k$) equals -0.0469 h^{-1}. Thus the rate constant (k) is 0.0469 h^{-1}.

Since

$$k t_{\frac{1}{2}} = \ln 2$$

$$t_{\frac{1}{2}} = \frac{\ln 2}{k}$$

$$= 14.8 \text{ h}$$

i.e. the half life = 14.8 h.

Second-order Processes

Consider the reaction below

$$A + B \rightarrow \text{products.}$$

Let the initial concentrations of A and B be $[A]_0$ and $[B]_0$, and after time t there are x moles of products formed:

Then

$$\frac{dx}{dt} = k_2[A][B]$$

$$= k_2([A_0] - x)([B]_0 - x).$$

On integration

$$\frac{1}{([A]_0 - [B]_0)} \cdot \ln \frac{[B]_0 ([A]_0 - x)}{[A]_0 ([B]_0 - x)} = k_2 t. \quad (7.2)$$

So a plot of

$$\ln \frac{[B]_0 ([A]_0 - x)}{[A]_0 ([B]_0 - x)} \text{ vs. } t$$

gives a straight line of slope $k_2([A]_0 - [B]_0)$.

If the initial concentrations of A and B are equal we cannot use eqn. (7.2). In this case

$$\frac{dx}{dt} = k_2([A]_0 - x)^2 \quad (\text{since } [A]_0 = [B]_0).$$

On integration

$$\frac{1}{([A]_0 - x)} = k_2 t + c.$$

Since $x = 0$ when $t = 0$, then

$$c = \frac{1}{[A]_0}.$$

$$\therefore \qquad \frac{x}{[A]_0\,([A]_0 - x)} = k_2 t.$$

The half-time for the reaction is found by putting $x = \frac{1}{2}[A]_0$.

$$\boxed{t_{\frac{1}{2}} = \frac{1}{k_2[A]_0}}. \tag{7.3}$$

Thus in second-order reactions the half-time is inversely proportional to the initial concentration of A, i.e. it takes twice as long for the concentration of A to fall from $0.5[A]_0$ to $0.25[A]_0$ as it does to fall from $[A]_0$ to $0.5[A]_0$.

EXAMPLE 7.2 *The rate constant for the dissociation of NADH from liver alcohol dehydrogenase is $3\,s^{-1}$. If the dissociation constant of the enzyme-NADH complex is 0.25×10^{-6} (mol dm^{-3}) what is the rate constant of the association reaction? What is the half-time of the reaction if the initial concentrations of NADH and enzyme are both 100 μmol dm^{-3}.*

Solution

For the reaction

$$\text{NADH} + \text{enzyme} \underset{k_{-1}}{\overset{k_1}{\rightleftharpoons}} \text{enzyme-NADH}$$

$$K_d = \frac{k_{-1}}{k_1}.$$

Now $K_d = 0.25 \times 10^{-6}$ (mol dm^{-3}), $k_{-1} = 3\text{s}^{-1}$.

$$\therefore \qquad k_1 = 1.2 \times 10^7 \text{ (mol dm}^{-3})^{-1}\,\text{s}^{-1}.$$

The half time for the reaction of NADH and enzyme (second order overall) can be calculated from the formula (eqn. (9.3))

$$t_{\frac{1}{2}} = \frac{1}{k[A]_0}$$

where $[A]_0$ is the initial concentration of the reactants, and k is the rate constant.

In this case

$$t_{\frac{1}{2}} = \frac{1}{1.2 \times 10^7 \times 10^{-4}}\ \text{s}$$

$$t_{\frac{1}{2}} = 8.3 \times 10^{-4}\ \text{s}.$$

Note that the rate of the back reaction is negligible.

Pseudo-first-order Processes

A reaction of the type

$$A + B \rightarrow \text{products}$$

is often of second order overall, with a rate law $-d[A]/dt = k_2[A][B]$. If, in such a case, the concentration of one of the reactants (say B) is much greater than that of the other, then it is clear that during the reaction [B] remains essentially constant. The rate law now becomes

$$-\frac{d[A]}{dt} = k'[A]$$

where k' is a new constant, known as the *pseudo-first-order rate constant*. (Note that $k' = k_2[B]$.)

That is, the process is now *apparently* first order in A (and zeroth order in B). Pseudo-first-order conditions provide a special and useful way of measuring the rate constant of a second-order reaction, since if k' and [B] are known, k_2 can be calculated. In practice, it is found that if [B] is in a 20 fold or greater excess over [A], then pseudo-first-order kinetics will be observed. These conditions often apply in biochemical studies, as for instance when a large concentration of some reagent is used to modify the functional groups of a macromolecule (which is present at fairly low concentrations).

EXAMPLE 7.3 *Sheep liver pyruvate carboxylase at a concentration of 1 μ mol dm^{-3} was modified with cyanate, and the extent of modification was followed by measuring the loss of the acetyl-CoA-dependent activity. From the data given below, determine the pseudo-first order rate constants (k') and the second-order rate constant (k_2) for the modification.*

	[Cyanate] (mmol dm^{-3})			
	50	100	200	300
Time (min)	(% initial activity)			
0	100	100	100	100
10	97	94	89	84
20	94	89	79	71
40	89	79	63	50
60	84	71	50	35
80	79	63	40	25
100	75	56	31	18
150	64	42	18	7.5
200	56	31	10	3

Solution

Because the cyanate concentration is many orders of magnitude greater than the enzyme concentration, then clearly the cyanate concentration remains effectively constant during the modification.

Thus, the rate expression for the modification is:

$$\frac{dA}{dt} = -k'A$$

where A is the enzyme concentration, and k' is the pseudo-first-order rate constant. (Note that $k' = k_2$ [cyanate].)

This expression can be integrated to give

$$\ln(A/A_{\text{initial}}) = -k't.$$

Thus k' may be obtained by plotting $\ln(A/A_{\text{initial}})$ against time: k' values may be obtained from this graph either from the slope or by noting that:

$$k' = \frac{0.693}{t_{\frac{1}{2}}}$$

where $t_{\frac{1}{2}}$ is the half-time of the reaction.

In this way, the following k' values may be obtained:

[Cyanate] (mmol dm^{-3})	$t_{\frac{1}{2}}$ (min)	k' (min^{-1})
50	236	2.94×10^{-3}
100	120	5.78×10^{-3}
200	60	1.16×10^{-2}
300	40	1.73×10^{-2}

Now, the second-order rate constant (k_2) may be obtained by noting that:

$$k' = k_2 \text{ [cyanate]}.$$

Hence, k_2 is the slope of the line obtained by plotting k' against [cyanate].

Hence

$$k_2 = 5.75 \times 10^{-5} \text{ (mmol dm}^{-3})^{-1} \text{ min}^{-1}$$

or

$$= 5.75 \times 10^{-2} \text{ (mol dm}^{-3})^{-1} \text{ min}^{-1}$$

Third-order and Higher Reactions

These reactions are extremely rare and are generally of the form

$$2A + B \rightarrow \text{products}.$$

An example would be the combination of two atoms or radicals which occurs in the presence of a third species.

Determination of the Order of a Reaction

The order of a reaction can be determined by comparing the experimental data on the concentrations of reactants or products as a function of time with the integrated forms of the various rate laws discussed above. It is worthwhile mentioning that data over as large a percentage of the total reaction as possible should be used, in order to decide the order unequivocally. In the initial phase of a reaction it is difficult to distinguish, for example, between first- and second-order reactions.

Often the determination of reaction orders can be simplified by use of the method of half-times.† It can be shown that if a reaction is of nth order, then

† The same equation applies for *any* fractional time, e.g. $t_{0.8}$, $t_{0.6}$, or $t_{0.25}$, etc.

$$\boxed{t_{\frac{1}{2}} \propto \frac{1}{[A]_0^{n-1}}}\,.$$

This expression will also apply for those reactions involving more than one reactant, provided that the initial concentrations of the reactants are in their stoichiometric proportions.

Thus we can look at the successive half-times for a reaction. In a first-order reaction they are the same; in a second-order reaction they increase in geometric progression.

The initial concentration of [A] is taken to be 100 in both cases. In the first-order reaction it is clear that the successive half-times are the same. In the second-order reaction they are in the proportion 1 : 2 : 4, etc.

The overall order of the reaction can be determined by this method. There are two main ways to determine the order with respect to any one reactant (e.g. A). The first method involves keeping all the other reactants in a large excess, and then studying the reaction rate as a function of A. This procedure is then varied so as to examine each of the reactants in turn.

In the second method, studies of the initial rate of the reaction are made.

$$\text{Initial rate} = -\frac{\mathrm{d}[A]}{\mathrm{d}t} = k[A]_0^n \cdot [B]_0^m \cdot [C]_0^p$$

where $[A]_0$, $[B]_0$, and $[C]_0$ are the initial concentrations of A, B and C; and n, m, and p are the orders with respect to these reactants.

The concentrations of B and C are kept constant, and $[A]_0$ is varied. Analysis of the variation of initial rate with $[A]_0$ will then given the order with respect to A. The procedure is again varied to determine the orders with respect to B and C in turn.

EXAMPLE 7.4 *The rate of formation of a product P in a reaction was studied.*

Concentration of P formed/μmol dm^{-3}	*0*	*34.3*	*56.8*	*71.6*	*81.7*	*88.0*	*92.1*	*100*
Time/min	*0*	*2.5*	*5.0*	*7.5*	*10.0*	*12.5*	*15.0*	*∞*

What is the order of the reaction?

Solution

If the concentration of P is plotted against time, we can deduce successive half times for the reaction of 4.25 min, 4 min, and 4.25 min. Since these are effectively constant the reaction is first order. The rate constant k can be evaluated from $kt_{\frac{1}{2}} = \ln 2$, i.e.

$$k = \frac{0.693}{4.25}\ \text{min}^{-1}$$
$$= 0.163\ \text{min}^{-1}.$$

k could also be evaluated from the appropriate first-order plot.

EXAMPLE 7.5 *The important biochemical intermediate, acetyl CoA, may be prepared by reacting CoASH with acetyl chloride, the reaction being followed by measuring the CoASH concentration at various times. The following data were obtained when the initial concentrations of both reagents were 10 mmol dm^{-3}*

Time (min)	*0*	*1*	*1.5*	*2*	*2.75*	*4.41*	*5*
[Acetyl CoA] (mmol dm^{-3})	*0*	*3.3*	*4.2*	*4.8*	*5.8*	*6.9*	*7.0*

Determine the order of the reaction, and evaluate the relevant rate constant.

Solution

First, we shall assume that the reaction is a first order process. If this is the case, then a graph of ln(CoA/CoA$_{initial}$) vs. time should give a straight line. From the data, we have:

Time (min)	0	1	1.5	2	2.75	4.41	5
[CoA] (mmol dm^{-3})	10	6.7	5.8	5.2	4.2	3.1	3.0
CoA/CoA$_{init}$	1	0.67	0.58	0.52	0.42	0.31	0.30
ln(CoA/CoA$_{init}$)	0	−0.40	−0.54	−0.65	−0.87	−1.17	−1.20

A straight line is not obtained, and hence it may be concluded that the reaction is not first order.

We shall now assume that the reaction is second order. Because the initial [CoA] and [acetyl chloride] are equal, we can use the equation:

$$\frac{[\text{acetyl CoA}]}{[\text{CoA}]_{init}\,([\text{CoA}_{init} - \text{acetyl CoA}])} = k_2t$$

From the data we obtain:

Time (min)	0	1	1.5	2	2.75	4.41	5
$\frac{[\text{acetyl CoA}]}{[\text{CoA}]_{init}\,[\text{CoA}_{init} - \text{acetyl CoA}]}$	0	0.049	0.072	0.092	0.138	0.223	0.233

In this case, a good straight line is obtained, which indicates that the reaction is second order. From the slope of the line, the second order rate constant is 4.83×10^{-2} (mmol dm^{-3})$^{-1}$ min^{-1}.

Units

It is clear that the first-order rate constants contain no unit of concentration. The units are simply (time)$^{-1}$ (e.g., s^{-1}, min^{-1}, h^{-1}, etc.)

Second-order rate constants contain concentration and time units. From the rate law, the units are seen to be (concentration)$^{-1}$ (time)$^{-1}$, (e.g. (mol dm^{-3})$^{-1}$ s^{-1}, (mol dm^{-3})$^{-1}$ h^{-1} etc.).

Pseudo-first-order rate constants are clearly in units of (time)$^{-1}$. Since they 'include' the concentration of the reactant which is in excess, they can be corrected so as to give true second-order rate constants by dividing by this concentration. Thus if the concentration of B is 1 mmol dm^{-3} (B is in excess) and the pseudo-first-order rate constant obtained is 10^{-2} min^{-1}, then the true second order rate constant is ($10^{-2}/10^{-3}$) (mol dm^{-3})$^{-1}$ min^{-1}; i.e. 10 (mol dm^{-3})$^{-1}$ min^{-1}.

The Kinetics of Some Other Types of Processes

In this section we will examine briefly various other types.

Parallel Reactions

A compound A could give rise to two distinct products:

$$A \xrightarrow{k_1} B$$

$$A \xrightarrow{k_2} C$$

with respective rate constants k_1 and k_2. If both reactions are first-order with respect to A, we have

$$-\frac{d[A]}{dt} = (k_1 + k_2)[A]$$

$$\frac{d[B]}{dt} = k_1[A]$$

$$\frac{d[C]}{dt} = k_2[A].$$

Thus the disappearance of A is still first order, but the overall rate constant is now the sum of the rate constants for the individual processes.

Examples of this type of reaction are known in organic chemistry, e.g. at 1000 K acetic acid gives CH_4 and CO_2 in one reaction and ketene and H_2O in a parallel reaction. Many examples can be found in which a biochemical compound is converted into two other compounds, e.g. AMP can be converted to ATP or to IMP. However these are generally enzyme catalysed reactions and the kinetics of these reactions are more complicated than those discussed here. Other examples can be found in isotope work. When an isotopically-labelled compound is injected into an animal, the amount of isotope remaining generally decreases in a first-order fashion with time. There are two processes responsible, namely the radioactive decay of the isotope and the loss of the compound from animal by excretion. Both of these, then, are first order.

Reversible Reactions

If the rate of the back reaction becomes significant, then the rate laws are more complex.

Consider the reaction:

$$A \underset{k_{-1}}{\overset{k_1}{\rightleftharpoons}} B$$

where k_1 and k_{-1} are the rate constants for the forward and back reactions. If x moles of A have been converted to B at time t then:

$$\frac{dx}{dt} = k_1(A_0 - x) - k_{-1}x$$

where A_0 is the initial concentration of A, and the initial concentration of B is assumed to be zero.

Because the reaction is reversible, at some time it will reach equilibrium, with no net formation of A or B, i.e. the reaction rates in both directions will be equal.

Hence, if x_e is the equilibrium concentration of B

$$k_1\,(A_0 - x_e) = k_{-1}\,x_e \tag{7.4}$$

$$\therefore \quad k_{-1} = \frac{k_1\,(A_0 - x_e)}{x_e}.$$

Hence

$$\frac{dx}{dt} = k_1(A_0 - x) - \frac{k_1\,(A_0 - x_e)}{x_e}.\,x$$

i.e.
$$\frac{dx}{dt} = \frac{k_1\,A_0}{x_e}(x_e - x).$$

As A_0, k_1, and x_e are constants, this expression may be integrated noting that $x = 0$ when $t = 0$, to give

$$\ln \frac{x_e}{x_e - x} = \frac{k_1\,A_0}{x_e}.\,t$$

This expression can be expressed in an alternative form, using the equilibrium (7.4) to give

$$\ln \frac{x_e}{x_e - x} = (k_1 + k_{-1})\,t \tag{7.5}$$

Note that the rate constant for the approach to equilibrium is equal to the sum of the first-order rate constants k_1 and k_{-1}.

A fuller treatment of this important case can be found in the text by Fersht.

Eqn (7.5) is sometimes written in the form:

$$\ln \frac{\Delta C_0}{\Delta C_e} = (k_1 + k_{-1})\,t$$

where ΔC_0 is the difference between the initial and equilibrium concentrations of A (equal to x_e if there is no B present at $t = 0$) and ΔC_e is the difference between the concentration of A at time t, and at equilibrium. In this form, the equation will apply regardless of whether there is any B present at the start.

This equation is the basis of *relaxation techniques* for the kinetic study of reversible reactions. If the reaction is at equilibrium and the equilibrium is suddenly altered (e.g. by altering the temperature or pressure) then the equilibrium will 'relax' to a new state, with a time course defined by the above equation, if we have opposed first-order reactions. This technique has great advantages for studying *rapid reactions* (such as the reversible binding of a substrate to an enzyme) because it avoids the need to start the reaction by mixing the reactants (which takes a finite time of the order of 1 ms). The temperature can be changed very rapidly by 5 or 10 K in about 10 μs by discharging a condenser between electrodes in the solution. By suitable measuring devices the change in concentration in A and B resulting from the change in the position of equilibrium† can be assessed. This is the *temperature jump* method for studying rapid reactions in solution—both inorganic (e.g. $H^+ + OH^- \rightarrow H_2O$, $k = 1.5 \rightarrow 10^{11}$ (mol $dm^{-3})^{-1} s^{-1}$) and also elementary steps involved in enzyme catalysis. Even if there are second-order reactions involved in the relaxation of an equilibrium, the time course will follow a first-order course provided that the perturbation of the equilibrium is not too great. This greatly simplifies the mathematical analysis of more complex reactions.

Illustration

One interesting biochemical system to which reversible first order equations apply is the carbonic acid system:

$$CO_2 + H_2O \underset{k_{-1}}{\overset{k_1}{\rightleftharpoons}} H_2CO_3 \underset{\text{very fast}}{\overset{k_a}{\rightleftharpoons}} H^+ + HCO_3^- \qquad (7.6)$$

The forward reaction rate r_f is given by the expression

$$r_f = k_1 [CO_2][H_2O]$$

since the $[H_2O]$ is essentially constant, this can be written as:

$$r_f = k_f [CO_2].$$

† The position of equilibrium will only be changed by altering the temperature or pressure if for the reaction $\Delta H°$ or $\Delta V°$ (the change in volume) are non zero respectively.

The reverse reaction rate r_b is given by the expression:

$$r_b = k_{-1}[H_2CO_3].$$

Now, the dissociation of H_2CO_3 is very fast, and will always be at equilibrium.

Hence

$$[H_2CO_3] = \frac{[H^+][HCO_3^-]}{K_a}.$$

Therefore

$$r_b = \frac{k_{-1}[H^+][HCO_3^-]}{K_a}$$

This means that eqn (7.6) reduces to

$$CO_2 \underset{k_b}{\overset{k_f}{\rightleftharpoons}} HCO_3^-$$

where

$$k_b = \frac{k_{-1}[H^+]}{K_a}.$$

Therefore, the rate at which CO_2 dissolves in a buffer solution of fixed pH can be simplified to two opposing first-order reactions.

Consecutive Reactions

An important class of reactions are those in which the product of a reaction is subsequently converted to a second product.

Consider the process.

$$A \xrightarrow{k_1} B \xrightarrow{k_2} C.$$

The rate equations can be written down easily:

$$-\frac{d[A]}{dt} = k_1[A]$$

$$\frac{d[B]}{dt} = k_1[A] - k_2[B]$$

$$\frac{d[C]}{dt} = k_2[B].$$

In this example it is seen that the concentration of the ***short lived*** intermediate B rises at first, reaches a maximum value, and then falls slowly to zero. This type of result forms the basis of the so-called '*steady-state approximation*', which states that the concentration of any intermediate reactive species can be considered to remain constant (i.e. the rate of its formation equals the rate of its breakdown). The time during which this is true is known as the '*steady state*' period. As is clear from the above diagram, there is a time in which the concentration of B rises rapidly; this is known as the *pre-steady state period. The subsequent rate of change of* [B] *is very small compared with the rates of change of* [A] *and* [C] *and under these conditions the steady state approximation is valid.*

If the concentrations of B remains constant, this would be expressed mathematically as

$$\boxed{\frac{d[B]}{dt} = 0}$$

The importance of the steady state approximation is that it greatly simplifies the solution of the rate equations in complex consecutive reaction pathways.

Consider the interaction of an enzyme E with its substrate S to form an enzyme-substrate complex, which can then break down to give the product P and regenerate E.

$$E + S \underset{k_{-1}}{\overset{k_1}{\rightleftharpoons}} ES \xrightarrow{k_2} E + P$$

According to the steady state approximation the rate of production of ES (i.e. $k_1[E][S]$) equals the rate of its breakdown (i.e. k_{-1} [ES] + k_2[ES]),

$$k_1[E][S] = k_{-1}[ES] + k_2[ES]$$

$$[ES] = \frac{k_1[E][S]}{k_{-1} + k_2}.$$

The rate of product formation† is then given by

† Provided [ES] is constant and small compared with S, the rate of product formation equals the rate of disappearance of substrate.

$$\frac{d[P]}{dt} = k_2[ES]$$

and by substituting for [ES] from the previous equation, the rate of product formation can be written down. It should be pointed out that this equation is in terms of the concentration of free enzyme [E] and not in terms of the total concentration of enzyme. This modification is referred to in Chapter 10 on enzyme kinetics.

The steady state approximation thus allows us to set up equations for each reactive intermediate in a reaction. These equations can then be solved to find the (steady state) concentrations of these intermediates. Derivation of the overall rate of the reaction is then a relatively simple task.

Having dealt with the rate laws for various types of processes we can now proceed to examine the effects of some of the variables which can give information on the mechanism of a reaction. The effect of temperature will be considered first.

EFFECT OF TEMPERATURE ON THE RATE OF A REACTION

It has long been observed that the rate of most reactions increases dramatically with temperature. The equation proposed by Arrhenius was of the form:[†]

$$\boxed{\frac{d\ln k}{dT} = E_a/RT^2}$$

where k is the rate constant, E_a is the 'activation energy' for the reaction, and R is the gas constant.

If E_a is assumed to be independent of temperature, this equation can be integrated to give

[†] We might note the 'analogy' between the Arrhenius equation and the van't Hoff isochore, eqn

$$\frac{d\ln K}{dT} = \frac{\Delta H^\circ}{RT^2},$$

and remember that

$$K = \frac{k_1}{k_{-1}}.$$

$$\ln\left(\frac{k_{T_1}}{k_{T_2}}\right) = -\frac{E_a}{R}\left(\frac{1}{T_1} - \frac{1}{T_2}\right)$$

or

$$\boxed{k = A\exp(-E_a/RT)}; \qquad (7.7)$$

A is known as the *pre-exponential* factor.

A plot of ln k vs. $1/T$, which is known as the *Arrhenius plot*, gives a straight line, from the slope of which E_a can be obtained. The value of A is obtained by extrapolation of the graph to $1/T = 0$.

EXAMPLE 7.6 *A reaction rate doubles on going from 293 K to 303 K. What is the activation energy of the reaction? ($R = 8.31$ J K^{-1} mol^{-1}.)*

Solution

In eqn (7.7)

$$\ln\frac{k_{T_1}}{k_{T_2}} = -\frac{E_a}{R}\left(\frac{1}{T_1} - \frac{1}{T_2}\right)$$

we put

$$\frac{k_{T_1}}{k_{T_2}} = \frac{1}{2}, \qquad T_1 = 293 \text{ K}, \qquad T_2 = 303 \text{ K}$$

from which

$$E_a = 51.4 \text{ kJ mol}^{-1}.$$

This result provides a handy 'rule of thumb'. If a reaction rate doubles for an increase of 10 K around room temperature, the activation energy is about 51.0 kJ mol^{-1}. A reaction whose E_a is less than 51.0 kJ mol^{-1} is less sensitive to temperature. Compare the example in Chapter 3 (p. 55) dealing with the temperature variation of the equilibrium constant of a reaction.

Significance of the Parameters in the Arrhenius Equation

Reactions are presumed to proceed via collisions between molecules. In gaseous systems, kinetic theory can be used to calculate the number of such collisions occurring under a given set of conditions. Not all collisions, however, will lead to products since only a certain fraction of the molecules will possess the necessary 'activation energy' so that collision can lead to reaction. We might then expect that the rate of the reaction would be given by

$$\text{Rate} = Z_{12} \exp(-E_a/RT)$$

where Z_{12} is the collision number (in suitable units) and $\exp(-E_a/RT)$ is the Boltzmann factor giving the fraction of the molecules which possess the necessary activation energy E_a. The thermal energy available is RT at T K.

The Z_{12} (collision) term would appear to correspond to the pre-exponential factor A in the Arrhenius equation. However it was found in practice that for most second-order gas reactions the calculated value of Z_{12} is very different from the observed value of A. An arbitrary 'steric factor' p was then included in the Arrhenius expression:

$$k = pA \exp(-E_a/RT).$$

A better insight into the significance of these parameters is afforded by a brief account of the *transition state* theory of reaction rates.

Transition State Theory

The collision theory assumes that molecules are hard spheres and only considers whether the molecules have enough energy to overcome the energy barrier. It takes no account of the nature of the reacting species. The transition state theory concentrates on the configuration of the reactants just as they are about to pass over an energy barrier and become products. This configuration is thought of as being at the highest energy point of the energy profile describing the reaction. It is called the transition state or activated complex.

In the most easily understood form of the theory, it is assumed that an equilibrium exists between the reactants and the activated complex, characterized by an equilibrium constant, K

$$K^{\ddagger} = \frac{[AB^{\ddagger}]}{[A][B]} \tag{7.8}$$

For this equilibrium we can assign values of $\Delta G^{\circ\ddagger}$, $\Delta H^{\circ\ddagger}$, and $\Delta S^{\circ\ddagger}$. (These are known as the *free energy*, *enthalpy*, and *entropy of activation*, respectively.)

The great conceptual advantage of this treatment is that it allows the application to reaction rates of all the basic thermodynamic considerations of reaction equilibria—thus forming the link between thermodynamics and kinetics. The values of $\Delta H^{\circ\ddagger}$ and $\Delta S^{\circ\ddagger}$ will also reflect the nature of the reacting species.

The rate at which the activated complexes are converted to products is then assumed to be slow enough so as not to disturb this equilibrium significantly. From the theory, the rate of such conversion (by passage over the energy barrier) is kT/h, where k is the Boltzmann constant and h is Planck's constant.

The overall rate is then given by:

$$-\frac{\mathrm{d[A]}}{\mathrm{d}t} = [\mathrm{AB}^{\ddagger}]\cdot\frac{kT}{h}$$

and since

$$-\frac{\mathrm{d[A]}}{\mathrm{d}t} = k'[\mathrm{A}][\mathrm{B}]$$

the rate constant k' is given by

$$k' = \frac{kT}{h}\cdot\frac{[\mathrm{AB}^{\ddagger}]}{[\mathrm{A}][\mathrm{B}]}$$

$$= \frac{kT}{h}\cdot K^{\ddagger} \qquad \text{(from eqn (7.8))}$$

$$= \frac{kT}{h}\exp\left(-\frac{\Delta G^{\circ\ddagger}}{RT}\right)$$

$$= \frac{kT}{h}\exp\left(+\frac{\Delta S^{\circ\ddagger}}{R}\right)\cdot\exp\left(-\frac{\Delta H^{\circ\ddagger}}{RT}\right)$$

which is of the form

$$\boxed{k' = A\exp(-E_{\mathrm{a}}/RT)}\,.$$

The $\Delta H^{\circ\ddagger}$ is more accurately related to the Arrhenius activation energy (E_{a}) (which is an internal energy change) by the equation

$$\Delta H^{\circ\ddagger} = E_{\mathrm{a}} + \Delta(PV^{\circ\ddagger})$$

The pre-exponential factor in the Arrhenius expression involves the entropy of activation $\Delta S^{\circ\ddagger}$. The values of $\Delta H^{\circ\ddagger}$ and $\Delta S^{\circ\ddagger}$ allow some insight into the factors governing the rates of reactions.

Meaning of $\Delta H^{\circ\ddagger}$ and $\Delta S^{\circ\ddagger}$

In general, $\Delta H^{\circ\ddagger}$ is a measure of the energy barrier which must be overcome by reacting molecules and is related to the strength of the bonds which are broken and made in the formation of the transition state from the reactants.

$\Delta S^{\circ\ddagger}$ is relate to how many molecules with the appropriate energy can actually react. The value of $\Delta S^{\circ\ddagger}$ includes steric and orientation requirements and also solvent effects, and provides a better insight into the roles of the reactants than the much less definite probability factor of the collision theory.

For example, *bimolecular* reactions would be expected to have more negative $\Delta S^{\circ\ddagger}$ values than monomolecular reactions since bringing two molecules together results in more ordering. *Denaturation*, by contrast has large positive $\Delta S^{\circ\ddagger}$ values since the organized structure of a macromolecule becomes disordered. In the case of boiling an egg for instance the value of $\Delta S^{\circ\ddagger}$ is the dominant factor in the rate equation.

EXAMPLE 7.7 *Comment on the following relative rates of intramolecular hydrolysis of esters by the carboxylate anion.*

COOR / COO^-	Me, Me — COOR / COO^-	COOR / COO^-	COOR / COO^-
1	20	230	10 000

Solution

Transition states of reactions have strict requirements of orientation of the participating species. As the initial orientation of the COO^- and COOR groupings approach that in the transition state, the molecules require less ordering, $\Delta S^{\circ\ddagger}$ decreases, and the rate increases. However, many factors determine $\Delta S^{\circ\ddagger}$ and so care should be used in any quantitative interpretation of it. For instance the reaction between ions in aqueous solution

$$Ce^{4+} + EDTA^{4-} \rightarrow (CeEDTA)$$

has a large positive value of $\Delta S^{\circ\ddagger}$. This is because both reactants are highly solvated and the activated complex has a much lower charge and is therefore much less solvated. The positive entropy change associated with the release of water easily overcomes the negative value associated with the reaction of two species to give a complex.

We might also see how the well-known S_N1 and S_N2 reactions in organic chemistry would have different values of $\Delta S^{\circ\ddagger}$.

$$S_N1 \; AB \rightarrow A^+ \ldots B^-$$

(separation of ions in transition state, hence a positive $\Delta S^{\circ\ddagger}$ expected)

$$S_N2 \; AB + C^- \rightarrow (C \ldots A \ldots B)^-$$

(decrease in the number of species in the transition state, hence a negative $\Delta S^{\circ\ddagger}$ expected).

This shows how the useful concept of the 'entropy of activation' replaces the much less definite 'steric factor' in the old collision theory of reaction rates.

The second variable to be considered in this section is the effect of pH.

THE EFFECT OF pH ON THE RATE OF A REACTION

We can divide this section into two principal categories.

(1) Those reactions in which the ionization of groups directly affects the reactivity of the reactants.

(2) Reactions subject to acid or base catalysis.

The Effect of Ionizing Groups

Many examples can be found in which, say, a basic form of a compound is a powerful nucleophile, whereas its conjugate acid is unreactive, e.g.

$$\underset{\text{unreactive}}{RNH_3^+} + H_2O \rightleftharpoons \underset{\text{reactive}}{RNH_2} + H_3O^+.$$

Clearly as the pH is raised, $[RNH_2]$ is increased and hence the reaction rate increases. Now

$$K_a = \frac{[RNH_2][H_3O^+]}{[RNH_3^+]}$$

$$\therefore \qquad [RNH_2] = \frac{[RNH_3^+] \cdot K_a}{[H_3O^+]}.$$

The fraction in the unprotonated form (F) is:

$$F = \frac{[RNH_2]}{[RNH_2] + [RNH_3^+]}$$

$$= \frac{K_a / [H_3O^+]}{\left(K_a / [H_3O^+]\right) + 1}$$

$$= \frac{K_a}{K_a + [H_3O^+]}.$$

The rate constant k is given by

$$k = k_0 F$$

where k_0 is the intrinsic rate constant for the unprotonated form;

$$\therefore \qquad k = k_0 \frac{K_a}{K_a + [H_3O^+]}.$$

Now if $[H_3O^+] >> K_a$ (i.e. pH < pK_a) then

$$k \approx k_0 \frac{K_a}{[H_3O^+]}$$

$$\therefore \qquad \log_{10} k = \log_{10} k_0 - pK_a + pH$$

so that a plot of $\log_{10} k$ vs. pH will be linear with a slope of unity. (This is observed provided pH ≤ pK_a – 1.5.). It should be noted that if such behaviour is observed it can be concluded that the unprotonated form is the reactive species, and that the conjugate acid is unreactive.

Note that when $[H_3O^+] << K_a$ (i.e. pH > pK_a)

$$k = k_0.$$

From the equation it is easily seen that at the intersection point $[H_3O^+] - K_a$, i.e. this method can be used to find the pK_a of an ionizing group.

Acid-base Catalysis

This is a very large subject and will only be treated in outline here.

Acid–base catalysis can be subdivided into general or specific catalysis. For instance ***general acid catalysis*** refers to contributions from all the acidic species present in the solution (e.g. H^+, CH_3CO_2H, H_2O etc.). The overall rate equation thus includes terms representing each of these species. In ***specific acid catalysis*** we consider only the term in H^+ (H_3O^+ in aqueous solution), since the undissociated acids do not contribute. An experimental distinction between the two types could be made by varying the concentration of undissociated acid at constant pH (i.e. by varying the concentration of the conjugate base). As the concentration of the undissociated acid is increased, the rate of a reaction subject to general acid catalysis will rise, whereas the reaction subject to specific acid catalysis will be unaffected.

In order to understand more clearly why some reactions are subject to *specific* and others to *general* acid–base catalysis, we must examine the various steps involved.

We can consider the general scheme

$$A + HX \underset{k_{-1}}{\overset{k_1}{\rightleftharpoons}} AH^+ + X^- \tag{7.9}$$

$$AH^+ + B \xrightarrow{k_2} \text{products.} \tag{7.10}$$

Rate of reaction $= k_2[AH^+][B]$.

Treating AH^+ as a reactive intermediate, and using the steady state approximation, then

$$\frac{d[AH^+]}{dt} = 0 = k_1[A][HX] - k_{-1}[AH^+][X^-] - k_2[AH^+][B]$$

$$\therefore \quad \boxed{[AH^+] = k_1[A][HX]/(k_{-1}[X^-] + k_2[B])} \tag{7.11}$$

We can consider *two limiting cases* of eqn (7.11). In the first, $k_{-1}[X] >> k_2[B]$—this corresponds to the equilibrium in eqn (7.9), being rapidly established and then being disturbed only slightly by a subsequent slow step (9.10). We obtain

$$[AH^+] = k_1[A][HX]/k_{-1}[X^-]$$

substituting for $[HX]/[X^-]$ using the expression

$$K_{HX} = [HX]/[H^+][X^-]$$

then

$$\boxed{\text{Rate} = \frac{k_1}{k_{-1}} \cdot K_{HX}[H^+][A]}$$

The rate law is of the form corresponding to *specific acid catalysis*. The second limiting case of eqn (7.11) is when $k_2[B] >> k_{-1}(X^-)$— i.e. the rate is determined by the forward reaction in eqn (7.9) and we obtain

$$\text{Rate} = k_2[B][AH^+] = \boxed{k_1[A][HX]}.$$

If other acids, e.g. HX′, HX″ are present, the rate would be:

$$\text{Rate} = k_1[A][HX] + k_1'[A][HX'] + k_1''[A][HX''] + \ldots,$$

i.e. this is a rate equation corresponding to *general acid catalysis*.

An example of the difference between *general* and *specific* catalysis is afforded by the aldol condensation reactions of acetone and acetaldehyde. The reaction of acetone to give diacetone alcohol is subject to *specific base* catalysis, indicating that a rapid proton abstraction occurs before the slow step:

$$CH_3COCH_3 + OH^- \rightleftharpoons \overline{C}H_2COCH_3 + H_2O \qquad \text{(fast)}$$

$$\overline{C}H_2COCH_3 + CH_3COCH_3 \xrightarrow{k_2} \text{diacetone alcohol} \qquad \text{(slow)}$$

$$\begin{aligned}\text{Rate} &= k_2[\overline{C}H_2COCH_3][CH_3COCH_3],\\ &= k_2 K'_{eq}[CH_3COCH_3]^2[OH^-],\end{aligned}$$

where

$$K'_{eq} = \frac{[\overline{C}H_2COCH_3]}{[CH_3COCH_3][OH^-]}.$$

The corresponding reaction of acetaldehyde is subject to *general base* catalysis, implying that proton abstraction is the slow step.

$$CH_3CHO + OH^- \xrightarrow{k_1} \overline{C}H_2CHO + H_2O \qquad \text{(slow)},$$

$$\overline{C}H_2CHO + CH_3CHO \xrightarrow{k_2} \text{aldol} \qquad \text{(fast)}.$$

The rate law contains contributions from all the bases (e.g. OH^-, H_2O, or added $CH_3CO_2^-$) in the solution

$$\text{Rate} = k_1[CH_3CHO][OH^-] + k_1'[CH_3CHO][H_2O] + \ldots$$

Presumably the difference between the two reactions can be attributed to the fact that the carbonyl group in acetone is less reactive than that in acetaldehyde (a result of the deactivating effect of a methyl group), and thus nucleophilic attack on the group, which is the second step of the reaction, is slower in the case of acetone.

The study of the mechanisms of acid–base-catalysed reactions is of considerable value in providing models for enzyme catalysis, where amino acid side chains are often postulated to act as general acids or bases. The following reactions depict imidazole acting as a general base in a model reaction, and part of a proposed mechanism of action of ribonuclease in which histidine acts as a general base.

The Effect of Ionic Strength on the Rate of a Reaction

A third parameter which can give some information on the mechanism of the reaction is the *ionic strength* of the solution.

This effect (known as the *salt effect*) on the reaction arises because changes in the ionic strength will change the activity coefficients of the reactants (if they are charged). Thus in the transition state theory expression:

$$k' = \frac{kT}{h} \cdot K^{\ddagger}.$$

We now have to express $K^{\ddagger}$ in terms of activities (rather than concentrations). The relationship between activity coefficients and ionic strength is given by the Debye–Hückel theory. For aqueous solutions of low ionic strength (at 298 K), the final equation derived is:

$$\boxed{\log_{10} \frac{k'}{k_0} = Z_A \cdot Z_B \sqrt{(I)}}$$

Where k' is the rate constant at an ionic strength I.

k_0 is the rate constant at zero ionic strength (obtained by extrapolation).

Z_A, Z_B are the charges (with signs) of the ions taking part in the reaction A + B → products.

The behaviour predicted by this equation has been observed experimentally in many cases (provided that the ionic strength is low enough for the Debye–Hückel theory to be valid). Note that if one of the reactants is uncharged, the product $Z_A . Z_B$ is zero and the rate is independent of the ionic strength.

This effect is known as the *primary salt effect.* A second type of salt effect is observed in acid–base catalysis when changes in the ionic strength can affect the degree of dissociation of weak acids and bases and hence the concentrations of the catalytic species. This indirect effect on the reaction rate is known as the *secondary salt effect.*

A final parameter to be mentioned is the effect of isotopic substitution on reaction rates.

THE EFFECT OF ISOTOPE CHANGES ON REACTION RATES

If an atom at a known position in a molecule is replaced by a different isotope of the atom, then observation of an effect on the rate of reaction of the molecule gives a very good indication that a bond to that atom is being broken in the rate determining step of the reaction.

Consider the replacement of H in a C—H bond by D (deuterium). Because of the heavier mass of D, the zero point energy of the C—D bond is less than that of C—H.

The dissociation energy of the C—D bond is thus approximately 5.0 kJ mol^{-1} greater than that of the C—H bond. From the $\exp(-E_a/RT)$ term in the Arrhenius expression, this would be expected to decrease the rate of a reaction whose rate determining step involves breaking the bond by a factor of 7.6 when H is replaced by D. Several 'isotope effects' of this magnitude are known, e.g. in the oxidation of benzaldehyde by $KMnO_4$ at pH = 7, but generally the values for the ratio $k_{C—H}/k_{C—D}$ are rather lower (~2).

The lower value could mean that in the activated complex the C—H bond may not be completely broken, or a new bond from the H to some other atom may be substantially formed. In some reactions (e.g. the nitration of benzene) no isotope effect is observed when H is replaced by D, and thus it is clear that breaking of a C—H bond is not involved in the rate determining step of the reaction.

Thus we see how kinetic investigations (including the form of the rate law, effects of temperature, pH, etc.) can be used to provide an insight into the pathway of a reaction. When this information is combined with, for example, data on the stereochemistry of the reaction, or isolation of intermediates it is often possible to build up a fairly detailed picture of the mechanism. Several notable examples are known in organic chemistry (especially substitution and elimination reactions) and more recently in inorganic systems (e.g. ligand-exchange reactions among transition metal ions).

PROBLEMS

1. Distinguish between the molecularity and order of a chemical reaction.

   An organic acid decarboxylates spontaneously at 298 K and at pH = 6. The evolution of CO_2 from 10 cm^3 of a 10 mmol dm^{-3} solution was measured at various times as follows:

Time/min	25	50	75	100	125	150	200
CO_2 evolved/cm^3	0.64	1.10	1.45	1.70	1.90	2.05	2.23

   Determine the order and rate constant for this reaction. Assume that no CO_2 remains in solution.

2. The compound A can give two alternative products B and C.

$$A \rightarrow B$$
$$A \rightarrow C$$

   The first-order rate constants are 0.15 min^{-1} and 0.06 min^{-1} respectively. What is the half-life of A?

   If the initial concentration of A is 0.1 mol dm^{-3}, at what time does the concentration of B = 0.05 mol dm^{-3}?

   What is the maximum concentration of B?

3. The following results were obtained for the reaction of 1 mmol dm^{-3} *N*-acetylcysteine with 1 mmol dm^{-3} iodoacetamide.

N-acetylcysteine concentration/mmol dm^{-3}	0.770	0.580	0.410	0.315	0.210	0.155	0.115
Time/s	10	20	40	60	100	150	200

   Determine the overall order of the reaction and the rate constant. You may assume that 1 mol of *N*-acetyl cysteine reacts with 1 mol iodoacetamide.

4. In the same reaction as in Question 3 under different conditions the following results were obtained:

   Initial concentrations of *N*-acetyl cysteine and iodoacetamide are 1 mmol dm^{-3} and 2 mmol dm^{-3} respectively.

N-acetyl cysteine concentration/mmol dm^{-3}	0.74	0.58	0.33	0.21	0.12	0.09
Time/s	5	10	25	35	50	60

   Assume as before that the stoichiometry is 1:1 and calculate the rate constant under these conditions.

5. The reaction between two compounds, A and B in solution is first order in B. The following results were obtained in a kinetic experiment at 300 K.

Initial concentration of B = 1.0 mol dm^{-3}.

Concentration of A/mmol dm^{-3}	1.000	0.692	0.478	0.29	0.158	0.110
Time/s	0	20	40	70	100	120

Determine the order in A and calculate the rate constant for the reaction.

Would the half life of A be different if the initial concentration of B = 0.5 mol dm^{-3}?

If the activation energy is 83.6 kJ mol^{-1} calculate the rate of reaction at of B = 0.5 mol dm^{-3}?

If the activation energy is 83.6 kJ mol^{-1} calculate the *rate* of reaction at 323 K when the concentration of A and B are both 0.1 mol dm^{-3}.

6. Explain what is meant by the steady approximation. Consider the scheme below for interaction of an enzyme and its substrate.

$$E + S \underset{k_{-1}}{\overset{k_1}{\rightleftharpoons}} ES$$

$$ES \xrightarrow{k_2} E + P$$

Assuming steady state conditions, derive an expression for the rate of formation of product P.

What conditions are necessary for the steady-state approximation to be valid?

What is the *maximum* rate of production of P, if the concentration of enzyme is E_0?

7. The rate constant of reaction is 15 (mol dm^{-3})$^{-1}$ min^{-1} at 298 K and 37 (mol dm^{-3})$^{-1}$ min^{-1} at 308 K. What is the activation energy for the reaction and the rate constant at 283 K?

8. The rate constant for the association of an inhibitor with carbonic anhydrase was studied as a function of temperature, with the results shown below. What is the activation energy for this reaction?

T/K	289.0	293.5	298.1	303.2	308.0	313.5
$10^{-6}\, k/(\text{mol dm}^{-3})^{-1}\, \text{s}^{-1}$	1.04	1.34	1.53	1.89	2.29	2.84

9. The variation of rate constant with ionic strength for reactions in water can be derived by considering the Debye–Hückel theory for dilute solutions.

$$\log_{10}(k/k_0) = Z_A Z_B \sqrt{(I)}$$

where k = rate constant, k_0 = rate constant at zero ionic strength, Z_A, Z_B are the charges (with signs) on the ionic species involved, and I = ionic strength.

For the reaction

$$Cr(H_2O)_6^{3+} + SCN^- \rightarrow Cr(H_2O)_5SCN^{2+} + H_2O$$

the following data were obtained:

k/k_0	0.87	0.81	0.76	0.71	0.62	0.50
$I\,(\times 10^3)$	0.4	0.9	1.6	2.5	4.9	10.0

Are these data consistent with this equation?

What is the dependence of the rate of the reverse reaction on ionic strength?

10. Nitramide (NH_2NO_2) decomposes in aqueous solution according to the equation

$$NH_2NO_2 \rightarrow N_2O + H_2O.$$

The rate of the reaction (as monitored by the release of N_2O) was studied in acetic acid/acetate buffers of varying composition. The following data were obtained:

pH	Total concentration of acetic acid + acetate (mmol dm^{-3})	$10^3\ k\ (min^{-1})$
4.09	18.2	2.12
4.11	20.3	2.46
4.40	20.3	3.82
4.46	26.5	5.26
5.01	20.3	7.26
4.77	28.4	8.00

Given that the pK_a of acetic acid is 4.7 deduce whether the rate constant depends on the concentration of H^+, acetate, or acetic acid.

11. The first-order decomposition of diacetyl ($CH_3COCOCH_3$) has values of $\Delta G^{\circ\ddagger}$ and $\Delta H^{\circ\ddagger}$ of 231.5 kJ mol^{-1} and 264.2 kJ mol^{-1} respectively at 558 K. Calculate the value of $\Delta S^{\circ\ddagger}$ and comment on its magnitude.

Solutions to Problems

1. *First order*, $k = 0.01205\ min^{-1}$.

The total CO_2 evolved from this solution would be 2.45 cm^3. From the successive half times, the reaction can be seen to be first order. k can be determined from the half time or by plotting

$$\ln\left(\frac{[P]_\infty}{[P]_\infty - [P]_t}\right) \text{ vs. t.}$$

($[P]_t$ is the amount of CO_2 evolved at time t.)

2. Half life of $A = 3.3$ min.

   B = 0.05 mol dm^{-3} after $t = 5.73$ min.

   Maximum concentration of B is 0.07 mol dm^{-3}.

   These results are derived by noting that $(-d[A]/dt = (k_1 + k_2)[A]$ (where k_1 and k_2 are the rate constants for the formation of B and C respectively), and that $d[B]/dt = k_1[A]$. Since $[A]_t = [A]_0 e^{-(k_1+k_2)t}$ we can substitute in the second equation to obtain:

$$[B]_t = \frac{k_1 [A]_0}{k_1 + k_2}[1 - \exp\{-(k_1 + k_2)t\}].$$

   In this example $k_1 = 0.15$ min^{-1}, $k_2 = 0.06$ min^{-1}.

   The maximum concentration of B is found by evaluating [B] at $t = \infty$.

3. The successive half times for the reaction are 27 s, 54 s and 120 s, and the reaction is thus *second order*. We can evaluate the rate constant from this directly ($k = 37$ (mol dm^{-3})$^{-1}$ s^{-1}) or by making an appropriate linear plot.

   For a second order reaction where the concentrations of the components are equal this is

$$\left(\frac{([A]_0 - [A]_t)}{([A]_0)([A]_t)}\right) \text{ against } t.$$

   ($[A]_t$ is the concentration of A remaining at time t, i.e. $([A]_0 - [A]_t)$ is the amount of product formed.) The slope of this line gives k directly.

4. Since the reaction is second order (from Question 3) we plot the appropriate second order expression against time (eqn (7.2)). From this the value of k is found to be 30 (mol dm^{-3})$^{-1}$ s^{-1}. Note that we obtain the amount of product formed at time t by subtraction of the *N*-acetylcysteine concentration at time t from its initial value.

5. By inspection of the decay curve the reaction is seen to be *first order* in A (the successive half times are 39 s, 41 s, and 37 s). Thus k = 0.018 s^{-1}.

   This reaction is carried out under pseudo-first-order conditions ([B] >> [A]). If the initial concentration of B is halved the pseudo first order rate constant is halved (i.e. $k = 0.009$ s^{-1}) and the new half life of A is 78 s.

   The second order rate constant, k_2, is obtained by dividing the pseudo first order rate constant by the concentration of the compound in excess. Thus $k_2 = 0.018$ (mol dm^{-3})$^{-1}$ s^{-1} at 300 K. The rate constant at 323 K is found by using the Arrhenius equation, k_2 at 323 K = 0.193 (mol $dm^{-3})^{-1}$ s^{-1}.

Thus when the concentrations of A and B are both 0.1 mol dm^{-3} the rate of the reaction is 0.00193 mol dm^{-3} s^{-1}. (Rate = k_2 [A] [B].)

6.

$$\frac{d[P]}{dt} = \frac{k_2 [S]}{\{(k_{-1} + k_2)/k_1\} + [S]} ([E]_{total}).$$

The outline of the derivation of this equation is given in Chapter 8.

For the steady state approximation to hold, the rate of breakdown of ES to give P must be comparable with or slower than the breakdown to regenerate E + S and [E] must be much smaller than [S]. This will then keep [ES] small and reasonably constant, as required by the approximation. The maximum rate of production of P will occur when [S] is large compared with the term $\{(k_{-1} + k_2)/k_1\}$, i.e.

$$\left(\frac{d[P]}{dt}\right)_{max} = k_2[E]_{total}.$$

7. E_a = 69.3 kJ mol^{-1}

   k_{283} = 3.43 (mol $dm^{-3})^{-1}$ min^{-1}.

8. From the graph of ln k vs. $1/T$

   E_a = 30.9 kJ mol^{-1}.

9. The plot of $\log_{10} (k/k_0)$ vs. $\sqrt{I}$ is a straight line of slope – 3. Since $Z_A . Z_B$ = – 3, *the data are consistent with the equation*. In the reverse reaction, one of the species is uncharged, and there is *no dependence of the rate constant* on the ionic strength.

10. For each buffer we can calculate the concentrations of H^+, acetate$^-$, and acetic acid (see Chapter 7). From this it can be seen that the rate constant does not depend on the concentrations of either H^+ or acetic acid, but is linearly dependent on the acetate$^-$ concentration according to the equation; rate constant = 0.38×10^{-3} + 0.5 [acetate$^-$]. This reaction is therefore an example of general base catalysis. (Catalysis by OH^- can be detected at higher pH values).

11. $\Delta S^{\circ\ddagger}$ = 58.6 J K^{-1} mol^{-1}. This indicates that the transition state is less ordered than the reactant and this situation is typical for unimolecular decompositions.

8

The Kinetics of Enzyme Catalysed Reactions

INTRODUCTION

At first sight the kinetics of enzyme catalysed reactions might appear to be vastly more complex than those of the relatively simple reactions considered in the previous chapter. Enzymes are much more efficient catalysts than any model catalysts yet devised, and in addition the catalytic activity of enzymes is often very sensitive to experimental conditions, such as temperature, pH, and ionic strength. Nevertheless, provided care is taken to obtain reliable and reproducible data (e.g. by the use of buffer solutions at defined temperatures) the kinetics of enzyme catalysed reactions can usually be interpreted in terms of the simple ideas developed so far.

There are at least four reasons why a study of enzyme kinetics is important for the biochemist. Firstly, such a study is observing the enzyme 'doing its job' as a biological catalyst. Secondly, a kinetic study provides us with valuable information about the mechanism of an enzyme catalysed reaction. Thirdly, knowledge of the kinetic parameters (Michaelis constant and maximum velocity) allows us to estimate the importance of a particular enzyme catalysed reaction under the conditions of enzyme and substrate concentration which might prevail in the cell. Fourthly, from measurements of how the rate of an enzyme catalysed reaction is affected by changes in variables such as pH and ionic strength, or by the binding of ligands which may regulate the enzyme activity, we can gain further insight into the likely properties of an enzyme in its environment *in vivo*.

In this chapter we shall mainly be discussing the steady-state behaviour of enzymes under conditions where the enzyme is present in very small concentrations (usually 1 per cent or less) compared to the concentrations of substrates or other ligands. This makes the algebraic treatment of the various equilibria rather easier, since it is then possible to set the *free* substrate concentration equal to the *total* added substrate concentration.

It should be mentioned, however, that many important kinetic studies are now being performed with concentrations of enzyme comparable with those of substrates. Using fast-recording or other special techniques, it then becomes possible to identify intermediate complexes in the overall reaction and to determine the rate constants of some of the individuals steps. This type of study will be dealt with in more detail later. In this connection we should note that there are many examples known where enzymes are present *in vivo* in quite high concentrations, and thus it is becoming increasingly clear that for a complete understanding of the rates of enzyme catalysed reactions in the cell we must accumulate kinetic data over as wide a range of enzyme concentrations as possible.

Steady-state Kinetics

Initially we shall confine the discussion to one-substrate reactions (i.e. those reactions in which only one substrate is acted on). This will allow us to introduce the basic concepts and definitions which can then be used in the study of enzymes with more than one substrate.

The treatment of enzyme kinetics assumes that a complex between enzyme and substrate is rapidly and reversibly formed. This complex then breaks down in a slow step to give the product and regenerate enzyme:

$$\mathrm{E} + \mathrm{S} \underset{k_{-1}}{\overset{k_1}{\rightleftharpoons}} \mathrm{ES}$$

$$\mathrm{ES} \xrightarrow{k_2} \mathrm{E} + \mathrm{P}.$$

It has been possible in a number of cases to observe the ES complex directly by using high concentrations of enzyme. Sometimes a reasonably stable intermediate can be isolated (e.g. acyl-chymotrypsin) and this must be incorporated into the overall kinetic scheme.

We can derive the kinetic expression to deal with this pathway in one of two ways.

(1) *The equilibrium approximation*

Here we assume that the $\mathrm{E} + \mathrm{S} \rightleftharpoons \mathrm{ES}$ equilibrium is only slightly disturbed by the breakdown of ES to give the product. This will clearly be a better assumption the lower the value of k_2 relative to k_{-1}.

$$K = \frac{[\text{E}][\text{S}]}{[\text{ES}]}$$

where [E] is the concentration of free enzyme and [S] is the concentration of free substrate. However since $[\text{S}_{\text{total}}] >> [\text{E}_{\text{total}}]$, $[\text{S}_{\text{free}}] = [\text{S}_{\text{total}}]$

Now the fraction F of enzyme in the form of ES is given by:

$$F = \frac{[\text{ES}]}{[\text{E}] + [\text{ES}]}$$

$$= \frac{[\text{S}]}{K + [\text{S}]}$$

Let V_{max} be the maximum rate of product formation, i.e. the rate when all the enzyme is in the form of ES, then

$$\text{Rate } v = V_{max} \cdot F$$

$$\therefore \qquad \boxed{v = V_{max} \cdot \frac{[\text{S}]}{K + [\text{S}]}} \qquad (8.1)$$

When [S] is low compared with K, the reaction is first order in S. At very high [S], the reaction rate tends to V_{max}, and is of zero order in S. In the intermediate range, the reaction is of a fractional order in S.

From eqn (10.1) it is clear that when $v = V_{max}/2$, $[\text{S}] = K$. K (which thus corresponds to the concentration of substrate when the velocity is half maximal) is known as the Michaelis constant (usually written K_m). In this case we can equate K with the dissociation constant of the ES complex.

(2) *The steady-state approximation*

We now abandon the assumption that the E + S $\rightleftharpoons$ ES equilibrium is not disturbed by the breakdown of ES, and use instead the steady-state approximation. After an initial phase (the pre-steady state) it is assumed that the concentration of ES remains constant or very nearly so. In cases where the ES complex can be observed, the experimental data generally resembles and the rate at which [ES] falls is very low, i.e. the approximation is a valid one for most purposes.

Now in the steady-state, the rate of production of ES (i.e. k_1 [E][S]) must equal the rate of its breakdown (i.e. k_{-1} [ES] + k_2 [ES])

$$\therefore \qquad k_1[E][S] = k_{-1}[ES] + k_2[ES]$$

$$\therefore \qquad [ES] = \frac{k_1 [E][S]}{k_{-1} + k_2}$$

therefore, as before, the fraction (F) of enzyme in the form of the ES complex is given by

$$F = \frac{[ES]}{[E] + [ES]}$$

$$= \frac{[S]}{[(k_{-1} + k_2)/k_1] + [S]}$$

and hence the rate of the reaction v is given by

$$v = V_{max} \cdot F$$

$$\therefore \qquad v = V_{max} \cdot \frac{[S]}{[(k_{-1} + k_2)/k_1] + [S]}. \qquad (8.2)$$

Again it should be remembered that $[S] = [S_{total}]$ since the enzyme is assumed to be present in very small amounts.

In the steady-state treatment, the term K in the denominator of eqn (8.1) has been replaced by $(k_{-1} + k_2)/k_1$. Only if $k_2 << k_{-1}$, does the term $(k_{-1} + k_2/k_1)$ become equal to k_{-1}/k_1, i.e. to the dissociation constant for the ES complex (K_s). We should therefore find that the Michaelis constant (the concentration of substrate where the velocity is half maximum) is not generally equal to K_s.

From both treatments of the kinetic scheme, we obtain the basic equation.

$$\boxed{v = \frac{V_{max}\,[S]}{K_m + [S]}} \qquad (8.3)$$

where K_m is the Michaelis constant.

EXAMPLE 8.1 *Using eqn (10.3), calculate the change in [S] required to increase the rate of a reaction from 10 to 90 per cent of the maximum rate. What further change is required to increase the rate to 95 per cent of the maximum rate?*

Solution

Put $v = 0.1\ V_{max}$.

Then $\frac{[S]}{K_m + [S]} = 0.1$,

i.e. $[S] = \frac{K_m}{9}$

Putting $v = 0.9\ V_{max}$ we find $[S] = 9\ K_m$,

i.e. an 81-fold change in [S] is required.

Putting $v = 0.95\ V_{max}$

we find $[S] = 19\ K_m$,

i.e. a further 2.11-fold change in [S] is required.

TREATMENT OF KINETIC DATA

From the result of the worked example we see that it is difficult to achieve truly saturating conditions and thus determine V_{max} (and hence K_m) from the plot of v against [S]. To overcome this difficulty we can rearrange eqn (8.3) in a number of ways. Three of the best known rearranged equations are given below.

The Lineweaver-Burk Method

This uses the rearranged equation

$$\boxed{\frac{1}{v} = \frac{K_m}{[S]} \cdot \frac{1}{V_{max}} + \frac{1}{V_{max}}}$$

Thus a plot of $1/v$ against $1/[S]$ gives a straight line whose intercepts on the x and y axes are $(-1/K_m)$ and $1/V_{max}$ respectively, and whose slope is (K_m/V_{max}).

It will be noticed that this plot is rather similar to that used in the analysis of binding data.

The Eadie-Hofstee Method

The basic equation is rearranged to give

$$\boxed{\frac{v}{[S]} = \frac{V_{max}}{K_m} - \frac{v}{K_m}}$$

A plot of $v/[S]$ against v gives a straight line with a slope of $-1/K_m$ and an intercept on the x-axis of V_{max}.

The Hanes Method

Eqn (8.3) is rearranged to give

$$\frac{[S]}{v} = \frac{[S]}{V_{max}} + \frac{K_m}{V_{max}}$$

A plot of $[S]/v$ against $[S]$ is linear with a slope of $1/V_{max}$ and an intercept on the x-axis of $-K_m$.

Which Plot to Use?

This question (i.e. which plot should be used to derive kinetic parameters from a set of data?) has generated a good deal of discussion among biochemists. Undoubtedly the Lineweaver–Burk plot is the most commonly used and it has the merit that the variables (v and [S]) are plotted on separate axes. On the other hand an analysis of the errors involved in the determination of the experimental data and the subsequent extraction of the parameters K_m and V_{max} has shown that the distribution of errors is highly non-uniform over the range of values of v and [S] in the Lineweaver–Burk plot. The error distribution is more uniform in the other two plots mentioned and so the use of these plots has been advocated by a number of workers, since statistical analysis of the data is more reliable. (In particular, the use of the Hanes plot has been recommended, since in this plot the error present in the x-axis is likely to be very small.) What should be remembered, however, is that any serious kineticist uses a computer method of fitting the data to the original rectangular hyperbola (eqn 8.3) to obtain the 'best fit' values of K_m and V_{max}, and that the linear plots are then merely representations of the data in an 'easy to comprehend' form. A fuller discussion of the merits of various plots is given in the books by Cornish-Bowden mentioned in the reading list.

Units

The units of K_m are those of concentration (i.e. mol dm^{-3}) as K_m is the *concentration* of substrate at which half maximal velocity is observed.

V_{max} can be expressed in a variety of units depending on what information is available. If we are working with an impure preparation of enzyme or if a purified enzyme has not been characterized as far as its molecular weight is concerned, the velocity (v) will generally be given in units of moles substrate consumed (or product formed) per weight of protein (or enzyme) per unit time (e.g. 1 mmol (g protein)$^{-1}$ s^{-1}). The SI unit of enzyme activity is the *katal* which is defined as: One katal is that amount of enzyme which, under the specified conditions, catalyses the production of 1 mole product (or the consumption of 1 mole substrate) per second.

The *specific activity* of an enzyme would then be expressed in units of katal (kg enzyme)$^{-1}$ (or katal (kg protein)$^{-1}$ for an impure enzyme preparation).

If, however, we know the concentration of enzyme active sites (in molar terms), we can evaluate k_{cat} for the reaction ($k_{cat} = V_{max}/[E]_0$; which is also known as the *turnover number* of the enzyme (moles substrate consumed or product formed per mole enzyme per unit time).

Some Practical Aspects

In order to obtain reliable results in enzyme kinetic work, it is necessary to take a number of experimental precautions of which we might mention the following:

(i) The substrates, buffers, etc. should be of as high a purity as possible, since contaminating substances may affect the activity of the enzyme. (Commercial preparations of NAD^+ sometimes contain inhibitors of dehydrogenases, for example.)

(ii) It must be ascertained that the enzyme preparation does not contain any substances (or other enzymes) which interfere with the assays of activity. Also the possibility of non-enzyme catalysed reactions should be tested by appropriate control experiments (e.g. by addition of heat-inactivated enzyme).

(iii) The enzyme should be stable (i.e. not lose any significant activity) under the conditions chosen, for at least the length of time required to perform the assays.

(iv) It should be checked that the measured rate is proportional to the amount of enzyme added and that the rate of product formation (or substrate consumption) is linear over the period of interest.

(v) The rate can be measured more accurately (and conveniently) by making a continuous record of product formation or substrate consumption rather than by performing analysis of the reaction mixture at fixed time intervals. A continuous recording is easily made, for instance, using a spectrophotometer if the reaction involves a change in light absorption.

If, however, the reaction of interest does not involve a convenient change in light absorption or some other readily measurable property this difficulty can sometimes be overcome by using a *coupled-assay procedure*, e.g. the reaction catalysed by pyruvate kinase:

$$\text{phosphoenolpyruvate} + \text{ADP} \xrightarrow{Mg^{2+}} \text{pyruvate} + \text{ATP}$$

is not easily monitored spectrophotometrically, but if NADH and lactate dehydrogenase are added to the reaction mixture, the pyruvate formed is reduced:

$$\text{pyruvate} + \text{NADH} \rightarrow \text{lactate} + \text{NAD}^+$$

and this latter reaction is easily followed since NADH absorbs radiation of wavelength 340 nm but NAD^+ does not.

It is essential to add the coupling enzyme(s) and substrate(s) in sufficient quantities to ensure that the overall measured rate is the rate of the first (pyruvate kinase) reaction, i.e. that the pyruvate as soon as it is formed is converted to lactate with concomitant oxidation of NADH.

EXAMPLE 8.2 *Under certain conditions the maximum specific activity of a purified preparation of triosephosphate isomerase was quoted as 10 000 μmoles glyceraldehyde-3-phosphate transformed per mg protein per minute. Express this in terms of katal* kg^{-1} *and evaluate the turnover number of the enzyme given that its active site molecular weight is 26,500.*

Solution

$10\,000\ \mu\text{mol mg}^{-1}\ \text{min}^{-1}$ is equivalent to $\dfrac{10\,000}{60}\ \mu\text{mol mg}^{-1}\ \text{s}^{-1}$,

i.e. to $\dfrac{10\,000}{60} \times 10^6\ \mu\text{mol kg}^{-1}\ \text{s}^{-1}$,

i.e. to $\dfrac{10\,000}{60}\ \text{mol kg}^{-1}\ \text{s}^{-1}$, i.e. 166.7 katal kg^{-1}.

Now 1 mole enzyme (active sites) = 26.5 kg
therefore turnover number $= 166.7 \times 26.5\ s^{-1}$
$= 4420\ s^{-1}$.

EXAMPLE 8.3 The activity of the enzyme urease, which catalyses the reaction

$$CO(NH_2)_2 + H_2O \rightleftharpoons CO_2 + 2NH_3$$

was studied as a function of urea concentration with the following results:

Urea concentration ($mmol\ dm^{-3}$)	30	60	100	150	250	400
Velocity (mmol urea consumed (mg enzyme)$^{-1}$ min^{-1})	3.37	5.53	7.42	8.94	10.70	12.04

What are the values of K_m and V_{max} for this reaction?

Solution

The data are rearranged into an appropriate form for a linear plot (i.e. $1/v$ vs. $1/[S]$, $v/[S]$ vs. v, $[S]/v$ vs. $[S]$).

From all three plots we deduce that the K_m is 105 mmol dm^{-3} and the V_{max} is 15.2 mmol urea consumed mg^{-1} min^{-1} (this is equivalent to 253.3 katal kg^{-1}). Of course we would not normally do all three plots; this was done here merely for illustrative purposes. It might be noted that the distribution of points are markedly different in the three plots and this fact should be borne in mind when suitable values for substrate concentration are being chosen for such an experiment.

Complications to the Basic Equation

We should note that an equation such as (8.3) is an *initial-rate equation* (i.e. the concentration of S is assumed to remain at its starting value). As the concentration of S falls during the course of a reaction, the rate will, of course, decline too. Also we do not have to consider the reverse reaction or any possible inhibition by product. However, deviations from these initial rate equations can occur for a number of reasons of which we might mention two of the most important.

(a) *Substrate Inhibition*

This occurs with some enzymes at high substrate concentrations and makes the linear plots curved.

An example of this type of effect is seen in the reaction catalysed by insect acetylcholinesterase (acetylcholine + water → acetate + choline) where marked inhibition occurs at acetylcholine concentrations above about 1 mmol dm^{-3}.

(b) *Multiple Binding*

Complex kinetics are often observed when an enzyme is composed of multiple subunits (and therefore possesses a number of active sites). We have already discussed this type of system in connection with ligand binding (Chapter 4) and, although the equations involved are complex, we should note that the treatment of enzyme kinetics has many analogies with the treatment of ligand binding data. The effects of interaction between subunits (which can be manifested as either positive or negative cooperativity) are often of considerable importance in the regulation of enzyme activity, since they render the enzyme activity more sensitive or less sensitive to changes in [S] than is the case with hyperbolic binding (eqn (8.3)).

Enzyme Inhibition

A very important aspect of the study of enzyme catalysed reactions is a study of the effects of inhibitors since this can give information on the active site of an enzyme, its mechanism of action, and possible regulation of physiological significance. Enzyme inhibition can also be discussed in terms of other, more complex schemes but we shall confine ourselves to the simplest cases only.

We shall assume that all the enzyme-containing complexes are in equilibrium with each other (i.e. that the k_2 step does not significantly disturb these equilibria). The general equation for this situation is

$$\frac{1}{v} = \frac{1}{V_{\max}}\left(1 + \frac{[\mathrm{I}]}{K_{\mathrm{ESI}}}\right) + \frac{K_{\mathrm{s}}}{V_{\max}}\left(1 + \frac{[\mathrm{I}]}{K_{\mathrm{EI}}}\right)\frac{1}{[\mathrm{S}]} \tag{8.4}$$

where [S] and [1] refer to substrate and inhibitor respectively. (If we use the steady-state approximation to treat this scheme, the equations derived are more complex, and of little use experimentally.)

The general expression (eqn 8.4) can be simplified by making certain assumptions. These will be discussed as three limiting cases, but it should be noted that several other cases are possible.

Competitive Inhibition

If, in eqn (8.4), $K_{ESI} = \infty$ (i.e. the ES complex cannot combine with I, nor the EI complex with S), the general equation simplifies to:

$$\frac{1}{v} = \frac{1}{V_{max}} + \frac{K_s}{V_{max}}\left(1 + \frac{[I]}{K_{EI}}\right)\frac{1}{[S]} \tag{8.5}$$

This is the situation known as ***competitive inhibition*** (the effect on the Hanes plot is left as exercise in Problem 1).

Effectively, the inhibitor 'pulls' some of the enzyme over into the form of the EI complex. When the concentration of S is increased sufficiently, this effect can be overcome. Thus V_{max} remains the same, but K_m is increased by the factor $(1 + [I]/K_{EI})$.

Many examples of competitive inhibition are known, e.g. carbamylcholine

$$(CH_3)_3\overset{\oplus}{N}—CH_2CH_2—O—\overset{\overset{\displaystyle O}{\|}}{C}—NH_2$$

and several other compounds containing a quaternary N atom act as competitive inhibitors with respect to acetylcholine

$$(CH_3)_3\overset{\oplus}{N}—CH_2CH_2—O—\overset{\overset{\displaystyle O}{\|}}{C}—CH_3$$

in the reaction catalysed by acetylcholinesterase from bovine erythrocytes. It might be thought that, if competitive inhibition were observed, it would prove that the inhibitor and substrate bind to the same site. While in many cases the structural analogy between the competitive inhibitor and the substrate may make this very likely, the conclusion is not necessarily justified. However, it can be said that if (from other data) S and I are known to bind to the same site on the enzyme, then competitive inhibition will be observed.

Non-competitive Inhibition

If, in eqn (8.4) $K_{ESI} = K_{EI}$ (i.e. the bind of S to the enzyme does not affect the binding of I) then

$$\frac{1}{v} = \frac{1}{V_{max}}\left(1 + \frac{[I]}{K_{EI}}\right) + \frac{K_s}{V_{max}}\left(1 + \frac{[I]}{K_{EI}}\right) \cdot \frac{1}{[S]} \tag{8.6}$$

This is known as *non-competitive inhibition*. K_m remains unaffected, whereas V_{max} is decreased by a factor $1/\{1 + ([I]/K_{EI})\}$.

Examples of non-competitive inhibition are rather less common than examples of competitive inhibition in the case of one substrate reactions (although there are many examples of non-competitive inhibition in the case of multi-substrate reactions). In one substrate reactions an example is the inhibition of the chymotrypsin-catalyzed hydrolysis of *N*-acetyl-L-tyrosine ethylester by indole (which behaves as a non-competitive inhibitor towards the substrate).

Uncompetitive Inhibition

If, in eqn (8.4) $K_{EI} = \infty$ (i.e. E cannot combine with I) then

$$\frac{1}{v} = \frac{1}{V_{max}}\left(1 + \frac{[I]}{K_{EI}}\right) + \frac{K_s}{V_{max}} \frac{1}{[S]} \tag{8.7}$$

This is known as simple *uncompetitive inhibition*. Both K_m and V_{max} are affected. Examples of simple uncompetitive inhibition are extremely rare in one-substrate reactions (an example of uncompetitive inhibition is the inhibition by Cl^- of hydration of phenylalanine) but are more often found in multi-substrate reactions (e.g. *S*-adenosylmethionine behaves as an uncompetitive inhibitor towards ATP in the reaction catalysed by yeast *S*-adenosylmethionine synthase). Another example of uncompetitive inhibition is the inhibition by Cl^- of hydration of Co_2 catalysed by carbonic anhydrase.

It should perhaps be noted that the net result of a non-competitive inhibitor (i.e. a decrease in V_{max}, but the same K_m) is *equivalent* to converting some of the enzyme present to an inactive form. Thus, for instance, diisopropylfluorophosphate behaves as a non-competitive inhibitor towards the substrates of chymotrypsin since it inactivates the enzyme by reacting with an essential serine group. We shall restrict the terms 'competitive', 'non-competitive', and 'uncompetitive' to those inhibitors which bind *reversibly* to any enzyme. It is also more useful, in our view, to use these terms to describe the effects that an inhibitor shows on the kinetic plots (since these can be readily determined), rather than to use them to describe the probable relationship between inhibitor and substrate binding sites on the enzyme (since this can be very difficult to establish).

It is worth re-emphasizing that in many cases (especially in the cases of multisubstrate reactions or multiple binding sites) the effects of an inhibitor do not conform exactly to any of the limiting situations discussed here. Such 'mixed' inhibition patterns clearly require different assumptions and equations for their description.

EXAMPLE 8.4 *The effect of choline on the reaction catalysed by insect acetylcholinesterase was studied with the following results:*

[choline] (mmol dm^{-3})	[Acetylcholine] (mmol dm^{-3})				
	0.1	0.15	0.25	0.40	0.70
	(relative velocity (arbitrary units))				
0	28.5	37.5	50.0	61.5	73.7
20	11.9	15.6	20.9	25.7	30.7
40	7.5	9.9	13.2	16.2	19.4

What type of inhibition is being observed in this case?

Solution

Plotting the data in a suitable form we note that K_m is unchanged (0.25 mmol dm^{-3}) in the presence of choline whereas V_{max} is lowered (the values in the presence of 0, 20 and 40 mmol dm^{-3} choline are 100.0, 41.7, and 26.3 respectively. From eqn (10.6) we might note that $V_{max}/(V_{max})_I = 1 + [I]/K_{EI}$, i.e. a plot of $V_{max}/(V_{max})_I$ against [I] should be linear with an intercept on the *y*-axis of 1. This is found to be the case here and the value of K_{EI} can be evaluated as 14.3 mmol dm^{-3}.

The non-competitive inhibition can be interpreted in terms of a second site for choline (i.e. distinct from the active site) and the note on substrate inhibition with this enzyme should be consulted.

TWO SUBSTRATE KINETICS

One substrate kinetics, which we have been discussing up to now, is only of limited applicability as can be seen by considering the types of reactions which enzymes catalyse. Any given enzyme can be placed into one of six categories according to the reaction type.

(1) *Oxidoreductases.*	Catalyse redox reactions in which one substrate is reduced at the expense of a second which is oxidized.
(2) *Transferases.*	Catalyse reactions in which a group is transferred from one substrate to another.
(3) *Hydrolases.*	Catalyse reactions in which a substrate is hydrolysed.
(4) *Lyases.*	Catalyse reactions in which a group is eliminated from a substrate to form a double bond.
(5) *Isomerases.*	Catalyse isomerization reactions.
(6) *Ligases.*	Catalyse the joining together of two molecules at the expense of ATP, or some other 'energy source'.

Enzymes belonging to categories (4) and (5) can be considered as one substrate enzymes, and hydrolases (category (3)) can also be included under this heading, since the second substrate (water) is present in vast excess. However, enzymes belonging to the other three categories clearly catalyse reactions of more than one substrate and it is these reactions that we shall turn to now. We shall indicate the various types of mechanism which can occur, discuss the resulting equations and outline how distinctions between different mechanisms can be achieved.

TYPES OF MECHANISMS

A basic division in two substrate reactions can be made into the following categories:

Those Involving a Ternary Complex

These reactions proceed via EAB and EPQ complexes (A and B are the substrates, and P and Q are the products)

$$E + A + B \rightarrow EAB$$

$$EAB \rightarrow EPQ$$

$$EPQ \rightarrow E + P + Q$$

This category can be subdivided into

(i) Those reactions in which the ternary complex (EAB) is formed in an *ordered* manner, i.e.

$$E + A \rightarrow EA$$

$$EA + B \rightarrow EAB \quad (\text{but } E + B \not\leftrightarrow EB).$$

(ii) Those reactions in which the ternary complex is formed in a *random* manner, i.e.

$$E + A \rightarrow EA \qquad E + B \rightarrow EB$$

or

$$EA + B \rightarrow EAB \qquad EB + A \rightarrow EAB$$

Those Not Involving a Ternary Complex

This category can be subdivided into

(i) Those reactions in which the first product is formed before the second substrate is bound. These cases involve a modification of the enzyme and are known as *enzyme substitution* or *ping-pong* mechanisms.

$$E + A \rightarrow E' + P$$

$$E + B \rightarrow E + Q$$

(ii) Those reactions in which a ternary complex is presumably formed, but its breakdown to yield the first product is very fast (so that the ternary complex is kinetically insignificant). This is the 'Theorell-Chance' mechanism and holds for the oxidation of ethanol catalysed by horse-liver alcohol dehydrogenase.

Kinetic Equations for Two Substrate Reactions

We shall not go into the details of the derivation of the various equations for these possible mechanisms (these are covered in the various books already mentioned). However, these derivations do not involve any new fundamental principles. We apply the steady-state approximation to evaluate the concentrations of the various enzyme containing complexes in the reaction pathway, and then the velocity of the overall reaction is set equal to the concentration of the complex preceding enzyme-regeneration multiplied by the rate constant for the step which regenerates enzyme.

The equations which result from this treatment are of the following form.

For the Ternary Complex Mechanisms

$$v = \frac{V_{max}[A][B]}{K'_A K_B + K_B[A] + K_A[B] + [A][B]} \tag{8.8}$$

where K'_A, K_A, and K_B are constants, the meaning of which we shall discuss shortly.

For the ping-pong Mechanism

$$v = \frac{V_{max}[A][B]}{K_B[A] + K_A[B] + [A][B]} \tag{8.9}$$

Significance of the Parameters in the Kinetic Equations

V_{max} in eqns (8.8) and (8.9) represents the maximum velocity at saturating levels of A and B.

In an *operational* sense, the parameters K_A and K_B in eqns (8.8) and (8.9) represent the Michaelis constants for each substrate (A and B respectively) in the presence of saturating concentrations of the other substrate. This can be readily shown, e.g. in eqn (8.8) by dividing numerator and denominator by [B] and then setting [B] $\to \infty$. The equation reduces to:

$$v = \frac{V_{max}[A]}{K_A + [A]},$$

i.e. K_A is the Michaelis constant for A at saturating levels of B.

A similar procedure, dividing eqn (8.8) through by [A] shows that K_B is the Michaelis constant for B at saturating levels of A. K'_A in eqn (8.8) does not have a similarly simple operational meaning.

In terms of the *mechanisms* of the reactions, K'_A, K_A and K_B represent combinations of rate constants of individual steps in the reactions. Their precise meaning varies according to the type of mechanism under discussion. In the case of the random-order ternary complex mechanisms (*a*) (ii), however, K'_A, K_A, and K_B have relatively simple meanings in terms of equilibrium constants:

$$\begin{array}{ccccc} & K'_A & EA & K_B & \\ E & \rightleftharpoons & & \rightleftharpoons & EAB \to E + \text{products} \\ & K'_B & EB & K_A & \end{array}$$

$$\left(\text{note } K'_{\mathrm{B}} = \frac{K'_{\mathrm{A}}\, K_{\mathrm{B}}}{K_{\mathrm{A}}}\right).$$

Derivation of the Kinetic Parameters from Experimental Data

The four parameters in eqn (8.8) can be derived in the following fashion. Values of the velocity, v, are measured at various values of [A] with the concentration of B constant. The procedure is then repeated at other fixed values of [B].

By taking the inverse of eqn (8.8), i.e.

$$\frac{1}{v} = \left(1 + \frac{K_{\mathrm{A}}}{[\mathrm{A}]} + \frac{K_{\mathrm{B}}}{[\mathrm{B}]} + \frac{K'_{\mathrm{A}}\, K_{\mathrm{B}}}{[\mathrm{A}][\mathrm{B}]}\right)\frac{1}{V_{\max}} \tag{8.10}$$

we see that *primary plots* of $1/v$ vs. $1/[\mathrm{A}]$ (at constant values of [B]) will be linear with

$$\text{a } \textit{slope} \text{ of } \quad \frac{1}{V_{\max}}\left[K_{\mathrm{A}} + \frac{K'_{\mathrm{A}}\, K_{\mathrm{B}}}{[\mathrm{B}]}\right] \tag{8.11}$$

$$\text{and an } \textit{intercept on the y-axis} \text{ of } \quad \frac{1}{V_{\max}}\left[1 + \frac{K_{\mathrm{B}}}{[\mathrm{B}]}\right] \tag{8.12}$$

Thus as [B] *increases,* both the *slope* and the *intercept* will *decrease.*

Secondary plots of the slopes and intercepts of the primary plot vs. $1/[\mathrm{B}]$ can then be made.

Reference to eqn (8.11) shows that a plot of the slopes vs. $1/[\mathrm{B}]$ is linear with a *slope* of $K'_{\mathrm{A}}\, K_{\mathrm{B}}/V_{\max}$ and an *intercept on the y-axis* of $K_{\mathrm{A}}/V_{\max}$. From eqn (8.12) it is seen that a plot of the intercepts vs. $1/[\mathrm{B}]$ is linear with a *slope* of $K_{\mathrm{B}}/V_{\max}$ and an *intercept on the y-axis* of $1/V_{\max}$. Thus all four parameters in eqn (8.8) can be derived from the slopes and intercepts of these secondary plots.

If a similar procedure is adopted for the ping-pong mechanism (for which eqn (8.9) holds) the resulting primary plot is seen to consist of a series of parallel lines.

This is readily seen by considering the inverse of eqn (8.9), i.e.

$$\frac{1}{v} = \left(1 + \frac{K_{\mathrm{A}}}{[\mathrm{A}]} + \frac{K_{\mathrm{B}}}{[\mathrm{B}]}\right)\frac{1}{V_{\max}}$$

Thus a plot $1/v$ vs. $1/[A]$ will have a slope of K_A/V_{max}. This slope is independent of [B], resulting in a set of parallel lines.

EXAMPLE 8.5 *The following data were derived from a study of the reaction catalysed by yeast alcohol dehydrogenase*

$$\text{ethanol} + NAD^+ \rightleftharpoons \text{acetaldehyde} + NADH$$

[Ethanol] (mmol dm^{-3})	[NAD^+] (mmol dm^{-3}) 0.05	0.1	0.25	1.0
	Velocity (katal kg^{-1})			
10	0.30	0.51	0.89	1.43
20	0.44	0.75	1.32	2.11
40	0.57	0.99	1.72	2.76
200	0.76	1.31	2.29	3.67

Determine the kinetic parameters for this reaction.

Solution

A primary plot $1/v$ vs. $1/[NAD^+]$ is made for the various values of [ethanol].

From the primary plot we can tabulate the values of slopes and intercepts:

[Ethanol] (mmol dm^{-3})	10	20	40	200
Slope	0.139	0.095	0.073	0.055
Intercept	0.55	0.38	0.28	0.22

The secondary plots (slopes and intercepts vs. $[\text{ethanol}]^{-1}$) (*b*) and (*c*).

From the *y*-axis intercept of (*c*) (0.2) we deduce that $V_{max} = 1/0.2 = 5$.

From the slope of (*c*) (3.5) we deduce that $K_{ethanol} = 5 \times 3.5 = 17.5$ (mmol dm^{-3}).

From the *y*-axis intercept of (*b*) (0.05) we deduce that $K_{NAD^+} = 5 \times 0.05 = 0.25$ (mmol dm^{-3}).

From the slope of (b) (0.89) we deduce that $K'_{NAD^+} = 5 \times 0.89/17.5 = 0.25$ (mmol dm^{-3}).

So the parameters are

$$\frac{V_{max} = 5 \text{ katal kg}^{-1},}{K'_{NAD^+} = 0.25 \text{ mmol dm}^{-3},} \qquad \frac{K_{NAD^+} = 0.25 \text{ mmol dm}^{-3}}{K_{ethanol} = 17.5 \text{ mmol dm}^{-3}.}$$

Distinction Between the Various Mechanisms

So far we have seen how to distinguish a 'ping-pong' mechanism from one involving a ternary complex (although it must be emphasized that care should be taken to ensure that lines in a primary plot such are truly parallel). To confirm that a particular reaction follows a 'ping-pong' mechanism we should look for (i) the occurrence of half-reactions, i.e. the formation of the first product in the absence of the second substrate, and (ii) the formation of a modified enzyme, on incubation of the enzyme with the first substrate. Classical examples of 'ping-pong' mechanisms are found among the transaminases in which the pyridoxal prosthetic group on the enzyme becomes modified during the course of the reaction:

$$\text{R—CH}(\overset{\oplus}{\text{N}}\text{H}_3)\text{—CO}_2^{\ominus} + \text{E—CHO} \rightleftharpoons \text{R—C(=O)—CO}_2^{\ominus} + \text{E—CH}_2\text{—NH}_3^{\oplus}$$

$$\text{R}'\text{—C(=O)—CO}_2^{\ominus} + \text{E—CH}_2\text{—}\overset{\oplus}{\text{N}}\text{H}_3 \rightleftharpoons \text{R}'\text{—CH}(\overset{\oplus}{\text{N}}\text{H}_3)\text{—CO}_2^{\ominus} + \text{E—CHO}$$

Overall

$$\text{R—CH}(\overset{\oplus}{\text{N}}\text{H}_3)\text{—CO}_2^{\ominus} + \text{R}'\text{—C(=O)—CO}_2^{\ominus} \rightleftharpoons \text{R—C(=O)—CO}_2^{\ominus} + \text{R}'\text{—CH}(\overset{\oplus}{\text{N}}\text{H}_3)\text{—CO}_2^{\ominus}$$

(E—CHO represents enzyme with pyridoxal phosphate prosthetic group.)

On the other hand, from an analysis of the steady-state kinetics we are unable to distinguish an ordered ternary complex mechanism from a random one, since both give rise to an equation of the form of eqn (8.8). We can make the distinction however on the basis of other data which might include the following:

(*a*) *product inhibition patterns.* The type of inhibition shown by products P and Q towards substrates A and B (i.e. is P competitive, non-competitive or uncompetitive with respect to A and/or B?) can be used to indicate a likely mechanism or exclude another.

(*b*) *substrate binding*. In a random order mechanism, substrates A and B should both be able to bind to the enzyme, whereas in the ordered mechanism the second substrate (say, B) cannot bind in the absence of the first substrate A. Determination of whether the enzyme can bind B or not can help to indicate which mechanism is correct.

(*c*) *isotope exchange at equilibrium*. This type of experiment involves measuring the influence of the concentration of substrates (and products) on the rate of exchange of an isotope between substrates and products when the reaction is at equilibrium. Ordered and random mechanisms can be distinguished on this basis.

More complete discussions on these aspects are given in the books by Engel and Ferdinand. However, we might mention that as a result of studies such as these, it has been shown that for instance muscle creatine kinase and yeast alcohol dehydrogenase proceed via random ternary complex mechanisms and malate dehydrogenase and lactate dehydrogenase proceed via ordered ternary complex mechanisms.

Pre-steady-state Kinetics

Up to now we have concentrated on the kinetic behaviour of enzymes when a steady-state has been achieved in reactions where [E] << [S]. However, a great deal of useful information can be obtained from studies of the pre-steady-state period. Usually in these experiments we make the concentrations of enzyme and substrate more nearly equal and special rapid mixing and detection apparatus is required to initiate and monitor the progress of the reaction. Processes with a half-time of the order of a few milliseconds can be followed using suitably designed apparatus (the limiting factor is usually the rate at which two solutions can be properly mixed). A full account of these types of study is given in the books by Fersht and Gutfreund.

Among the types of information available from these studies are the following:

(i) The participation of complexes and intermediates in the reaction pathway can be deduced and the rate constants of the individual steps in the pathway evaluated. Thus the participation of an enzyme substrate complex ($E\text{–}H_2O_2$) was demonstrated in the peroxidase catalysed oxidation of leucomalachite green by hydrogen peroxide. In the case of the reaction catalysed by pig-heart lactate dehydrogenase (where steady-state kinetics had indicated an ordered ternary complex mechanism with NAD^+ binding preceding that of lactate), the rapid reaction studies showed that the overall steady state rate corresponded to that of NADH release (step ④ in the scheme below)

$$E + NAD^+ \rightarrow E^{NAD^+} \xrightarrow[\text{①}]{\text{lactate}} E^{NAD^+}_{lactate}$$

$$\uparrow \qquad\qquad\qquad\qquad\qquad\qquad \text{②}\downarrow$$

$$E \xleftarrow[\text{NADH}]{\text{④}} E^{NADH} \xleftarrow[\text{pyruvate}]{\text{③}} E^{NADH}_{pyruvate}$$

The rates of the elementary steps in the reaction (e.g. steps ①, ②, ③ and ④) can also be deduced affording a more complete description of the kinetics of the enzyme catalysed reaction.

(ii) In a reaction scheme such as that shown above where an intermediate (in this case E^{NADH}) is formed rapidly but breaks down only slowly then initially this intermediate will accumulate. If the breakdown of this intermediate is very slow indeed then a measure of the amount of intermediate accumulated (or other product released as the intermediate is formed) will give a measure of the amount of enzyme undergoing the reaction. This data can be used to evaluate the purity of an enzyme preparation since in effect we are determining the concentration of 'active sites'.

ADDITIONAL STUDIES ON ENZYME MECHANISMS

The studies of steady-state and pre-steady-state kinetics allow us to deduce which enzyme-containing complexes are kinetically significant in the overall reaction pathway and to assign rate constants to some or all of the elementary steps involved. However, if we are going to describe the mechanism completely we need to determine the structure of these complexes and to evaluate what parts of the enzyme molecule are involved in the binding of substrates and the catalytic processes. To answer these questions data from techniques such as X-ray crystallography (which can afford detailed structures of an enzyme and its complexes with substrates and substrate analogues) and chemical modification of amino acid side chains (which can determine which groups on the enzyme are important for the catalysis) are invaluable. Another useful technique in this respect is the study of the variation of enzyme activity with pH since this can, in favourable cases, give a good indication of the importance of certain amino-acid side chains in an enzyme.

The Effect of pH on Enzyme Catalysed Reactions

Changes in pH can have a number of effects on enzyme catalysed reactions. There could be (i) unfolding and consequent inactivation of the protein outside a certain pH range, (ii) changes in the position of equilibrium of a reaction if H^+ appears as a reactant or product in the overall equation, or (iii) ionization of groups in the substrate(s). These possibilities could be checked by performing appropriate experiments. However, we shall confine ourselves to a fourth possibility, namely changes in the ionisation state of amino acid side chains in the enzyme. The equations derived for such processes are similar to those used in Chapter 7 to describe the variation in reaction rate with pH.

The parameters K_m and V_{max} can both change with pH and a full kinetic analysis is necessary at each pH value (to check, for instance, that the substrate is still saturating). It is easier to discuss the variation of V_{max} with pH, since this generally reflects the variation of a single rate constant, whereas K_m is usually a function of several rate constants. (Of course it is possible that the rate determining step could be different at different pH values and this would complicate the analysis considerably.)

If a plot of $\log_{10} V_{max}$ for an enzyme catalysed reaction against pH can be concluded that only the acidic form of the ionizing species is catalytically active. A pK_a for this group can be derived, as before, by extrapolating the linear portions of the curve.

It is tempting to try to assign this observed pK_a to a particular type of amino acid side chain, and hence to implicate this amino acid as being involved in the mechanism of action of the enzyme. However, there are difficulties in this procedure, since the pK_as of amino acid side chains in enzymes might well be very different from the pK_a for the free amino acid. (An example of this is seen in the case of pepsin where an aspartic acid group involved in the mechanism has a pK_a of 1.1 compared with the 'normal' pK_a of the side chain of free aspartic acid (approximately 4.0). A very rough guide to the pK_a of some types of amino acid side chains in proteins is given in Table 8.1.

Also included are some data on the enthalpies of ionization (ΔH_i) of these various groups, and it is possible that this could help in the assignment of pK_a values to specific groups (if the rate dependence on pH were known as a function of temperature).

TABLE 8.1

Ionization properties of some amino acid side chains

Group	pK_a(298 K)	ΔH_i(kJ mol^{-1})
β-Carboxyl (Asp) } γ-Carboxyl (Glu) }	~4	~ ±4
Imidazole (His)	~6	~29
ε-Amino (Lys)	~10	~46
Phenolic OH (Tyr)	~10	~25

Sometimes a second ionizing group may be present in the enzyme such as in the following scheme:

$$\underset{\text{inactive}}{HX\diagdown E \diagup YH} \underset{pK_{a1}}{\overset{-H^+}{\underset{+H^+}{\rightleftharpoons}}} \underset{\text{active}}{HX\diagdown E \diagup Y^-} \underset{pK_{a2}}{\overset{-H^+}{\underset{+H^+}{\rightleftharpoons}}} \underset{\text{inactive}}{X^-\diagdown E \diagup Y^-}$$

The plot of $\log_{10}(V_{max})$ against pH is then a curve of the type.

Again the values of pK_{a_1} and pK_{a_2} can be obtained by extrapolation of the appropriate linear portions of the plot (although if the two pK_as are close, say within 1.5 pH units, the ionizations will not be independent of each other and the pK_a values obtained by such procedures may need correction to obtain the true pK_a values). An example of this type of analysis is afforded by fumarase (in the fumarate → malate direction) in the example below.

EXAMPLE 8.6 *The data listed were obtain for the variation with pH of V_{max} for the reaction catalysed by fumarase at 298 K.*

pH	5.2	5.5	6.0	6.5	6.75	7.0[a]	7.5	8.0	8.5
V_{max} (arbitrary units)	54	105	184	224	236	230	162	86	33

Comment on this data and deduce appropriate pK_a values.

Solution

The plot of $\log_{10} V_{max}$ against pH suggests that two ionizing groups are involved in the catalytic activity of fumarase. From the points of intersection of the linear portions the pK_as can be determined as 5.9 and 7.5 for pK_{a_1} and pK_{a_2} respectively.

EXAMPLE 8.7 *It was found that at 308 K the lower pK_a was the same as at 298 K within 0.1 unit. What is the enthalpy of ionization of this group?*

Solution

Since the pK_a is less than 0.1 unit different.

$$\log_{10}\left(\frac{K_{308}}{K_{298}}\right) \leq 0.1$$

(where K refers to the ionization constant of the group of lower pK_a)

$$\therefore \qquad \ln\left(\frac{K_{308}}{K_{298}}\right) \leq 0.2303.$$

Now from eqn (3.9)

$$\ln\left(\frac{K_{T_2}}{K_{T_1}}\right) = -\frac{\Delta H^\circ}{R}\left(\frac{1}{T_2} - \frac{1}{T_1}\right)$$

we deduce that

$$-\frac{\Delta H^\circ}{R}\left(\frac{1}{308} - \frac{1}{298}\right) \leq 0.2303,$$

i.e. $\Delta H^\circ \leq 17.5$ kJ mol^{-1}.

The enthalpy of ionization of the group is thus 17.5 kJ mol^{-1} or less.

These data suggest that the group involves in the ionization may be a carboxyl group even though the pK_a value (5.9) is rather larger than normal. An imidazole group would be expected to possess a heat of ionization of about 29 kJ mol^{-1} (see Table 8.1).

THE EFFECT OF TEMPERATURE ON ENZYME-CATALYSED REACTIONS

Many enzyme-catalysed reactions follow the normal Arrhenius equation for dependence of reaction rate against temperature. An apparent activation energy for the reaction can thus be derived. This activation energy would be expected to be lower than that for the non-catalysed reaction. It has been shown that the overall rate equation has contributions from the various individual rate processes (characterized by k_1, k_{-1}, k_2, etc.) all of which we would expect to be temperature dependent. A detailed analysis of the system (such as studying the effect of temperature on the rate constants in the E + S $\rightleftharpoons$ ES equilibrium) would be required to derive the values of the activation energies of the various steps in the reaction.

We should note that there are several complications in the study of enzyme catalysed reactions. These include:

(i) Above a certain temperature, the enzyme molecule will become unfolded, so that the three-dimensional integrity of the catalytic site will become lost. The rate at which this process occurs is generally dependent upon the pH, the concentration of substrates or other ligands, ionic strength, etc. For many enzymes incubation at temperatures above about 323 K leads to fairly rapid denaturation, but some enzymes (e.g. those from thermophilic bacteria) remain stable at much higher temperatures. Generally, substrates protect an enzyme to some extent against heat inactivation.

In those cases where the rate of inactivation has been studied as a function of temperature, it has been found that the $\Delta S^{\circ\ddagger}$ for this process is normally very large and positive, as would be expected if the compact, active, structure of the enzyme has become substantially unfolded.

(ii) In a series of consecutive reactions (such as a multistep enzyme-catalysed reaction) with different activation energies, there can be a change in the rate-limiting step of the reaction with temperature. The rate of the reaction with the lower activation energy is less sensitive to temperature, and at a high enough temperature this will be the slow step of the reaction. This situation will result in a change in the slope of the Arrhenius plot at a certain temperature.

(iii) The enzyme may exist in two interconvertible, active forms which possess different activation energies. We would then expect a break in the Arrhenius plot around the temperature where the change-over between the two becomes significant.

Two examples of this type of behaviour are seen with the ATPase enzymes which are involved with the transport of (a) Na^+ and K^+ and (b) Ca^{2+}. In these cases the transition probably arises from structural changes within the tightly bound phospholipid molecules which are associated with these enzymes. The possibility of structural changes in enzymes around the transition temperature could be checked by various means (e.g. sedimentation data, circular dichroism, fluorescence spectra, etc.).

PROBLEMS

1. What are the effects on the Hanes plot ([S]/v vs. [S] of competitive, non-competitive, and uncompetitive inhibitors?
2. The liver contains two enzymes which can convert glucose to glucose-6-phosphate at the expense of ATP. One of these (hexokinase) has a K_m for glucose of 40 μmol dm^{-3} and a maximum catalytic activity of 0.7 μmol glucose transformed min^{-1} (g tissue)$^{-1}$. The other (glucokinase) has a K_m for glucose of 10 mmol dm^{-3} and a maximum catalytic activity of 4.3 μmol min^{-1} (g tissue)$^{-1}$. (The data refer to the enzymes from rat liver.)

   Use these results to calculate the proportion of glucose which is phosphorylated by glucokinase (as compared with hexokinase) in a fasting rat (glucose concentration = 3 mmol dm^{-3}) and in a well-fed rat (glucose concentration = 9.5 mmol dm^{-3}). Assume that ATP is present in saturating concentrations in each case.
3. Fructose biphosphatase (which catalyses the hydrolysis of fructose biphosphate to fructose-6-phosphate and phosphate) is inhibited by AMP. The following data were obtained from a study of the rat liver enzyme:

	[FBP] (μmol dm^{-3})				
	4	6	10	20	40
[AMP] (μmol dm^{-3})	Velocity (katal kg^{-1})				
0	0.059	0.076	0.101	0.125	0.150
8	0.034	0.043	0.056	0.071	0.083

   Comment on these data.
4. The enzyme nucleoside diphosphokinase will catalyse the following reaction:

$$GTP + dGDP \xrightarrow{Mg^{2+}} GDP + dGTP.$$

In an experiment with the enzyme isolated from erythrocytes, the following results were obtained:

	[GTP] (μmol dm^{-3})			
	22	30	50	200
[dGDP] (μmol dm^{-3})	Velocity (katal kg^{-1})			
20	0.095	0.112	0.141	0.196
25	0.102	0.120	0.155	0.223
40	0.112	0.136	0.180	0.284
100	0.125	0.156	0.218	0.385

What can you deduce from these data regarding a likely mechanism for the enzyme catalysed reaction? How could you check your conclusion?

5. **A study was made of the reaction catalysed by rabbit muscle creative kinase:**

$$\text{creatine} + \text{ATP} \xrightarrow{Mg^{2+}} \text{phosphocreatine} + \text{ADP}.$$

The following data were obtained:

	[ATP] (mmol dm^{-3})			
	0.46	0.62	1.23	3.68
[Creatine] (mmol dm^{-3})	Velocity (katal kg^{-1})			
6	0.377	0.463	0.660	0.968
10	0.555	0.678	0.950	1.308
20	0.845	1.005	1.338	1.803
40	1.180	1.378	1.718	2.295

Evaluate the kinetic parameters for this reaction. From other data it appears that the reaction proceeds via a random order ternary complex mechanism. What can you deduce about the binding of substrates to the enzyme?

6. **The effect of one of the products (pyruvate) on the reaction catalysed by rabbit muscle lactate dehydrogenase was studied, with the following results:**

$$NAD^+ + \text{lactate} \rightarrow \text{NADH} + \text{pyruvate}.$$

At a fixed NAD+ concentration (1.5 mmol dm^{-3})

[Pyruvate] (μmol dm^{-3})	[Lactate] (mmol dm^{-3}) 1.5	2.0	3.0	10.0
	Velocity (katal kg^{-1})			
0	1.88	2.36	3.10	5.81
40	1.05	1.34	1.88	4.19
80	0.73	0.94	1.34	3.27

At a fixed lactate concentration (15 mmol dm^{-3})

[Pyruvate] (μmol dm^{-3})	[NAD^+] (mmol dm^{-3}) 0.5	0.7	1.0	2.0
	Velocity (katal kg^{-1})			
0	3.33	3.91	4.50	5.42
30	2.65	3.13	3.60	4.33
60	1.97	2.30	2.66	3.21

What types of inhibition are being observed in these cases?

7. The velocity of the reaction catalysed by pyruvate kinase was studied as a function of phosphoenolpyruvate (PEP) concentration with the results shown in row A.

$$\text{phosphoenolpyruvate} + \text{ADP} \xrightarrow{Mg^{2+}} \text{pyruvate} + \text{ATP}.$$

(It is assumed that ADP is present in saturating concentrations.)

Also given are data on the effects of two inhibitors of this reaction. In row B the data refer to 10 mmol dm^{-3} 2-phosphoglycerate as inhibitor, and in row C, the data refer to 10 mmol dm^{-3} phenylalanine as inhibitor. Comment on these data.

Inhibitor	[PEP] (mmol dm^{-3}) 0.25	0.3	0.5	1.0	2.0
	Velocity (katal kg^{-1})				
A None	1.10	1.25	1.63	2.05	2.43
B 2-Phosphoglycerate	0.73	0.83	1.15	1.68	2.08
C Phenylalanine	0.60	0.68	0.90	1.25	1.50

8. The activity of the enzyme nitrogenase (which catalyses the reduction of nitrogen to ammonia) from *Azotobacter* was studied as a function of temperature with the following results:

T (K)	278	284	290	296	302	308	314	320	326	330
Velocity (arbitrary units)	1.65	11.7	72.2	284	483	781	1180	1260	200	9.5

Comment on these data.

9. The trypsin catalysed hydrolysis of *N*-benzoyl-L-arginine ethyl ester was studied (at 298 K) as a function of pH with the following results.

pH	4.0	4.5	5.0	5.5	6.0	6.5
V_{max} (katal kg^{-1})	0.0028	0.0087	0.027	0.076	0.180	0.320

pH	7.0	7.5	8.0	8.5	9.0
V_{max} (katal kg^{-1})	0.425	0.473	0.491	0.497	0.499

What is the pK_a of the ionizing group in this process? At 308 K the pKa of this group is 6.08. What is the enthalpy of ionization of this group?

Solutions to Problems

1. *Competitive inhibition.* K_m is raised, but V_{max} is unchanged, so the line will have the same slope but a point of intersection on the *x*-axis further to the left of the line in the absence of inhibitor.

   Non-competitive inhibition. K_m is unchanged, but V_{max} is decreased, so the line will have the same point of intersection on the *x*-axis but an increased slope.

   Uncompetitive inhibition. K_m is increased and V_{max} is decreased, so the line will intersect the *x*-axis further to the left and have an increased slope.

2. Applying the standard eqn (8.3) we see that for the *fasting rat*, v (hexokinase) = 0.691 μmol min^{-1}g^{-1} and v (glucokinase) = 0.992 μmol min^{-1} g^{-1}, i.e. the glucokinase contribution is 59 per cent of the total.

   For the *well-fed rat*, v (hexokinase) = 0.697 μmol min^{-1} g^{-1} and v (glucokinase) = 2.095 μmol min^{-1} g^{-1}, i.e. the glucokinase contribution is 75 per cent of the total.

   Thus glucokinase can alter its activity in response to changes in the levels of glucose, whereas hexokinase is effectively working at its maximum rate under these conditions. After a carbohydrate intake (e.g. a meal) the glucokinase in the liver can be used to phosphorylate a large proportion of the extra glucose.

3. From an appropriate plot we deduce that in the absence of AMP the K_m for FDP is 8 μmol dm^{-3} and the V_{max} is 0.185 katal kg^{-1}. AMP behaves

as a ***simple non-competitive inhibitor***, with V_{max} lowered to 0.102 katal kg^{-1}. From eqn (8.6) we can calculate that K_{EI} is 9.8 μmol dm^{-3}.

4. From a primary plot (1/v vs. 1/[dGDP]) we can deduce that the reaction apparently follows a 'ping-pong' mechanism. The kinetic parameters in eqn (8.9) are K_{dGDP} = 49 μmol dm^{-3}, K_{GTP} = 107 μmol dm^{-3}, and V_{max} = 0.79 katal kg^{-1}.

   A likely mechanism would be reaction of the enzyme with GTP to form a phosphorylenzyme and GDP. The phosphorylenzyme would then react with dGDP to give dGTP. Tests which could be done include isolation of the presumed phosphorylenzyme and examination of partial reactions.

5. From the primary and secondary plots we deduce values of K_{ATP} = 0.28 mmol dm^{-3}, K'_{ATP} = 1.7 mmol dm^{-3}, $K_{creatine}$ = 8.5 mmol dm^{-3}, and V_{max} = 3.03 katal kg^{-1}. Referring to the scheme for the random ternary complex mechanism this means that the binding of ATP (strictly speaking MgATP) to the enzyme is enhanced by the prior binding of creatine and vice versa. This is an example of ***substrate synergism***.

6. From the appropriate plots we deduce that pyruvate acts as a competitive inhibitor towards lactate and a non-competitive inhibitor towards NAD^+. Data such as this has been used to indicate that this enzyme obeys an ordered ternary complex mechanism with NAD^+ binding first.

7. The K_m for PEP is 0.4 mmol dm^{-3} and V_{max} is 2.95 katal kg^{-1}. 2-Phosphoglycerate acts as a ***competitive inhibitor*** towards PEP (K_{EI} = 11.5 mmol dm^{-3}). The close structural similarity between these two molecules makes it likely that they bind at the same site on the enzyme.

   Phenylalanine inhibition is more complex and does not belong to any of the three simple categories mentioned. It might be suspected that phenylalanine binds to a site distinct from the substrate site, since there is little structural similarity between the molecules. Present evidence supports this idea and suggests that the inhibition is mediated via changes in the three dimensional structure of the enzyme.

8. From the Arrhenius plot it is seen that there are three distinct phases. (*i*) T < 293 K—the activation energy is 220 kJ mol^{-1} (*ii*) 315 > T > 293 K—the activation energy is 65 kJ mol^{-1}; (*iii*) T > 315 K the enzyme loses activity, presumably because of thermal denaturation. The transition at 293 K may well be due (in this case) to changes in the phospholipids associated with this enzyme.

9. From a plot of log (V_{max}) vs. pH it is seen that the pK_a (298 K) of the ionizing group is 6.25. By applying the van't Hoff isochore the enthalpy of ionization can be calculated to be 29.9 kJ mol^{-1}. Reference to the data in Table 8.1 suggests that this group is likely to be a histidine (only the unprotonated form being active). It is now known that a histidine is involved in the mechanism of action of the enzyme.

9

Spectroscopy

INTRODUCTION

Spectroscopy is the study of the interaction between electromagnetic radiation and matter: it is a phenomenon of quantum mechanics. Detailed information regarding structure and bonding, as well as various intra- and intermolecular processes, can be obtained from the analysis of atomic and molecular spectra. In this chapter we discuss several important spectroscopic techniques.

DEFINITIONS

Absorption and Emission

All branches of spectroscopy can be divided into one of two categories: absorption or emission. The fundamental equation for both absorption and emission is

$$\Delta E = E_2 - E_1 = h\nu \tag{9.1}$$

where E_1 and E_2 are the energies of the two quantized levels involved in a transition. Transitions that result from external irradiation are described as *induced* absorption or emission. Further, an excited molecule may lose an amount of energy in the form of radiation without being irradiated. This is called *spontaneous emission.*

Line Width and Resolution

Every spectral line has a finite, nonzero width which is usually defined as the width at half-height of the peak. If the two levels involved in an absorption have precisely known energy values, then their difference must also be an exactly measurable quantity. In this case we would observe a line of no width, which would be a line of infinite intensity. But atoms and molecules are generally unstable in the excited state and will emit spontaneously. Consequently, the lifetime of an atom or molecule in an excited state is much shorter than it would be in a ground state.

The width of the absorption line is determined by ΔE. The lifetime for the ground state is usually very long and its level is well defined.

Beside "lifetime" broadening, spectral lines can be broadened by collision as well as by other processes, such as chemical exchange reactions. The electronic absorption spectrum of benzene in the vapour state and in solution. The liquid phase has a higher frequency of collision, so excited molecules are more readily deactivated and hence have a shorter lifetime. Consequently, greater line widths are observed.

Related to line width is a quantity called *resolution*, which measures how well the lines are separated from one another. In general, the resolution of any spectrum depends not only on the molecule and its environment but on the spectrometer design as well. We define the *resolving power R* of a spectrometer as its ability to separate radiations at wavelengths λ and $\lambda + \Delta\lambda$, that is,

$$R = \frac{\lambda}{\Delta \cdot \lambda} \tag{9.2}$$

Units

The position of a line corresponds to the difference in energy between two levels involved in a transition. The position can be measured in several different units.

1. *Wavelength.* Wavelength (λ) can be measured in

meter (m)
centimeter (cm)
micrometer (μm)
nanometer (nm)
angstrom (Å)

where

$$1 \text{ m} = 100 \text{ cm}$$
$$1\ \mu\text{m} = 10^{-4} \text{ cm} = 10^4 \text{ Å}$$
$$1 \text{ nm} = 10^{-3}\ \mu\text{m} = 10 \text{ Å}$$
$$1 \text{ Å} = 10^{-8} \text{ cm} = 0.1 \text{ nm}$$

Both Å and nm are commonly used for wavelength. In this text we usually use the nm unit.

2. *Frequency.* Frequency (ν) is given in cps or Hz (Hertz):

$$1 \text{ cycle per second (cps)} = 1 \text{ Hz}$$

3. *Wave Number.* Wave number ($\bar{\nu}$) is the number of waves per centimeter.

$$\bar{\nu} = \frac{1}{\lambda} = \frac{\nu}{c} \tag{9.3}$$

Depending on the particular branch of spectroscopy, any one of these units may be employed for labeling a spectrum. No confusion should arise as long as we remember the fundamental equation, $c = \nu\lambda$.

Intensity

The *intensity* of absorption lines depends on the difference in population (the number of molecules occupying a particular energy level) between the two levels. The ratio of populations in these two levels is given by the Boltzmann expression.

$$\frac{n_2}{n_1} = e^{-(E_2 - E_1)/kT} \tag{9.4}$$

where n_1 and n_2 are the populations in the lower and upper levels, k is Boltzmann's constant, and T is the absolute temperature. The greater the difference in population ($n_1 - n_2$), the more intense the absorption line will be.

The reader should keep in mind that any type of spectrum (absorption or emission) is actually a superposition of a very large number of spectra from individual molecules. It is normally impossible to detect the energy absorbed or emitted by a single molecule. Further, the interaction between a photon of electromagnetic radiation and a molecule can give rise to only one transition and hence one line. When we look at any spectrum containing more than one line, we see the statistical sum of all the spectra, each consisting of one line caused by one specific transition.

SELECTION RULES

Transitions may not take place between any two levels in an atom or molecule even if the frequency of radiation appropriate to the resonance condition ($E_2 - E_1 = h\nu$) is present. Generally, a transition has to obey certain *selection rules*, which are theoretical quantities obtained by quantum-mechanical calculations. Transitions, then, are classified as *allowed* (having a high probability) or *forbidden* (having a low probability), depending on how they occur according to the selection rules.

Theoretically, we predict two types of transitions as forbidden: spin-forbidden transitions and symmetry-forbidden transitions.

Spin-Forbidden Transitions. Spin-forbidden transitions involve a change of *spin multiplicity*. The spin multiplicity is given by the value $(2S + 1)$, as illustrated in Table 9.1. The numerical value of $(2S + 1)$ tells us the number of different ways the unpaired spins can line up in an external magnetic field. The selection rule is that in a transition there must not be a change in the spin multiplicity; that is, we must have $\Delta S = 0$. For example, a transition from a singlet to a triplet, or vice versa, is normally strongly forbidden.

Symmetry-Forbidden Transitions. A quantitative measure for the intensity of a transition is provided by the quantity called *transition dipole moment* μ_{ij}, where

$$\mu_{ij} = \int \psi_i \, \mu \psi_j \, d\tau \tag{9.5}$$

where ψ_i and ψ_j are the wave functions for states *i* and *j* and μ is the dipole moment vector connecting the two states. The integration is taken over all coordinates, and $d\tau$ represents the volume element ($d\tau = dx\, dy\, dz$). Unless the product ψ_i, $\mu\psi_j$ is of a certain symmetry, this integral will be zero and no transition can occur. Consider the electronic transition in a molecule as an example. The physical significance of the dipole-moment vector is that it is related to the difference

TABLE 9.1 **Spin Multiplicity of Atoms and Molecules**

Number of Unpaired Electrons	Electron Spin S	$2S + 1$	Multiplicity
0	0	1	Singlet
1	$\frac{1}{2}$	2	Doublet
2	1	3	Triplet
3	$\frac{3}{2}$	4	Quartet
.	.	.	.
.	.	.	.
.	.	.	.

in the electronic dipole moments of the initial and final state. This difference in electronic dipole moment arises as the result of a different electron distribution in the two states and can be thought of as representing charge migration during the transition. When the integral becomes zero, the transition is totally forbidden.

Complex mechanisms cause various degrees of breakdown in the selection rules. Consequently, some transitions predicted as forbidden may appear as weak lines.

Spectrophotometry

Spectrophotometry refers to the measurement of absorption of electromagnetic radiation (e.g. light) by compounds. The basic condition for absorption of radiation of a given frequency ν is that there are two *energy levels* of the compound separated by an amount of energy E where

$$\boxed{E = h\nu} \tag{9.6}$$

h is Planck's constant, 6.63×10^{-34} Js.

What are these energy levels? The results of quantum mechanics show that a molecule can possess different types of energy such as that associated with rotation of the molecule (or parts of the molecule) about certain axes, or that associated with vibration of atoms in the chemical bonds, etc. However, the type of energy with which we shall principally be concerned is known as *electronic energy*. This is determined by the distribution of electrons within the various available orbitals of the molecule. Absorption of radiation promotes electrons from a given energy level to one of higher energy (Fig. 9.1).

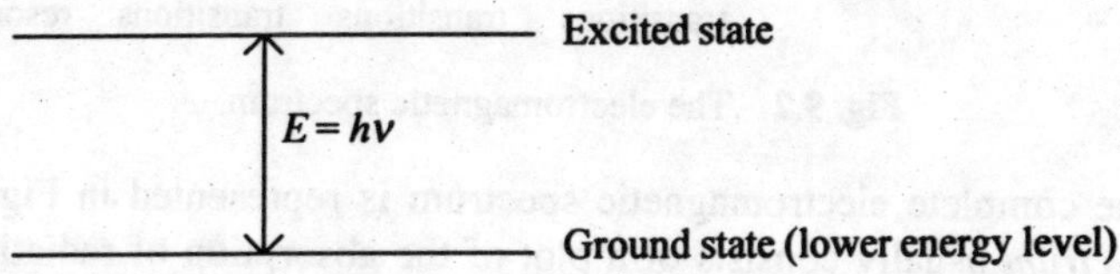

Fig. 9.1 Absorption of radiation of frequency ν.

EXAMPLE 9.1 *What is the energy corresponding to radiation of wavelength 589 nm? (This is the so called sodium D line—commonly observed by putting sodium salts into flames.)*

Solution

The frequency ν is related to wavelength λ by:

$$\nu = c/\lambda$$

where c is the velocity of light = 3×10^8 m s^{-1}.
Thus

$$\nu = \frac{3 \times 10^8}{589 \times 10^{-9}} = 5.1 \times 10^{14} \text{ Hz.}$$

(N.B. 1 nm = 10^{-9} m) From the formula $E = h\nu$ we obtain

$$E = 6.63 \times 10^{-34} \times 5.1 \times 10^{14} \text{ J}$$

$$E = 3.38 \times 10^{-19} \text{ J}$$

This is the energy associated with *one* atom. If we wished to express it as an energy per mole, we must multiply this number by Avogadro's number (6.03×10^{23}), whence

$$E = 2.04 \times 10^5 \text{ J mol}^{-1}$$

$$E = 204 \text{ kJ mol}^{-1}.$$

Increasing wavelength →

10^{-14} m ———————————————— 10^{+5} m

Decreasing energy →

Cosmic radiation	α-Rays	X-Rays	Ultra-violet visible	Infra-red	Microwave	Radio
			⟷ Electronic transitions	⟷ Vibrational transitions	⟷ Rotational transitions	⟷ (Magnetic resonance)

Fig. 9.2 The electromagnetic spectrum.

The complete electromagnetic spectrum is represented in Fig. 9.2. A *spectrum* usually consists of a plot of the absorption of radiation as a function of the wavelength or frequency. So far we have dealt with the wavelength (i.e. energy) of absorption. We now consider the amount of absorption. This is described by the *Beer-Lambert Law*.

The Beer-Lambert Law states that

$$\boxed{\log_{10}\left(\frac{I_0}{I_t}\right) = \varepsilon cl} \qquad (9.7)$$

where I_0 is the radiation incident on the compound, I_t is the radiation transmitted by the compound (i.e. $I_0 - I_t$ is the radiation absorbed) c is the concentration of the compound, and l is the length of cell through which radiation actually travels (i.e. the path length), ε is known as the *extinction coefficient* of the compound at the particular wavelength of the incident radiation and is thus a measure of the absorbing power of the compound. The units of ε will depend on those chosen for c and l. The term $\log_{10} (I_0/I_t)$ is known as either the *absorbance* (A) of the compound and is clearly dimensionless and hence the units of the product (εcl) must also be dimensionless.

Theoretically, A can vary from zero to infinity. In experimental designs, however, A is usually measured on a scale between zero and 2. (Note that A is dimensionless.) Beer's law applies to all types of absorption spectroscopy and holds only for relatively low concentrations. At high concentrations ($\gtrsim$ 0.5 M), deviations occur and A is no longer a linear function of concentration c.

Another useful quantity is the *transmittance* (T), defined as

$$T = \frac{I}{I_0} \tag{9.8}$$

$$-\log T = A = \varepsilon bc$$

The per cent transmittance is given by $(I/I_0) \times 100$.

EXAMPLE 9.2 *15.8 per cent of the 340 nm radiation incident on a certain solution of NADH is transmitted. Given that the extinction coefficient of NADH at this wavelength is 6.22×10^6 cm^2 mol^{-1}, what is the concentration of NADH in the solution? The path length is 1 cm.*

Solution

If

$$I_t = \frac{15.8}{100} I_0$$

the *absorbance* (A) is given by

$$A = \log_{10}\left(\frac{I_0}{I_t}\right)$$

$$= 0.801.$$

Now

$$A = \varepsilon Cl$$

and

$$\varepsilon = 6.22 \times 10^6 \text{ cm}^2 \text{ mol}^{-1},$$
$$l = 1 \text{ cm}.$$

Hence

$$C = \frac{0.801}{6.22 \times 10^6} \text{ mol cm}^{-3}$$

$$= \frac{0.801}{6.22 \times 10^3} \text{ mol dm}^{-3}$$

$$C = 1.29 \times 10^{-4} \text{ mol dm}^{-3}$$

Since the units of ε were given as cm^2 mol^{-1} and that of l was cm, the concentration is expressed as mol cm^{-3}. This is converted to mol dm^{-3} by multiplication by 10^3.

EXAMPLE 9.3 *Calculate the absorbance at 340 nm of (a) a 1 mmol dm^{-3} solution and (b) a 1 μmol dm^{-3} solution of NADH in a cell of path length 1 cm. Calculate the percentage transmitted radiation in these two cases.*

Solution

From the Beer–Lambert Law

$$A = \varepsilon Cl.$$

ε is in units of cm^2 mol^{-1}, so C is expressed in mol cm^{-3}.

Case (*a*)

$$A = (6.22 \times 10^6)\,(1 \times 10^{-3} \times 10^{-3})\,(1)$$

i.e.

$$A = 6.22$$

$$\log_{10}\left(\frac{I_0}{I_t}\right) = 6.22$$

i.e.

$$I_t = 6.01 \times 10^{-5}\%\ I_0.$$

Case (*b*)

$$A = (6.22 \times 10^6)\,(1 \times 10^{-6} \times 10^{-3})\,(1)$$

i.e. $A = 0.00622.$

In this case

$$I_t = 98.6\%\ I_0$$

Appropriate Concentration Ranges

Clearly it would be extremely difficult to measure the concentration of either of the solutions in the worked example accurately by spectrophotometry. In the first case the intensity of transmitted radiation is negligibly small, whereas in the second case the amount of radiation actually absorbed is very small. In practice I_t should be between 10 per cent and 90 per cent of I_0 (i.e. the absorbance should be between 1.0 and 0.05), but the most accurate measurements will be where $I_t \sim 50$ per cent of I_0 (i.e. absorbance = 0.3). These conditions can be realized by alteration of the concentration of the absorbing compound, the sample path length or by changing the wavelength of the radiation (i.e. by changing ε). The first procedure is most commonly employed.

EXAMPLE 9.4 *By what factor would you dilute 1 mmol dm^{-3} solution of NADH to measure its absorbance at 340 nm accurately? The path length can be assumed to be 1 cm.*

Solution

The absorbance can be most accurately measured when it is approximately equal to 0.3.

Now $A = \varepsilon cl$.

For a path length of 1 cm, $A = 0.3$ corresponds to a concentration of 4.8×10^{-5} mol dm^{-3} (i.e. 48 μmol dm^{-3}), i.e. we must dilute the 1 mmol dm^{-3} solution 20 fold to determine the absorbance accurately.

If it is undesirable to dilute the solution, we could use short pathlength cells (e.g. 1 mm) and measure the absorption of radiation at a different wavelength, e.g. 366 nm (where ε is known to be 3.3×10^6 cm^2 mol^{-1}).

The Beer–Lambert Law is very widely used to provide a simple means of estimating the concentration of an absorbing compound. Thus the concentrations of solutions of nucleotides and nucleic acids can be estimated by their absorption of 260 nm radiation. For proteins a suitable wavelength is 280 nm. Of course in each case we have to know the value of ε.

Two Absorbing Compounds.

In many systems of interest there is more than one absorbing compound, and in some cases the spectra of these overlap

The Beer–Lambert Law is, of course, applicable to each compound separately. However in the region marked X, the total absorbance contains contributions from both species. We can determine the concentrations of the individual compounds either by making measurements at a wavelength where there is no overlap, or by making measurements at two wavelengths and using simultaneous equations, as illustrated in the example below.

EXAMPLE 9.5 *The absorbance of a sample of (enzyme + AMP) is 0.46 at 280 nm and 0.58 at 260 nm. Calculate the concentration of each component given that for the enzyme*

$$\varepsilon_{280} = 2.96 \times 10^7 \text{ cm}^2 \text{ mol}^{-1} \text{ and } \varepsilon_{260} = 1.52 \times 10^7 \text{ cm}^2 \text{ mol}^{-1}$$

and for AMP

$$\varepsilon_{260} = 1.5 \times 10^7 \text{ cm}^2 \text{ mol}^{-1} \text{ and } \varepsilon_{280} = 2.4 \times 10^6 \text{ cm}^2 \text{ mol}^{-1}.$$

The pathlength is 1 cm.

Solution

Let the concentrations of enzyme and AMP be x and y mmol dm^{-3} (i.e. μmol cm^{-3}) respectively.

Since $A = \varepsilon c l$ for each component and

$$(\text{Absorbance})_A = (\text{Absorbance of enzyme})_A + (\text{Absorbance of AMP})_A,$$

where λ is the given wavelength, we obtain

at 260 nm: $0.58 = 15.2x + 15y$

and at 280 nm: $0.46 = 29.6x + 2.4y$.

Solution of these two equations gives $x = 0.0135$, $y = 0.0252$, i.e. the concentrations of enzyme and AMP are 13.5 μmol dm^{-3} and 25.2 μmol dm^{-3} respectively.

ISOSBESTIC POINTS

A special case of overlapping spectra occurs when the two compounds concerned are in equilibrium with each other. Consider the ionization of *ortho*-nitrotyrosine:

$$\underset{\text{(I)}}{\text{OH, } NO_2 \text{, } CH_2} + H_2O \rightleftharpoons \underset{\text{(II)}}{\text{O, } NO_2 \text{, } CH_2} + H_3O^+$$

The neutral molecule (I) and its anion (II) have maximum absorptions at 360 nm and 428 nm respectively. (The wavelength of maximum absorption is often abbreviated as λ_{max}.)

As we alter the relative contributions of the two compounds (in this case by changing the pH) there will be a shift in the maximum of the absorption spectrum.

Although the intensity of absorption varies in going from (I) to (II), there is one wavelength at which the intensity remains constant. This is known as an *isosbestic point*. If an isosbestic point is observed while some experimental parameter (e.g. pH) is varied in a titration, this means that *two* (and only two) absorbing species are in equilibrium with each other. Measurements of absorption at the isosbestic point are often used to obtain the *total* concentration of absorbing species. Of course, in order to obtain the individual concentrations, we must make measurements at two wavelengths as described previously.

INFRARED SPECTROSCOPY

Infrared (ir) *spectroscopy* deals with the vibrational motion of molecules. Again, the simplest system is a diatomic molecule that is treated as a *simple harmonic oscillator*.

Infrared spectroscopy is a highly useful technique for chemical analysis. The complexity of molecular vibrations virtually assures that two different molecules cannot produce identical ir spectra. Matching the IR spectrum of an unknown with that of a standard compound, a procedure called *fingerprinting*, is the unequivocal method of identification. Over 100,000 reference spectra have been recorded for fingerprinting. The details of an ir spectrum reveals much information about the structure and bonding of the molecule.

The IR technique is of great value in the study of intra- and intermolecular hydrogen bonding. Formation of a hydrogen bond,

$$A—H\cdots B$$

will cause the A—H stretching band to broaden and shift to lower frequencies. In addition, new bands might appear as a result of the torsional and bending motions of the system.

ELECTRONIC SPECTROSCOPY

Both visible and ultraviolet (uv) spectra involve electronic transitions. Within the framework of molecular orbital theory, electrons can be classified as σ, π, or n (nonbonding), depending on the orbitals in which they reside. A gross selection rule, which is based on symmetry considerations, requires that only transitions of the type $\sigma \rightarrow \sigma^*$, $\pi \rightarrow \pi^*$, and $n \rightarrow \pi^*$ are allowed. Since the energy difference between σ and σ^* orbitals is quite large, these transitions usually fall in the uv region. On the other hand, $\pi \rightarrow \pi^*$ and $n \rightarrow \pi^*$ transitions may fall in either the uv or visible region.

An electronic transition is always accompanied by simultaneous changes in rotational and vibrational states. Large moments of inertia in polyatomic molecules rule out the observation of rotational fine structures and vibrational transitions usually appear as broad, unresolved bands. Analogous to the group vibrational frequencies in IR, electronic spectra can often be characterized by functional groups such as $>C=C<$, $>C=O$, $—NO_2$, and —Ph, which are called *chromophores*. The actual maximum adsorption wavelength for these chromophores depends not only on the molecule involved but also on the environment, being affected by such variables as solvent and temperature.

The visible region is roughly between 400 and 750 nm. An object appears colored because it selectively absorbs certain wavelengths and transmits (or reflects) others in this region. For example, copper sulfate solution, which absorbs in the yellow region, appears as yellow's complementary color, blue. Table 9.2 lists the various wavelengths absorbed in the visible region and the complementary color observed.

Many transition-metal complexes are colored. A particularly simple system is $Ti(H_2O)_6^{3+}$. Since there is only one d electron present and hence only one transition, analysis is fairly straightforward. $Ti(H_2O)_6^{3+}$

is an example of *d-d* transition: that is, transition takes place from one *d* level to another.

TABLE 9.2 **Relationship of Wavelength to Color**

Wavelength Absorbed (nm)	Color Observed (transmitted)
400 (violet)	Greenish yellow
425 (dark blue)	Yellow
450 (blue)	Orange
490 (blue-green)	Red
510 (green)	Purple
530 (yellow-green)	Violet
550 (yellow)	Dark blue
590 (orange)	Blue
640 (red)	Bluish green
730 (purple)	Green

When several *d* electrons are present, analysis can be quite complex. In addition, we may also have charge-transfer bands as a result of metal-ligand interactions.

In saturated organic molecules, such as the alkanes, the transitions are of the $\sigma \rightarrow \sigma^*$ type. Aromatic molecules or compounds containing $>C=C<$ and $>C=O$ chromophores also have $\pi \rightarrow \pi^*$ and $n \rightarrow \pi^*$ transitions. Diphenylpolyenes $C_6H_5—(CH=CH)_n—C_6H_5$ are an interesting series as the chromophore $>C=C<$ has an important effect on their color. The transition from the highest filled π orbital to the lowest empty π^* orbital falls in the uv for $n = 1$ and 2. As n increases, there is a gradual shift toward the visible region, and the color changes from pale yellow for $n = 3$ to greenish black for $n = 15$. Similarly, the highly conjugated porphyrin system is green in the chlorophyll molecule (complexed with Mg^{2+}) and red in hemoglobin (complexed with Fe^{2+}). In these cases the color of the complex is greatly affected by the extent of electron delocalization and the metal ion present.

The electronic spectra of most amino acids arise from the $\sigma \rightarrow \sigma^*$ transitions that occur in the far uv, below 230 nm. Phenylalanine, tryptophan, and tyrosine are the only exceptions. These molecules contain the —Ph chromophore, which absorbs strongly above 250 nm. Absorbance at 280 nm, due mainly to the tryptophan and tyrosyl residues, is used to measure the concentration of protein solutions.

The optical properties of DNA and RNA are studied in terms of the purines (adenine and guanine) and pyrimidines (cytosine, thymine, and uracil). The concentration of nucleic acid solutions is determined by the absorbance at 260 nm. Both DNA and RNA exhibit an interesting phenomenon called *hypochromism*. In general, the molar absorptivity of intact DNA is some 20 to 40% lower than that which we would expect from a knowledge of the total number of nucleotides present. For example, molar absorptivity of calf thymus DNA at 260 nm increases from about 6500 to 9500 when the polymer undergoes thermal denaturation.

Charge-Transfer Spectra

A special type of electronic spectra arises from charge-transfer interaction between a pair of molecules. When tetracyanoethylene, an acceptor, is dissolved in carbon tetrachloride, a colorless solution is obtained. Upon the addition of a small amount of a donor aromatic hydrocarbon, such as benzene or toluene, the solution immediately turns yellow. Many similar reactions have been observed. In 1952, Mulliken proposed the following scheme to explain the observed spectra:

$$\mathrm{D + A} \rightleftharpoons \underbrace{[(\mathrm{D, A}); (\mathrm{D^+, A^-})]}_{\text{ground state}} \xrightarrow{h\nu} \underbrace{[(\mathrm{D, A}); (\mathrm{D^+, A^-})]^*}_{\text{excited state}}$$

where D and A are the donor and acceptor molecules and (D, A) and (D^+, A^-) represent the covalent and ionic resonance structures of the charge-transfer complex. In the ground state the normal van der Waals forces are responsible for holding the molecules together, and there is little if any actual transfer of charge from D to A. However, when the complex is excited by a suitable wavelength, a large charge transfer takes place, and the ionic structure will now make a major contribution to the excited state. If the exciting wavelength falls in the visible region, the solution will appear colored. There is an interesting difference between this electronic transition and the normal absorption. Here an electron is excited from a lower level (bonding molecular orbital) in the donor molecule to a higher level (antibonding molecular orbital) in the acceptor molecule.

The tendency for charge-transfer formation generally depends on the ionization potential of the donor and the electron affinity of the acceptor. Charge-transfer spectra have also been observed for many transition-metal complexes. There the interaction is between the various ligands and the mental ion, with the former as the donor and the latter the acceptor.

The visible and uv technique is not as reliable for compound identifications as is the IR technique, since an electronic spectrum generally does not possess the fine details of an ir spectrum. However, electronic spectroscopy is a useful tool in quantitative analysis. The concentration of a solution can be readily determined from Beer's law by measuring the absorbance (if the molar absorptivity is known). An interesting situation may arise if a solution contains two absorbing substances in equilibrium whose bands overlap. At some wavelength over the overlapped region, the molar absorptivities of the two species may be equal. If the sum of the concentrations of these two compounds in solution is held constant, there will be *no* change in absorbance at this wavelength as the ratio of these two compounds is varied. This invariant point is called the *isosbestic point*. The existence of one or more isosbestic points in a system is a good indication of chemical equilibrium between two compounds.

NUCLEAR MAGNETIC RESONANCE SPECTROSCOPY

The fundamental properties of a proton are (1) mass, (2) charge, and (3) spin, represented by the symbol I. I can be zero, an integer, or a half-integer. The actual value of I depends on the particular nucleus.

Since the spinning motion of a charge particle generates a magnetic field, nuclei that have spin $I \neq 0$ will possess a *magnetic dipole moment*, μ_I. When these nuclei are placed between the poles of an external magnet, the energy of interaction is given by

$$E = -\mu_I H \cos\theta \tag{9.10}$$

where H is the external magnetic field and θ the angle between the vectors μ_I and H. The minus sign indicates that this is an attractive interaction. Classically, θ can have any value so that E varies continuously. Quantum mechanics restricts μ_I in the magnetic field to certain allowed orientations characterized by the nuclear spin quantum number $M_I = \pm \frac{1}{2}$. The energy is now given by

$$E = -\frac{\mu_I}{I} H M_I \tag{9.11}$$

In the absence of the magnetic field, that is, when $H = 0$, these two orientations are equivalent in energy, but degeneracy is removed when an external field is turned on. The splitting of the energy levels, together with its classical analog. If the nucleus is now irradiated with the appropriate frequency such that

$$E_2 - E_1 = \Delta E = h\nu = 2\mu_I H \tag{9.12}$$

transition from the lower to higher level can be induced. Such a transition gives rise to the phenomenon known as *nuclear magnetic resonance* (*nmr*). Table 9.4 lists some nuclei for which the nmr technique is applicable.

The basic features of an nmr spectrometer are (1) a source of radiation, (2) a receiver coil to detect the absorption of energy, (3) a dc magnetic field, and (4) a recorder or an oscilloscope to display the signals. The sample tube is placed between the poles of the magnet operating at about 14,000 gauss (G), which for protons corresponds to a frequency of 6×10^7 Hz, or 60 MHz. This falls in the radio-frequency range. According to Eq. (9.12) there are two variables: ν and H. Experimentally, it is more convenient to hold the frequency constant and vary H until the resonance condition is met. The absorption of energy by the nuclei is then detected, amplified, and displayed on a recorder or oscilloscope.

TABLE 9.4 **Nuclei Suitable for Nmr Study**

Isotope	Nuclear Spin (I)	Natural Abundance (per cent)
^{1}H	$\frac{1}{2}$	99.985
^{13}C	$\frac{1}{2}$	1.5×10^{-2}
^{14}N	1	99.63
^{15}N	$\frac{1}{2}$	0.37
^{17}O	$\frac{5}{2}$	3.7×10^{-2}
^{19}F	$\frac{1}{2}$	100
^{31}P	$\frac{1}{2}$	100

Since many compounds contain hydrogen atoms. ^{1}H nmr or *proton magnetic resonance* (*pmr*) is the most studied case.

ELECTRON SPIN RESONANCE SPECTROSCOPY

Electron spin resonance (esr) is very similar to nmr in theory. The spinning motion of an electron also generates a magnetic field and the orientation of the electron magnetic moment in an external magnetic field is characterized by the electron spin quantum number $M_S = \pm \frac{1}{2}$. The resonance condition is usually expressed as

$$\Delta E = h\nu = g\beta H \tag{9.13}$$

where g is a dimensionless constant, called the Landé g factor, equal to 2.0023,* and β is the *Bohr magneton*, given by $eh/2\pi mc$, where e and m are the electronic charge and mass and c is the velocity of light.

Because the electron magnetic moment is about 600 times greater than that for a proton, the experimental setup for esr experiments is different from that for the nmr. Normally, an esr measurement is carried out in a magnetic field of about 3400 G and a frequency of 9.5×10^9 Hz or 9.5 GHz, which falls in the microwave region. Spectrometer designs usually require that esr lines be presented as the first derivative of the absorption lines.

In most molecules electrons regularly occur in pairs with opposite spins as required by the Pauli exclusion principle; hence esr experiments cannot be performed. A few molecules, such as O_2, NO, NO_2, and ClO_2, do contain one or more unpaired electrons in their stable ground state, and the esr spectrum for each molecule has been observed. Many transition-metal ions, such as Fe^{3+}, Mn^{2+}, and Cu^{2+}, also possess unpaired *d* electrons. This fact is of particular importance in biochemistry because these paramagnetic ions are required cofactors for many proteins and enzymes.

Although only one transition and therefore one line is observed for isolated electrons, or electrons trapped in a matrix, the esr spectrum of hydrogen atoms consists of two lines of equal intensity. This *hyperfine splitting* results from the magnetic interaction between the unpaired electron and the nucleus, analogous to the spin-spin interaction discussed earlier for the nmr spectrum of ethanol. Only two transitions are allowed, however, because of the selection rules, $\Delta M_I = 0$ and $\Delta M_S = \pm 1$. An interpretation of the selection rules is that the motion of a nucleus is much slower than that of an electron, so that during the time it takes for an electron to change its orientation, the nuclear spin has no time to reorient. The separation between these two lines gives the *hyperfine splitting constant* (*A*).

PROBLEMS

1. What energy (in kJ mol^{-1}) does electromagnetic radiation of the following wavelengths represent.
   (*a*) 250 nm (typical value for ultraviolet (u.v.) absorption).
   (*b*) 5000 nm (typical value for infrared (i.r.) absorption).
   (*c*) 1 cm (typical value for electron spin resonance (e.s.r.) absorption).
   (*d*) 500 cm (typical value for nuclear magnetic resonance (n.m.r.) absorption).
   (*e*) 247 m (typical value for B.B.C.).
2. The human eye (when completely adapted to darkness) is able to perceive a point source of light against a dark background when radiation falls on the retina at a rate of 2×10^{-16} J s^{-1}. What is the minimum rate of incidence of quanta of radiation (photons) on the retina which can be perceived, assuming an incident wavelength of 550 nm?
3. The absorbance at 280 nm of a 1 g dm^{-3} solution of adenylate kinase (M.W. = 21,000) is 0.53, in a 1 cm cell. Calculate ε, and state clearly the units.
4. The absorbance (in a cell of 1 cm path length) of a solution containing NAD^+ and NADH is 0.21 at 340 nm and 0.85 at 260 nm. The extinction coefficient of NAD^+ and of NADH at 260 nm is 1.8×10^7 cm^2 mol^{-1}, while the value at 340 nm for NADH is 6.22×10^6 cm^2 mol^{-1}. (NAD^+ does not absorb at 340 nm.)

   Calculate the concentrations of NAD^+ and NADH in the solution.
5. The enzyme creatine kinase (CK) is often assayed according to the following scheme:

$$\text{creatine} + \text{ATP} \xrightarrow{\text{CK}} \text{creatine}-\text{Ⓟ} + \text{ADP}$$

$$\text{ADP} + \text{Ⓟ-enolpyruvate} \xrightarrow[\text{kinase}]{\text{pyruvate}} \text{ATP} + \text{pyruvate}$$

$$\text{pyruvate} + \text{NADH} \xrightarrow[\text{dehydrogenase}]{\text{lactate}} \text{lactate} + \text{NAD}^+$$

   The oxidation of NADH is monitored at 340 nm (where NAD^+ does not absorb). In a cell of path length 1 cm the absorbance of a 1 mmol dm^{-3} solution of NADH at 340 nm is 6.22.

   (*a*) 0.02 cm^3 of a solution of creatine kinase whose concentration is 0.08 g dm^{-3} is added at 298 K to 1 cm^3 of assay mixture (containing all the appropriate substrates, cofactors and coupling enzymes in saturating amounts), in a cell of path length 1 cm. The decrease in absorbance at 340 nm is 0.52 per minute. What is the activity of the creatine kinase solution in μmol substrate consumed min^{-1} (mg enzyme)$^{-1}$. (These are known as International Units of enzyme activity.) What is this activity expressed as katal kg^{-1} (SI units; see Chapter 10).

(b) The above assay system can also be used to determine the concentration of a solution of creatine.

A stock solution of creatine was diluted 50 fold into an appropriate assay mixture. After addition of creatine kinase the absorbance at 340 nm had decreased by 0.8. Calculate the concentration of creatine in the stock solution. (It can be assumed that the creatine kinase equilibrium is driven well over to the right.)

6. In the reaction catalysed by glutamate dehydrogenase

$$\text{Glutamate}^- + NAD^+ + H_2O \rightleftharpoons \alpha\text{-ketoglutarate}^{2-} + NADH + NH_4^+ + H^+$$

the initial concentrations of glutamate and NAD^+ are 20 mmol dm^{-3} and 1 mmol dm^{-3} respectively. The reaction was carried out in phosphate buffer at pH = 7 and the concentration of NADH was measured spectrophotometrically. The absorbance at 340 nm (in a cell of path length 1 cm) increased until a constant value of 0.561 was reached. Calculate the equilibrium constant for the above reaction. (ε for NADH = 6.22×10^6 cm^2 mol^{-1} at 340 nm). Comment on the major sources of error.

7. The haemoglobin (Hb) content of a solution can be determined by treating the solution with a reagent solution containing an excess of ferricyanide and cyanide. This converts both Hb and oxy-Hb to cyanomet-Hb which can be determined from its absorbance at 540 nm. The following results were obtained with a standard haemoglobin solution (0.6 g dm^{-3}):

Volume (cm^3) of standard Hb solution added to reagent (total volume = 5 cm^3)	Absorbance at 540 nm
0	0.025
1	0.090
2	0.160
3	0.230
4	0.290

When 0.01 cm^3 of a blood sample was added to 5 cm^3 of the reagent solution, the absorbance at 540 nm was 0.19. What is the concentration of Hb in the blood sample?

Solutions to Problems

1. Using $E = h\nu$ and $s = c/\lambda$, we obtain the following energies.

(*a*) 479.7 kJ mol^{-1}.
(*b*) 24.0 kJ mol^{-1}.
(*c*) 1.2×10^{-2} kJ mol^{-1}.
(*d*) 2.4×10^{-5} kJ mol^{-1}.
(*e*) 4.85×10^{-7} kJ mol^{-1}.

2. The *incident light* of wavelength 550 nm corresponds to an *energy* of 3.6×10^{-18} J per photon.

   The *minimum detectable energy* is 2×10^{-16} J s^{-1}. Thus, the *minimum rate of incidence of photons* is $(2 \times 10^{-16})/(3.6 \times 10^{-18})$ s^{-1}, i.e. approx 550 photons s^{-1}.

   This illustrates the extreme sensitivity of the eye.

3. Using the equation

$$\text{Absorbance} = \varepsilon\ c.l.$$
$$\varepsilon_{280} = 1.11 \times 10^{7}\ \text{cm}^{2}\ \text{mol}^{-1}$$

4. The NADH can be calculated from the absorbance at 340 nm. The total (NAD^+ + NADH) can be calculated from the absorbance at 260 nm. This gives

$$[\text{NAD}^+] = 13.5\ \mu\text{mol dm}^{-3}$$
$$[\text{NADH}] = 33.8\ \mu\text{mol dm}^{-3}$$

5. (*a*) The decrease in absorbance at 340 nm corresponds to the oxidation of 83.6 nmol NADH min^{-1} cm^{-3}. The amount of enzyme added is 0.02×0.08 mg. Thus the activity of the enzyme is 52.3 μmol substrate consumed min^{-1} mg^{-1}. (0.87 katal kg^{-1}.)

   Note that 1 mol of NADH oxidized is equivalent to 1 mol of ATP and creatine consumed in this coupled assay.

   (*b*) The NADH consumed is equivalent to 128.6 μmol dm^{-3}. Thus the stock solution of creatine is 50×128.6 μmol dm^{-3} = 6.43 mmol dm^{-3}.

   Note that the creatine kinase equilibrium, which would normally lie well over to the left (Chapter 2), is driven over to the right by coupling to the pyruvate kinase and lactate dehydrogenase systems.

6. At equilibrium the concentration of NADH formed = 0.090 mmol dm^{-3} = $[NH_4^+]$ = [α-ketoglutarate^{2-}]. The concentrations of glutamate$^-$ and NAD^+ are obtained by difference. Noting that $[H^+] = 10^{-7}$ mol dm^{-3} and that H_2O is in its standard state, K is evaluated as 4.05×10^{-15} (mol dm^{-3}).

   The amount of NADH formed is very small, compared with the concentrations of glutamate$^-$ and NAD^+. Since the concentration of NADH formed is raised to the third power in the expression for K, any inaccuracies in estimating this concentration constitute a major source of error in determining K. For accurate work, the effects of ionic strength on the activity coefficients (and thus on K) should also be considered. In practice, the true equilibrium constant would be obtained by an extrapolation to zero ionic strength (infinite dilution).

7. By constructing a standard curve (absorbance vs. amount of Hb) it can be seen that the 0.01 cm^3 of blood contains 1.4 mg Hb; i.e. the Hb content is 140 g dm^{-3} (A doctor would express this as 14 g (100 $cm^3)^{-1}$.) Normal ranges for males are 130–180 g dm^{-3} and for females 110–160 g dm^{-3}.

   Note that it is important to have readings from a range of standards above and below the sample reading, since linearity of absorbance vs. amount cannot necessarily be assumed.

10

Macromolecules

BINDING OF LIGANDS TO MACROMOLECULES

Introduction

It is convenient to discuss the analysis of ligand binding data in a separate section, as several additional points are brought out although no new fundamental principles are involved.

Let us consider, initially, the case where one mole of A combines with one mole of P

$$\mathrm{P + A \rightleftharpoons PA.}$$

Then the equilibrium constant for this association reaction is

$$K = \frac{[\mathrm{PA}]}{[\mathrm{P}][\mathrm{A}]} \quad \text{where [ ] represent concentrations.}$$

Biochemists usually refer to a 'dissociation constant' (often given the symbol K_d) which is merely the reciprocal of the above association constant

i.e.
$$K_\mathrm{d} = \frac{[\mathrm{P}][\mathrm{A}]}{[\mathrm{PA}]} \tag{10.1}$$

The Binding Equation—Treatment of Binding Data

Let us define the number of moles of A bound per mole of P under a given set of conditions as r. Then

$$r = \frac{\text{concentration of A bound to P}}{\text{total concentration of all forms of P}} \tag{10.2}$$

$$= \frac{[\mathrm{PA}]}{[\mathrm{P}]+[\mathrm{PA}]}. \tag{10.3}$$

Now, from eqn (10.1)

$$[PA] = \frac{[P]\,[A]}{K_d}$$

So that

$$r = \frac{([P]\,[A])/K_d}{[P] + ([P]\,[A])/K_d}, \qquad (10.4)$$

i.e.

$$\boxed{r = \frac{[A]}{K_d + [A]}} \qquad (10.5)$$

Note that when $r = 0.5$ (i.e. P is half saturated with A) then $[A] = K_d$.

The fundamental eqn (10.5) can be rearranged into several forms suitable for graphical treatment. Two of the most common are dealt with below:

(1) The equation is rearranged to give

$$\frac{1}{r} = 1 + \frac{K_d}{[A]}. \qquad (10.6)$$

So that a graph of $1/r$ against $1/[A]$ gives a straight line of slope K_d.

(2) Alternativelv the equation can be rearranged to give

$$\frac{r}{[A]} = \frac{1}{K_d} - \frac{r}{K_d}. \qquad (10.7)$$

A graph of $r/[A]$ against r gives a straight line whose slope is $-1/K_d$.

Thus if, from experimental data the amount of A bound to P can be calculated (at given total concentrations of P and A), we can evaluate the dissociation constant. It is important to note that in these equations [A] refers to the concentration of A which is free in solution, i.e. not bound to P.

EXAMPLE 10.1 *Mg^{2+} and ADP form a 1 : 1 complex. In an experiment, the concentration of ADP was kept constant at 80 μmol dm^{-3} and the concentration of Mg^{2+} varied. The following results were obtained.*

Total Mg^{2+} (μmol dm^{-3})	20	50	100	150	200	400
Mg^{2+} bound to ADP (μmol dm^{-3})	11.6	26.0	42.7	52.8	59.0	69.5

Determine the dissociation constant for MgADP under these conditions.

Solution

At each value of the total Mg^{2+} concentration, the free Mg^{2+} concentration ([A] in the equations) can be evaluated simply by difference. The value of r is found by dividing the concentration of bound Mg^{2+} by the ADP concentration (i.e. 80 μmol dm^{-3}). We can convert the data into the correct from for graphical treatment.

Total Mg^{2+} (μmol dm^{-3})	20	50	100	150	200	400
Bound Mg^{2+} (μmol dm^{-3})	11.6	26.0	42.7	52.8	59.0	69.5
Free Mg^{2+} (μmol dm^{-3})	8.4	24.0	57.3	97.2	141.0	330.5
r	0.145	0.325	0.534	0.660	0.738	0.869
$\frac{1}{r}$	6.90	3.08	1.874	1.515	1.356	1.151
$\frac{1}{[Mg^{2+}]_{free}}$ (μ mol $dm^{-3})^{-1}$	0.1190	0.0417	0.0175	0.0103	0.0071	0.0030
$\frac{r}{[Mg^{2+}]_{free}}$ (μ mol $dm^{-3})^{-1}$	0.0173	0.0135	0.0093	0.0068	0.0052	0.0026

The appropriate plots ($1/r$ vs. $1/[Mg^{2+}]_{free}$ and $r/[Mg^{2+}]_{free}$ vs. r). Of course we would not normally do both, but this is done here for the sake of completeness.

From both plots we obtain the result that K_d = 50 μmol dm^{-3} or 50×10^{-6} (mol dm^{-3}). It is noticeable that in the 'double reciprocal plot' the experimental points are much more unevenly spaced than in the alternative plot. This has led many workers to prefer the type of plot for the analysis of binding data, since it is rather easier in this case to draw the best straight line through the experimental points. In any experiment it is important to make a proper analysis of the distribution of errors in the method of plotting the data. This is also true in the analogous plots which are used in the analysis of enzyme kinetic data and is discussed in the books by Cornish-Bowden mentioned in the reading list.

It is possible to simplify the experiment considerably if one component is present in a considerable excess over the other. For instance, suppose that P is present at a concentration of 1 μmol dm^{-3} and [A] is varied from 50 μmol dm^{-3} up to 500 μmol dm^{-3}. Then, throughout the titration very little of the total A is actually bound to P and it is a very good approximation to write $[A]_{free} = [A]_{total}$. The equations would then become

$$\frac{1}{r} = 1 + \frac{K_d}{[A]_{total}} \quad \text{and} \quad \frac{r}{[A]_{total}} = \frac{1}{K_d} - \frac{r}{K_d}.$$

We often use this simplification in enzyme kinetic work. The substrate (S) of the enzyme is almost always greatly in excess over the enzyme concentration. (i.e. $[S]_{free} = [S]_{total}$.) In this type of work, we use the velocity of the enzyme catalysed reaction (v) to give a measure of r (the amount of S bound to E) in the equations, since only the ES complex shows enzyme activity. We shall see in Chapter 10 that we do in fact plot $1/v$ vs. $1/[S]_{total}$ or $v/[S]_{total}$ vs. v to obtain the *Michaelis constant* which characterizes the interaction of the enzyme with its substrate.

The simplification of the algebra which is achieved by setting $[S]_{free}$ equal to $[S]_{total}$ is illustrated in the following example.

Example 10.2 *Consider the equilibrium* $E + S \rightleftharpoons ES$, *and let* K_s *be the dissociation constant of the ES complex.*

What is the fraction of enzyme present as ES if

(a) $K_s = 50 \times 10^{-6}$ (mol dm^{-3}), $[E]_{total} = 100$ μmol dm^{-3}, $[S]_{total} = 150$ μmol dm^{-3},

(b) $K_s = 50 \times 10^{-6}$ (mol dm^{-3}), $[E]_{total} = 1$ μmol dm^{-3}, $[S]_{total} = 150$ μmol dm^{-3}

Solution

In the case (a) $[E]_{total}$ is comparable with $[S]_{total}$. We must therefore solve a quadratic equation to calculate [ES].

Let x μmol dm^{-3} be the concentration of ES

then $(100 - x)$ μmol dm^{-3} is the concentration of free E

and $(150 - x)$ μmol dm^{-3} is the concentration of free S,

$$\therefore \qquad K_s = 50 \times 10^{-6} = \frac{(100 - x)(150 - x)}{x} \cdot 10^{-6}$$

$$50x = (100 - x)(150 - x)$$

$$\therefore \quad x^2 - 300x + 15\,000 = 0,$$

$$\therefore \quad x = 63.4$$

The total concentration of enzyme is 100 μmol dm^{-3} so in this case 63.4 per cent of the enzyme is present as ES.

In case (b) $[E]_{total} << [S]_{total}$ and thus $[S]_{free}$ can be set equal to $[S]_{total}$

$$\therefore \quad K_s = 50 \times 10^{-6} = \frac{[E]\,[S]_{total}}{[ES]}$$

$$= \frac{[E]}{[ES]} \cdot 150 \times 10^{-6}.$$

Thus $[E]/[ES] = \frac{1}{3}$, and hence [ES] is 75 per cent of the total enzyme.

Multiple Binding Site Equilibria

So far we have only considered the case where one mole of a ligand (A) is bound by a macromolecule (P). However, there are many examples known where several moles of A can be bound by one mole of P. We now wish to examine these cases and to discuss equations which will enable us to determine the number of binding sites in these cases.

If we have a situation where the binding of ligand to macromolecule is very tight we can determine the number of binding sites by a direct titration. Aliquots of ligand are added to a known concentration of the macromolecule. When the amount of ligand added exceeds the concentration of binding sites available, the added ligand remains free in solution.

In cases where the binding is not sufficiently tight we must consider the appropriate equilibrium equations.

For a single binding site, we derived the equation $r = [A]/(K_d + [A])$, where r is the number of moles of A bound per mole of P, [A] is the concentration of free A and K_d is the dissociation constant.

For the case where one mole of P can bind up to n moles of A this equation becomes modified:

$$\boxed{r = \frac{n[A]}{K_d + [A]}} \tag{10.8}$$

where the n sites are assumed to be equal and independent (i.e. the free energy of binding is the same for each site). K_d is now an average dissociation constant.

As before, we can rearrange this equation in two main ways

(1) By taking the reciprocal of each side of the equation we derive

$$\frac{1}{r}=\frac{1}{n}+\frac{K_d}{n[A]}. \tag{10.9}$$

Thus a plot of $1/r$ against $1/[A]$ gives a straight line of slope K_d/n and an intercept on the y axis of $1/n$. This is often referred to as the Hughes–Klotz equation

(2) The alternative rearrangement is

$$\frac{r}{[A]}=\frac{n}{K_d}-\frac{r}{K_d}, \tag{10.10}$$

so that a plot of $r/[A]$ against r gives a straight line of slope $-1/K_d$ and an intercept on the x-axis of n. This is often termed the Scatchard equation.

These two types of plots are illustrated in the worked example below.

EXAMPLE 10.3 *In an experiment the concentration of an enzyme is kept constant at 11 μmol dm^{-3}, and the concentration of inhibitor [I] varied. The following results were obtained*

$[I]_{total}$ (μmol dm^{-3})	5.2	10.4	15.6	20.8	31.2	41.6	62.4
$[I]_{free}$ (μmol dm^{-3})	2.3	4.8	7.95	11.3	18.9	27.4	45.8

Determine the dissociation constant for the enzyme-inhibitor complex and the number of inhibitor binding sites on the enzyme.

Solution

At each value of $[I]_{total}$ we can evaluate $[I]_{bound}$ by subtraction; r is obtained by dividing $[I]_{bond}$ by the concentration of enzyme (i.e. 11 μmol dm^{-3}). The following table can be constructed:

From the 'double reciprocal plot' we find that the intercept on the y axis is 0.5, so that $n = 2$. The slope of the line is 7.6 so that $K_d = 15.2 \times 10^{-6}$ (mol dm^{-3}).

$[I]_{total}$ (μmol dm^{-3})	5.2	10.4	15.6	20.8	31.2	41.6	62.4
$[I]_{free}$ (μmol dm^{-3})	2.3	4.8	7.95	11.3	18.9	27.4	45.8
$[I]_{bound}$ (μmol dm^{-3})	2.9	5.6	7.65	9.5	12.3	14.2	16.6
r	0.264	0.510	0.695	0.864	1.118	1.291	1.510
$\frac{1}{r}$	3.793	1.964	1.438	1.158	0.894	0.775	0.663
$\frac{r}{[I]_{free}}$ (μmol dm^{-3})$^{-1}$	0.115	0.106	0.087	0.076	0.059	0.047	0.033
$\frac{1}{[I]_{free}}$ (μmol dm^{-3})$^{-1}$	0.435	0.208	0.126	0.088	0.053	0.036	0.022

From the 'Scatchard' plot, again we find that $n = 2$ and the value of K_d is 15.2×10^{-6} (mol dm^{-3}). It is also clear that the sites are equivalent and independent, since, otherwise a curved plot would be expected.

As in the previous example (Mg^{2+} and ADP) we find that the Scatchard plot has a more even spacing of the experimental points, than does the 'double reciprocal'. (However, this need not always be the case.)

It is important to note that in order to determine the number of binding sites n accurately it is essential to cover as wide a range of the total saturation curve as possible. Roughly, the required range is the region between 25 per cent saturation and 75 per cent saturation of the sites on the macromolecule for the ligand.

Non-equivalent Ligand Sites on a Macromolecule

In the example discussed above, it is clear that there are two equivalent sites on the enzyme for the inhibitor. However, it often happens that the binding plots do not yield straight lines, and this reflects the fact that the ligand sites are not equivalent. An early example of this situation was found in the case of oxygen binding to haemoglobin (Hb) where the saturation curve is sigmoid in shape and should be compared with the curve for myoglobin (Mb) which is described as hyperbolic (i.e. follows a rectangular hyperbola). Many other examples of this type of effect have since been found, both for binding proteins and for enzymes.

Myoglobin consists of a single polypeptide chain, whereas haemoglobin consists of four subunits held together by non-covalent bonds, and it has become clear that the sigmoidal shapes of the saturation curves for haemoglobin and for other proteins are a consequence of a multisubunit structure. Haemoglobin would be said to show 'positive co-operativity' in binding of oxygen (i.e. binding of the first oxygen molecule enhances the binding of the subsequent ones). The 'advantage' of a sigmoid saturation curve is that over certain concentration ranges the fractional saturation of a molecule (or the rate of an enzyme catalysed reaction) can more readily respond to changes in ligand (or substrate) concentration, leading to greater possibilities of control of metabolic activity in the cell. Cases are also known of 'negative co-operativity' (in which the binding of the first ligand molecule discourages binding of subsequent ones). In this case, the fractional saturation or reaction rate would be less sensitive to changes in ligand or substrate concentration over certain concentration ranges.

The type of binding curve can be recognized from the saturation curve, the double reciprocal plot or the Scatchard plot.

Two principal theories to account for the sigmoid shape of saturation curves (e.g. for haemoglobin) have been proposed. Both stress the importance of changes in protein shape (or conformation) which can be associated with the binding of ligands to proteins. The 'concerted' theory of Monod, Wyman, and Changeux is based on the proposition that when a ligand binds to one subunit in a multisubunit protein all the subunits change their shape in a concerted fashion. In the other, 'sequential', model of Koshland, Nemethy, and Filmer, the conformational changes are assumed to be sequential, i.e. the ligand changes the conformation of only the subunit to which it binds. The conformational change will, of course, alter the interaction(s) between this subunit and its neighbouring subunit(s) making the binding of ligand to the latter stronger or weaker.

Although a distinction between the theories is not always easy (or valuable) to make, it is clear that the 'concerted' model cannot apparently account for the phenomenon of negative co-operativity (such as is apparently observed, for instance, in the binding of NAD^+ to rabbit muscle glyceraldehyde-3-phosphate dehydrogenase). However, alternative explanations to negative co-operativity can be involved (such as the binding sites being non-equivalent in the absence of ligands).

Experimental Methods for Obtaining Binding Data

This aspect of the subject will not be discussed in detail here either. There are two principal types of technique employed in these studies.

(1) The macromolecule with its bound ligand is separated from the free ligand (usually on the basis of size). The concentration of free ligand can be readily evaluated, and hence the concentration of bound ligand is determined by subtraction of this number from the known total ligand concentration.
(2) We can use an indirect method to monitor binding—e.g. by looking for a spectroscopic change in either the macromolecule or the ligand which distinguishes the free macromolecule (or ligand) from the complex between the two. Under any given set of conditions we can then determine the contributions of the free and bound forms.

Using either of these types of techniques we were able to evaluate the parameters ν and [A] in eqn (10.8). The data can then be plotted either in the 'double reciprocal' or in the 'Scatchard' form to determine the parameters *n* and *K*.

COLLIGATIVE PROPERTIES OF MACROMOLECULES

Introduction

A large number of biologically interesting molecules are of high molecular weight and these are termed macromolecules, e.g. proteins, nucleic acids, polysaccharides, etc. The discussion of the thermodynamic properties of these molecules introduces several new features. One of these new features is the idea that the solution can contain species of different molecular weights and so average molecular weights can be involved. Secondly, solutions of macromolecules would not be expected to be ideal because of the obvious chemical differences between solvent and solute. It is extremely difficult to calculate the activity coefficients in solutions of macromolecules, since there is no simple theory available (as there is, for instance, in the case of strong electrolytes). However, as before, in extremely dilute solutions ideal behaviour will be approached. Thus in order to apply the thermodynamic equations to solutions of macromolecules, it is necessary to extrapolate the results of any measurements to infinite dilution.

Osmotic Pressure

We perform the extrapolations using empirical equations based on the thermodynamic ones already described for ideal solutions. This is illustrated by considering the osmotic pressure of solutions of macromolecules.

For dilute ideal solutions we derived the relationship

$$\pi = RTc$$

where c is the molar concentration of solute. If the concentration of solute is x g dm^{-3} and the molecular weight is M, then

$$c = x/M$$

hence

$$\frac{\pi}{xRT} = \frac{1}{M}.$$

Now for solutions of macromolecules the equation above is inadequate and is replaced by a more complex one

$$\frac{\pi}{RTx} = \frac{1}{M} + \frac{B(x)}{M} + \frac{C(x^2)}{M^2} + \ldots$$

where B, C, etc. are constants known as *virial coefficients*. Since M and M^2 are also constants we can write:

$$\frac{\pi}{RTx} = \frac{1}{M} + B'(x) + C'(x)^2 + \ldots$$

where B', C' are new constants.

If x/M is small (i.e. the solution is dilute) then $(x/M)^2$ and higher terms are very small compared with x/M, and so we obtain

$$\boxed{\frac{\pi}{RTx} = \frac{1}{M} + B'(x)}. \qquad (10.11)$$

Since the first term on the right hand side of this equation is a constant, a plot of π/RTx against x is a straight line. The intercept on the y-axis can be used to calculate the molecular weight of the solute:

$$\left(\frac{\pi}{RTx}\right)_{x=0} = \frac{1}{M}$$

$$\therefore \qquad M = \frac{1}{(\text{Intercept on } y\text{-axis})}.$$

EXAMPLE 10.4 *The following osmotic pressure data were obtained for solutions of bovine serum albumen at 298 K.*

Protein concentration (g dm^{-3})	18	30	50	56
Osmotic pressure (atm)	0.0074	0.0134	0.0253	0.0292

Calculate the molecular weight of bovine serum albumen.

Solution

We evaluate π/xRT for each value of x. Note that π is in atmospheres and $R = 0.082$ dm^3 atm mol^{-1} K^{-1}.

x (g dm^{-3})	18	30	50	56
$\pi/xRT\,(\times 10^5)$	1.68	1.83	2.07	2.13

From the plot of π/xRT against x, the intercept on the y-axis is 1.47×10^{-5}. Hence the molecular weight of bovine serum albumen is 68000.

We should note that a plot of π against x is not linear, thus showing the necessity to use the expanded equation for osmotic pressure and performing the extrapolation to infinite dilution. For the case of an ideal solution there would be no change in the value of the term π/xRT as x varies.

The determination of the molecular weight of a macromolecule from the other colligative properties of the solution (i.e. vapour pressure lowering and freezing point depression) is not feasible because the effects observed are so small.

Significant exceptions to this statement have been found in the cases of the sera of fish living in cold waters (e.g. Arctic or Antarctic regions). These sera contain (in addition to small solutes) special types of proteins, known as 'anti-freeze proteins', which can exert a much larger effect on the freezing point of the sera than would be expected on the basis of their molecular weights. For instance, the average temperature in McMurdo Sound in Antarctica is 1.85 K below the freezing point of pure water and sera from fish living in temperate zones would freeze at this temperature, since they do not contain 'anti-freeze' proteins. However, in two species of Antarctic fish the serum freezing point was 2 K below that of pure water, enabling these fish to survive in these conditions. Investigations of the structures of these anti-freeze proteins have shown that they are based on relatively simple repeating units of small numbers of amino acids (often containing large quantities of attached carbohydrate). The mechanism by which this type of structure exerts its effect is under current investigation.

Apart from the non-ideality of solutions of macromolecules, we often have to consider the effects associated with macromolecules which carry a net charge. A very important example of this is the *Donnan* effect.

THE DONNAN EFFECT

Consider two solutions separated by a membrane permeable to small ions but not to charged macromolecules (e.g. a dialysis membrane is not permeable to proteins). If the charged protein is present on only one side of the membrane, there will be an imbalance of ions across the membrane at equilibrium. This is known as the *Donnan effect*. (Sometimes it is also called the *Gibbs-Donnan effect*.)

For simplicity consider two compartments of equal volume, separated by the membrane. Suppose we add an x molar solution of protein (with a charge of Z) and its associated ion (e.g. Na^+) to one side of the membrane, and that we add x_2 molar NaCl to the other side. Movement of ions will occur across the membrane but since electrical neutrality must be preserved at all times on both sides of the membrane, it follows that if y g ions of Na^+ move from side 2 to side 1, y g ions of Cl^- must also migrate in this direction. Equilibrium is achieved when the free energy of all species, to which the membrane is permeable is the same on either side of the membrane. In this case we consider NaCl and so, at *equilibrium*

$$(G_{NaCl})_1 = (G_{NaCl})_2$$

(the subscripts refer to the appropriate side of the membrane). Using the definition

$$G = G^\circ + RT \ln a \qquad (a = \text{activity})$$

we obtain

$$(a_{NaCl})_1 = (a_{NaCl})_2$$

or

$$(a_{Na^+} \cdot a_{Cl^-})_1 = (a_{Na^+} \cdot a_{Cl^-})_2$$

at equilibrium. If y moles of NaCl have been transferred when equilibrium is reached, then this condition becomes

$$(Zx_1 + y)\,(y) = (x_2 - y)\,(x_2 - y)$$

provided the activity can be replaced by concentration. From this equation

$$y = \frac{x_2^2}{Zx_1 + 2x_2} \tag{10.12}$$

and we can calculate the final composition of the solution on either side, if the values of x_1, x_2, and Z are known.

The ratio of the Cl^- ion concentration on side 2 to that on side 1 at equilibrium is $(x_2 - y)/y$, and using the above value for y, this ratio becomes $(Z_1x_1 + x_2)/x_2$ or more explicitly:

$$\left[\frac{(Cl^-)_2}{(Cl^-)_1}\right]_{(final)} = 1 + \left[\frac{(Na^+)_1}{(Na^+)_2}\right]_{(initial)} \tag{10.13}$$

EXAMPLE 10.5 · *At a certain pH, ribonuclease carries a net negative charge of 3. If 0.1 dm³ of a solution of 3 mmol dm⁻³ ribonuclease (as its sodium salt) is added to one side of a membrane, and 0.1 dm³ of 50 mmol dm⁻³ NaCl is added to the other side, what are the final concentrations of ions on the two sides of the membrane?*

Solution

The concentration of Na^+ ions initially present on the protein side is 9 mmol dm^{-3}.

Let y mmol dm^{-3} NaCl be transferred to the protein side at equilibrium. Then using the formula (eqn 10.12)

$$y = \frac{x_2^2}{Zx_1 + 2x_2}$$

$$= \frac{50 \times 50}{9 + 100} \text{ mmol dm}^{-3}$$

(since $x_2 = 50$ mmol dm^{-3}, $Z = 3$, $x_1 = 3$ mmol dm^{-3})

hence $y = 22.94$ mmol dm^{-3}

so that on the protein side we have (at equilibrium)

$$[\text{Protein}^{3-}] = 3 \text{ mmol dm}^{-3}, [Na^+] = 31.94 \text{ mmol dm}^{-3},$$

$$[Cl^-] = 22.94 \text{ mmol dm}^{-3}$$

and on the non-protein side,

$$[Na^+] = [Cl^-] = (50 - 22.94) \text{ mmol dm}^{-3}$$
$$= 27.06 \text{ mmol dm}^{-3}.$$

EXAMPLE 10.6 *How would the result be different if we initially had 0.2 mol dm^{-3} NaCl on the non protein side?*

Solution

Protein side: [Protein^{3-}] = 3 mmol dm^{-3}, [Na^+] = 106.8 mmol dm^{-3}, [Cl^-] = 97.8 mmol dm^{-3}; non-protein side: [Na^+] = [Cl^-] = 102.2 mmol dm^{-3}.

Studies of the binding of small charged ligands to macromolecules are often performed by a technique known as *equilibrium dialysis*. A solution containing the macromolecule is placed on one side of a membrane, and a solution of the ligand is added to the other side. When equilibrium has been attained, the concentrations of ligand on the two sides of the membrane are measured. On the side containing the macromolecule the ligand concentration is the sum of free ligand and bound ligand.

In any such study we should consider the Donnan effect, since this can also lead to differences in the concentrations of charged ligands between the two sides of the membrane. It is thus important, in binding studies to minimize the Donnan effect. For instance, in our example, we see from eqn. (6.3) that at equilibrium the ratio $(Cl^-)_2/(Cl^-)_1$ approaches unity when the term $[(Na^+)_1/(Na^+)_2]_{\text{initial}}$ is small. Thus Donnan effects can be minimized by: (1) lowering the macromolecular concentration (so that $(Na^+)_1$ initial is small), (2) working at high ionic strengths (so that $(Na^+)_2$ initial is large), or (3) adjusting the pH so that the protein carries no net charge (i.e. it is at its *isoelectric* point). We should also take these precautions when doing osmotic pressure measurements on solutions of macromolecules, since the differences in ionic concentrations themselves give rise to an osmotic pressure.

THE MOLECULAR WEIGHTS OF MACROMOLECULES

So far, we have made the assumption that there is a unique molecular weight for a macromolecule. This assumption is justified if there is only one type of species present as in the case of most purified proteins. Such solutions are termed *monodisperse* and the definition of molecular weight is self explanatory. However, in many instances solutions of biological macromolecules contain species with various molecular weights. If the molecular weight distribution is continuous the system is termed *polydisperse*. (This might be the case, for example, for nucleic acid fragments.) If the distribution of molecular weight is discrete, the system is termed *paucidisperse*. An association-dissociation equilibrium would be an example of a paucidisperse system.

It is useful to define certain average molecular weights to describe *polydisperse* and *paucidisperse* systems.

Number-average Molecular Weight ($\overline{M}_n$)

In this average, the species present in solution contribute according to their relative populations so that the average is defined by

$$\boxed{\overline{M}_n = \frac{\Sigma n_i m_i}{\Sigma n_i}} \tag{10.14}$$

where n_i is the relative number of the ith species, an m_i is its molecular weight.

Consider the example of enolase. At a certain concentration suppose that 60 per cent of the species are monomers of molecular weight 41 000 and 40 per cent are dimers of molecular weight 82 000.

The number-average molecular weight is then given by:

$$\overline{M}_n = \frac{\frac{60}{100}(41\,000) + \frac{40}{100}(82\,000)}{\frac{60}{100} + \frac{40}{100}} = 57\,400.$$

This molecular weight would be obtained from measurements of colligative properties (e.g. osmotic pressure) since these properties depend only on the concentration of the species present, irrespective of size. We shall see later that it can also be derived from sedimentation equilibrium.

Weight-average Molecular Weight ($\overline{M}_w$)

In this average the species present in solution contribute according to their relative weights, so that this average is defined by:

$$\boxed{\overline{M}_w = \frac{\Sigma n_i m_i^2}{\Sigma n_i m_i}}. \qquad (10.15)$$

Thus in the above case of enolase, the weight average molecular weight is given by:

$$\overline{M}_w = \frac{\frac{60}{100}(41\,000)^2 + \frac{40}{100}(82\,000)^2}{\frac{60}{100}(41\,000) + \frac{40}{100}(82\,000)} = 64\,430.$$

This average molecular weight is larger than the number average because it gives increased importance to the heavier species. We could obtain $\overline{M}_w$ from studies of light scattering, sedimentation equilibrium, etc.

Z-Average Molecular Weight

In this average the species present in solution contribute according to the square of their relative weights, so that this average is defined by:

$$\boxed{\overline{M}_z = \frac{\Sigma n_i m_i^3}{\Sigma n_i m_i^2}} \qquad (10.16)$$

Thus in the above example of enolase, the z-average molecular weight can be shown to be 70 820.

The z-average molecular weight is easily obtained from sedimentation equilibrium measurements. Although the z-average molecular weight does not have a simple physical meaning, it can be used in conjunction with the other molecular weight averages to provide information on the composition of a *poly-* or *paucidisperse* system. (Clearly for a *monodisperse* system all the molecular weight averages will, of course, be the same.)

DETERMINATION OF MOLECULAR WEIGHTS

Of the many methods that can be used to determine molecular weights, we shall concentrate on the three that are probably the most widely used. Two of these (electrophoresis and gel filtration) are *semi-empirical* in that they are based on calibrations with standards of known molecular weight. The third method (analytical ultracentrifugation) is an *absolute* method, i.e. does not depend on comparisons with molecular weight standards. Because the analytical ultracentrifuge provides the only direct measure of molecular weights, we shall discuss its use in some detail. The section dealing with this is more advanced than much of this chapter and may be omitted at a first reading. Nevertheless the importance of this technique justifies its inclusion here.

Semi-Empirical Methods for The Estimation of Molecular Weights

Two of the most popular techniques currently in use for molecular weight determination are electrophoresis and gel filtration. (Their popularity may be due in large measure to the fact that no expensive apparatus is required and that no complex mathematical equations are involved!)

Electrophoresis

Electrophoresis is the name given to the movement of a charged molecule under the influence of an applied electric field. In order to minimize the effects of convection, electrophoresis is normally conducted on a support material such as paper or more commonly a cross-linked gel composed of polyacrylamide. The rate of movement, or *mobility*, of a molecule in an electric field of given strength will depend on a number of factors such as the charge, size, and shape of the molecule and the nature of the support medium. Clearly, the method cannot be used to estimate the molecular weight of a molecule unless comparison is being made with some other species which is known to be similar. However, this limitation can be neatly overcome in the case of proteins by performing the gel electrophoresis in the presence of a detergent, sodium dodecyl sulphate (SDS—$CH_3(CH_2)_{11}OSO_3^- Na^+$) which breaks the non-covalent bonds maintaining the shape of the protein. It has been found that (i) nearly all proteins bind a constant amount of SDS per g protein (the negative charge of SDS overwhelms the intrinsic charge of the protein), and (ii) protein-SDS complexes are rod-shaped: the length of the rod depending on the molecular weight of the protein. Since the charge and hydrodynamic properties are thus merely functions of the molecular weight of the protein, it is not difficult to see why a relationship between mobility and molecular weight should exist. It is found in practice that a graph of log (molecular weight) vs. mobility is linear over a wide range of molecular weights of standard proteins. The molecular weight of the protein of interest can then be read off the standard line. Using gels of different acrylamide concentrations the range of molecular weights can be varied: low per cent acrylamide gels are used in the higher molecular weight ranges.

An analogous approach can be used for determination of the molecular weight of a sample of RNA. In this case formamide is added to destroy the secondary structure of the molecule (base pairing and stacking) and a graph of log (molecular weight) against mobility is linear over a wide range of molecular weights. Using this approach, RNA samples of known molecular weight were employed to construct a standard curve and then it was shown that haemoglobin messenger RNA had a molecular weight of about 200 000.

Gel Filtration

In this method a cross-linked polymer in bead form is used to separate molecules according to size. If a solution is passed through a column (or over a plate) containing the polymer small molecules enter the pores of the bead and are retarded relative to larger molecules which cannot enter the pores and thus pass straight through the column. Molecules of intermediate size will be retarded to an intermediate extent. The usual procedure is to determine the elution volumes for a number of proteins of known molecular weight and construct a standard curve (elution volume against log (molecular weight)) from which the molecular weight of the molecule of interest can be determine. By varying the degree of cross-linking of the polymer, the range of molecular weights in the 'fractionation range' can be altered, e.g. Sephadex G-200 (a cross-linked polysaccharide) is useful in the range of molecular weights from 20 000 to 300 000.

The success of the method depends on the various molecules having roughly similar shapes. Globular proteins are very suitable in this connection, but proteins of a different shape (e.g. myosin which is rod-shaped) would not be expected to fall on the standard curve. Recent work, for instance, has shown that glycoproteins give anomalously high molecular weights in this method. The gel filtration technique is suitable for nucleic acid work only with the smaller nucleic acids (e.g. 4S and 5S RNA).

The Analytical Ultracentrifuge

In a closed isolated system the temperature, pressure, and the concentration of all components are the same in all parts of the system. For a two component system we may designate one component as solvent and the concentration, *c*, of the solute will be the only composition variable.

If we introduce an external force (gravity, centrifugal) which acts only on the mass of any part of the system and is directly proportional to the mass, the uniform concentration distribution will be disturbed. Those macromolecules which are more dense than the solution will sediment and those which are less dense will move upwards until a new equilibrium is reached.

In the analytical ultracentrifuge the sample is enclosed in a transparent cell and can be considered as a closed, isolated system in a strong and well defined external, centrifugal field. The temperature is kept at a fixed, constant value and the concentration distribution can be monitored by optical methods.

The optical methods are normally based on the measurement of the refractive index as a function of position within the cell (schlieren and interferometer optics) or on the absorption properties of the solute (photoelectric scanner).

Sedimentation Velocity

If the ultracentrifuge is operated at relatively high speeds (up to 65 000 r.p.m.), the uniformly distributed macromolecules will migrate toward the bottom of the cell leaving a region which contains only solvent molecules, and creating a boundary between the solvent and the solution. *The sedimentation velocity method is essentially based on the measurement of the velocity of movement v of this boundary,* and the sedimentation coefficient of the macromolecule can be determined. The sedimentation coefficient, *s*, is defined by:

$$s = \frac{dr/dt}{\omega^2 r} = \frac{v}{\omega^2 r}$$

where ω is the angular velocity (radians s^{-1}) of the rotor and *r* is the distance of the boundary, measured from the axis of rotation.

Numerically, the *sedimentation coefficient* is the distance in cm, covered by a macromolecule during 1 s under the effect of a 'force' of 10^{-2} N kg^{-1}, in H_2O solvent at 293 K. Sedimentation coefficients are usually given in Svedberg, S, units; $1/S = 10^{-13}$s.

The sedimenting macromolecule in the cell of an ultracentrifuge moves under the effect of three forces.

(i) The centrifugal force, F_c

$$F_c = \omega^2 rM$$

where M is the molecular mass.

(ii) The buoyant force, F_b

$$F_b = -\omega^2 rM\bar{v}\rho$$

where $\bar{v}$ is the partial specific volume of the molecule and ρ is the density of the solvent. This force (which is analogous to the 'upthrust' of the Archimedes principle) opposes the centrifugal force because of displacement of solvent by the particle.

(iii) the frictional force, F

$$F = -fv$$

where f is the frictional coefficient (which depends on the shape of the particle and on the viscosity of the solvent) and v is the velocity of the particle relative to the solvent. This force opposes the motion.

Initially the particle will accelerate and reach a certain velocity at which the total force is zero. The particle will then sediment with this velocity v, given by

$$\omega^2 rM - \omega^2 rM\bar{v}\rho - fv = 0,$$

rearranging and substituting

$$\boxed{\frac{M(1-\bar{v}\rho)}{f}} = \boxed{\frac{v}{\omega^2 r}} = s. \tag{10.17}$$

Molecular parameters — Measurable parameters

Eqn (10.17) expresses the sedimentation coefficient in terms of molecular parameters on the one hand and in terms of measurable quantities on the other. It is apparent from eqn (5.7) that the sedimentation coefficient depends on

(a) molecular properties of the sedimenting particle, such as: molecular weight, M, partial specific volume v and shape, f.

(b) properties of the solution: density, ρ and viscosity η (since f depends on this).

Because of the dependence of the sedimentation coefficient on the properties of the solution, the measured s values should be corrected to standard conditions to obtain comparable values for different molecules. The standard solvent is H_2O at 293 K and the s value under these conditions is denoted by $s^{\circ}_{20,w}$.

For macromolecular solutions which are generally non-ideal, s should be measured at four or five different concentrations, and the corrected $s_{20,w}$ values should be extrapolated to zero concentration, to obtain the standard $s^{\circ}_{20,w}$ value, a molecular parameter, which reflects the molecular weight, shape, and density of a macromolecule. An experimentally more convenient form of eqn (10.17) can be obtained by integration (noting that $v = \mathrm{d}r/\mathrm{d}t$)

$$\boxed{s = \frac{1}{\omega^2}\frac{\mathrm{d}\ln r}{\mathrm{d}t}} \quad (10.18)$$

The radial distance of the boundary is measured at various values of elapsed time during the run and a plot of ln r vs. t gives a straight line, the slope of which can be used to compute the sedimentation coefficient. ω can be calculated from the rotor speed

$$\omega = \frac{2\pi \text{ r.p.m.}}{60} = 0.1047 \text{ r.p.m.}$$

If r is measured in cm and t in seconds s will be obtained in seconds(s). The value of s from eqn (10.18) is divided by 10^{-13} to obtain sedimentation coefficients in Svedberg units, S.

If should be stressed that the value of the sedimentation coefficient of a molecule is not sufficient to estimate its molecular weight. It can be used to identify macromolecular components of known molecular weight and conformation or to check the homogeneity of the sample. If more than one macromolecular component is present, this will be reflected in the sedimentation velocity schlieren patterns either as separate peaks or as an asymmetric single peak.

Table 10.1 presents sedimentation coefficient and molecular weight data for some macromolecules. Note the differences in molecular weight of molecules with similar sedimentation coefficients.

TABLE 10.1
Sedimentation coefficient and molecular weight data for selected macromolecules

	$s^{\circ}_{20,\mathrm{w}}$ (S)	Molecular weight
Myosin	6.4	540 000
α-Amylase	4.5	52 000
Fibrinogen	7.9	330 000
Glyceraldehyde-3-phosphate dehydrogenase	7.9	146 000
Ribonuclease	1.64	12 400
Bushy stunt virus	132	10 700 000

In sedimentation velocity experiments a boundary is created between the solution and the solvent. The concentration gradient is the driving force for diffusion which tends to spread the boundary. The flow of matter through a unit surface, driven by a concentration gradient (given by Fick's first law), is equal to the concentration gradient multiplied by the diffusion coefficient, D. D is in units of $\mathrm{cm^2\,s^{-1}}$ (or $\mathrm{m^2\,s^{-1}}$).

For ideal solutions $D = RT/f$, where f is the frictional coefficient. Substituting this into eqn (10.17) the molecular weight can be expressed in terms of the sedimentation (s) and diffusion (D) coefficients:

$$\boxed{M = \frac{RTs}{D(1-\bar{v}\rho)}}. \qquad (10.19)$$

This is the *Svedberg equation*, which is commonly used to calculate the molecular weight from the experimentally determined parameters s and D. It should be noted that in a *poly-* or *paucidisperse* system the molecular weight obtained from eqn (10.19) is not any of the averages discussed earlier (M_n, M_w, M_z). It is, in fact, a *mixed average* (denoted by M_{so}).

The validity of the Svedberg equation is restricted to two component systems at *infinite dilutions*. For many globular molecules $s^{\circ}_{20,\mathrm{w}}$ can be directly measured in very dilute solutions ($c < 0.1$ per cent) at 293 K in dilute aqueous buffers. If the values of both the diffusion coefficients and partial specific volume at 293 K are also known, the molecular weight can be calculated.

EXAMPLE 10.7 *Calculate the molecular weight of D-glyceraldehyde-3-phosphate dehydrogenase from the following data:*

$s^{\circ}_{20,w} = 7.865$ S; $D^{\circ}_{293,w} = 4.75 \times 10^{-11}$ m^{2}s^{-1} $\bar{v}_{293} = 0.729$ cm^3 g^{-1}.

Solution

Use the Svedberg equation

$$M_{sD} = \frac{RTs}{D(1-\bar{v}\rho)}$$

$R = 8.31$ J K^{-1} mol^{-1}; $\rho = 0.9982$ kg dm^{-3} (water at 293 K) $T = 293$ K

$$s = 7.86 \times 10^{-13} \text{ s}; \qquad D = 4.75 \times 10^{-11} \text{ m}^2\text{s}^{-1}$$

$$M_{sD} = \frac{8.31 \times 293 \times 7.86 \times 10^{-13}}{4.75 \times 10^{-11}\,(1 - 0.729 \times 0.9982)} \text{ kg mol}^{-1}$$

$$= 148.0 \text{ kg mol}^{-1},$$

i.e.

$$M_{sD} = 148\ 000 \text{ g mol}^{-1}$$

The molecular weight is thus 148000.

In certain cases the experiments to determine the sedimentation coefficient cannot be conducted at 293 K or near infinite dilution. In such cases the obtained *apparent sedimentation coefficient values have to be corrected to standard conditions* (293 K in water) and extrapolated to zero concentration before using them either for comparison or to calculate the molecular weight.

EXAMPLE 10.8 *In order to preserve the enzyme an ultracentrifuge run was carried out on D-glyceraldehyde-3-phosphate dehydrogenase at 278 K. Only schlieren optics were available on the ultracentrifuge (concentration range 0.1–1.0 per cent) and the rotor speed was 60 000 r.p.m. The following boundary position vs. time data were obtained for an 0.4 per cent enzyme solution. (The time of the first photograph has been taken as zero.)*

Determine the $s^{\circ}_{20,w}$ sedimentation coefficient of the enzyme.

The partial specific volume at 293 K is 0.729 cm^3 g^{-1}; the temperature increment of $\bar{v}$ is 6×10^{-4} cm^3 g^{-1} K^{-1}; the density of the

Number of photograph	Time (s)	r (cm)	ln r
1	0	6.2215	1.8280
2	360	6.2680	1.8355
3	720	6.3140	1.8428
4	1080	6.3610	1.8502
5	1440	6.4085	1.8576
6	1800	6.4560	1.8650

buffer at 278 K and 293 K is 1.0000 kg dm^{-3} and 0.9982 kg dm^{-3} respectively, while the viscosity of the buffer at the corresponding temperatures is 1.5188×10^{-3} and 1.0087×10^{-3} kg m^{-1} s^{-1}.

Solution

(a) Calculation of apparent sedimentation coefficient

The apparent sedimentation coefficient can be calculated from the slope of the ln r vs. t plot.

From eqn (10.18)

$$s_{\text{app}} = \frac{\text{d} \ln r}{\omega^2 \, \text{d}t} = \frac{\text{slope}}{\omega^2} = \frac{2.05 \times 10^{-5}}{\omega^2}$$

$$\omega^2 = \left(\frac{2\pi \, \text{r.p.m.}}{60}\right)^2 = \left(\frac{2\pi \times 6 \times 10^4}{60}\right)^2 = 3.94 \times 10^7 \ \text{s}^{-2}$$

$$s_{\text{app}} = \frac{2.05 \times 10^{-5}}{3.94 \times 10^7} = 5.2 \times 10^{-13} \ \text{s}$$

$$s_{\text{app}} = 5.2 \ \text{S (Svedberg).}$$

(b) Correction to standard conditions

This sedimentation coefficient has to be corrected to standard conditions (293 *K in water).*

The correction takes into account the variations of η, $\bar{v}$ and ρ with temperature. A further correction will still be needed to give the value at infinite dilution. To remember that this value of s is for a 0.4 per cent enzyme solution we shall use the notation $s_{\text{app}}^{0.4\%}$. (By convention for any concentration c, the term is written s_{app}^{c} etc.) If the actual measurements are performed in a buffer solution (b) at temperature *T*, then we can calculate an expression for the value in water (w) at 293 K as

$$s^{c}_{20,w} = s^{c}_{app\,(T,b)} \cdot \frac{\eta_{T,b}\,(1-\bar{v}\rho)_{293,w}}{\eta_{293,w}\,(1-\bar{v}\rho)_{T,b}}$$

where the subscript *T*, b represents the value of the parameter in buffer at *T*. From the data given, we can calculate $\bar{v}_{278.b} = 0.720$

$$s^{0.4\%}_{20,w} = 5.2 \times \frac{(1.5188\times10^{-3})\times(1-0.729\times0.9982)}{(1.0087\times10^{-3})\times(1-0.720\times1.0000)}\,\text{S}$$

$$= 7.565\ \text{S}$$

Note the *large difference* between the real and approximate values and *the necessity for the correction.*

To calculate the $s^{\circ}_{20,w}$ value one further correction is required as illustrated in example. 10.9.

EXAMPLE 10.9 *In the above example on D-glyceraldehyde-3-phosphate dehydrogenase the following* $s^{c}_{20,w}$ *values were also obtained at 0.2, 0.3, and 0.5 per cent protein concentrations:*

$$s^{0.2\%}_{20,w} = 7.72\ \text{S};\quad s^{0.3\%}_{20,w} = 7.65\ \text{S};\quad s^{0.5\%}_{20,w} = 7.48\ \text{S}.$$

Calculate the $s^{\circ}_{20,w}$ *value.*

Solution

Plot these values of sedimentation coefficient vs. concentration and extrapolate to zero concentration. This gives $s^{\circ}_{20,w}$ = 7.86 S.

This corrected and extrapolated $s^{\circ}_{20,w}$ can now be used as a new molecular parameter of the dissolved macromolecule. Its value is related to the molecular weight, shape and density of the particle.

From the shape of the boundary during the sedimentation run, additional qualitative information can be obtained. A single symmetrical boundary reflects a homogeneous sample. A sharp, slowly spreading boundary is characteristic of particles with loose extended structures.

It is possible to determine the molecular weight from the sedimentation and diffusion coefficients (or viscosities of the solution).

However, obtaining the accurate corrected values of these parameters involves lengthy and tedious procedures and numerous conditions must be fulfilled (homogeneity, ideality, etc.) as we have seen. Most of these difficulties can be avoided by using sedimentation equilibrium.

Sedimentation Equilibrium

If the rotor speed is not great enough to pile up all the material at the bottom of the cell, after a certain time a concentration gradient will be formed by the sedimentation transport. This gradient is large enough to drive a *diffusion transport which balances the sedimentation transport* in all parts of the cell. This state is the sedimentation equilibrium. *Because this is an equilibrium state, there is no dependence on the shape of the solute or the viscosity of the solution.* An exact thermodynamic treatment of sedimentation equilibrium is therefore possible (this type of treatment is not possible in the case of sedimentation velocity where the system is not at equilibrium).

By considering the distribution of molecules in an external gravitational field, a simple equation can be derived to describe the equilibrium concentration gradient for a single solute component in an ideal two component solution

$$\frac{1}{c}\frac{\mathrm{d}c}{\mathrm{d}r} = \frac{\omega^2 r M (1-\bar{v}\rho)}{RT} \tag{10.20}$$

where c, the concentration, is a function of r, the radial position, but not a function of time (since the system is at equilibrium). This equation relates the term $M(1 - \bar{v}\rho)$, the effective molecular weight after correction for buoyancy, to the equilibrium concentration distribution in a cell spinning with angular velocity ω.

The validity of eqn (10.20) can be extended to solutions containing more than one solute component. Eqn. (10.20) treated in various ways will yield the number, weight or z-average molecular weight. The most commonly used equation yields the weight-average molecular weight $\overline{M}_w$

$$\boxed{\overline{M}_w = \frac{2RT}{(1-\bar{v}\rho)\omega^2}\frac{\mathrm{d}\ln c}{\mathrm{d}r^2}} \tag{10.21}$$

To calculate the molecular weight the concentration as a function of radial distance has to be known. This can be directly obtained from equilibrium runs monitored by a photoelectric scanner. However a more widely used method in sedimentation equilibrium because of its precision and sensitivity is the Rayleigh interferometer optics system. There is a disadvantage in this method in that the fringe displacements are proportional not to the concentration, but to the concentration difference between the meniscus and the point of interest, $(c_r - c_m)$. This difficulty can be eliminated in two ways.

(i) Using the meniscus depletion method. In this method a high rotor speed is applied to reduce the solute concentration at the meniscus to zero $c_m = 0$. In this way the fringe displacement throughout the cell (ΔI) will be proportional to the concentration c. The molecular weight can be calculated by plotting ln ΔI vs. r^2, and using the slope of this plot in place of d ln $c/\mathrm{d}r^2$ in eqn. (10.21)

(ii) By rewriting eqn (10.20) and extending to a multicomponent system, we can obtain $\overline{M}_w$ as:

$$\overline{M}_w = \frac{2RT}{(1-\overline{v}\rho)\omega^2} \frac{c_r - c_\mathrm{m}}{c_0\,(r^2 - r_\mathrm{m}^2)}$$

where c_0 is the initial concentration of the solute, $c_r - c_m$ can be directly obtained in ΔI fringe displacement units from the equilibrium run. c_0 can be determined (also in ΔI fringe displacements) from an additional synthetic boundary run, (where solvent is layered over the solution in an appropriate cell, thus making the junction between the liquids the boundary).

In the case of polydisperse or reversibly associating systems the molecular weight distribution of the sample will depend on the radial position in the cell. A steeper concentration gradient will be established for the larger than for the smaller species in polydisperse systems. Thus more associated molecules will be present near the bottom of the cell than near the meniscus. In such cases the actual value of then n^-, w^-, and z-average molecular weights *at each point* of the cell can be calculated, and the molecular weight of the components, their relative abundance, and the dissociation constant can be estimated.

The following equations can be used for calculations of the molecular weights in the meniscus depletion method.

$$\overline{M}_n = \frac{M_1 c}{\displaystyle\int_0^c \frac{M_1}{M_w\,(c)}\,\mathrm{d}c}$$

$$\overline{M}_z = \frac{2RT}{(1-\bar{v}\rho)\omega^2} \frac{d^2c/d(r^2)^2}{dc/d(r^2)}$$

$\overline{M}_w$ can be calculated from eqn (6.11). M_1 is the molecular weight of the monomer in the case of reversibly associating systems.

The analysis of interacting systems is complicated. However, a monomer-dimer-tetramer equilibrium can be described in terms of dissociation constant; by the analysis of the concentration distribution at sedimentation equilibrium.

To summarize, the sedimentation equilibrium method is a means of avoiding most of the difficulties encountered with the sedimentation velocity method. Using interferometer optics the experiments can be carried out at low concentrations (0.01–0.03 per cent) such that no correction for concentration dependence is necessary. Using a short solution column (0.15–0.30 cm) the time requirement for the experiment (i.e. to reach equilibrium) can be drastically reduced to 4–8 hours. An added advantage of the meniscus depletion method is that the molecular weight of the sample can be determined even in the presence of small amounts of large aggregates as shown in the worked example below.

A single meniscus depletion sedimentation equilibrium experiment is sufficient to determine the molecular weight of a macromolecule and to judge the homogeneity of the sample. The experiment requires 20–30 μg of material and can be carried out in 5–6 hours.

Note on the measurement of $\bar{v}$

The partial specific volume of most proteins is about 0.7 $cm^3 g^{-1}$. The Svedberg equation contains the term $(1 - \bar{v}\rho)$ and so an error of 1 per cent in $\bar{v}$ (in water at 293 K) will result in a 3 per cent error in M. This is in contrast to an error of 1 per cent in either s or D which result in a 1 per cent error in M—because of the linear dependence. This illustrates the requirement to know $\bar{v}$ must more precisely than s or D. The most accurate method of measuring $\bar{v}$ uses a pycnometer—but this requires large amounts of sample. $\bar{v}$ can also be estimated for proteins in a particular solvent from the chemical composition of the solution. This estimation is based on the partial specific volume of each type of amino acid present. The most accurate method of calculating $\bar{v}$ is from density measurements, and by use of the mechanical oscillator densimeter a precision of 10^{-6} kg dm^{-3} can be achieved.

The use of two solvents of different densities in the analytical ultracentrifuge avoids the requirement for an independent determination of $\bar{v}$. For instance one ultracentrifuge run in H_2O and one in a mixture of H_2O/D_2O leads to two simultaneous equations which can be solved for $\bar{v}$ and M. This is illustrated in example 10.10.

EXAMPLE 10.10 *Two meniscus depletion sedimentation equilibrium experiments were performed with an immunoglobulin sample. In the first run H_2O was used as solvent and in the second a 20 per cent H_2O–80 per cent D_2O mixture.*

The initial protein concentration was 0.03 per cent and a 16 000 r.p.m. rotor speed was selected, both runs were conducted at 298 K.

Rayleigh interferometer optics were used to take photographs after reaching the sedimentation equilibrium.

The fringe displacements ΔI_i were measured as a function of radial distance r_i. The data are given in Tables 10.2 and 10.3 and represent typical experimental measurements.

The densities of H_2O and D_2O at 298 K are 0.997 and 1.105 kg dm^{-3} respectively (R = 8.31 J K^{-1} mol^{-1}). Calculate (a) the partial specific volume of the immunoglobulin and (b) its molecular weight, and (c) comment on the homogeneity of the sample, and on the necessity of corrections.

Solution

Calculate the r^2 and the corresponding ln ΔI_i values. These are given in Tables 10.2 and 10.3. Then plot ln ΔI vs. r_i^2.

From the graphs we can determine the slope of the linear part of the plots. (The increase in the slope near the cell bottom reflects the presence of higher molecular weight components, possibly aggregates). We then use eqn (10.21) for the calculation as follows:

$$M = \frac{2RT}{(1-\bar{v}\rho)\omega^2}\frac{\mathrm{d}\ln c}{\mathrm{d}r^2} = \frac{2RT}{(1-\bar{v}\rho)\omega^2}\underbrace{\frac{\ln \Delta I}{r^2}}_{\text{slope}}$$

from which by rearranging

$$M(1-\bar{v}\rho) = \frac{2RT}{\omega^2}\cdot \text{slope.}$$

TABLE 10.2

Data for worked example (experiment in H_2O)

No.	r_i(cm)	r_i^2 (cm^2)	ΔI_i(μm)	ln ΔI_i
r_m	6.7054			
2	6.7100		Effectively −2	
3	6.7250		no 0	
4	6.7400		change, 3 $\Sigma \Delta I_i = 0$	
5	6.7550		just −1	
6	6.7700		random 0	
7	6.7850		scatter	
8	6.8000			
9	6.8150			
10	6.8300	46.649	26	3.266
11	6.8450	46.854	72	3.749
12	6.8600	47.060	69	4.233
13	6.8750	47.266	122	4.718
14	6.8900	47.472	182	5.204
15	6.9050	47.679	296	5.691
16	6.9200	47.886	483	6.179
17	6.9350	48.094	788	6.669
18	6.9500	48.303	1733	7.458
19	6.9650	48.511	3837	8.253
20	6.9800	48.720	9740	9.184
21	6.9950	48.930	—	—
r_b	7.0250	49.351	—	—

We have two parameters M and $\bar{v}$ to determine and two equations from the two runs:

$$M(1-\bar{v}\rho_1) = \frac{2RT}{\omega^2}\text{(slope 1)} = A$$

$$M(1-\bar{v}\rho_2) = \frac{2RT}{\omega^2}\text{(slope) } 2 = B.$$

Now A and B can be calculated from the experimental results.

The above equations can be rearranged to make $\bar{v}$ and M the subjects.

TABLE 10.3

Data for worked example (experiment in H_2O/D_2O)

No.	r_i (cm)	r^2 (cm^2)	ΔI_i (μm)	$\ln \Delta I_i$
r_m	6.7125	Effectively	0	$\Delta I_i = 0$
2	6.7250	no change	5	
3	6.7400	just random	−4	
4	6.7550	scatter	0	
5	6.7700			
6	6.7850			
7	6.8000			
8	6.8150			
9	6.8300	46.649	24	3.165
10	6.8450	46.854	35	3.540
11	6.8600	47.060	50	3.916
12	6.8750	47.266	73	4.293
13	6.8900	47.472	107	4.671
14	6.9050	47.679	156	5.050
15	6.9200	47.886	228	5.430
16	6.9350	48.094	334	5.811
17	6.9500	48.303	687	6.532
18	6.9650	48.511	1468	7.292
19	6.9800	48.720	3133	8.050
20	6.9950	48.930	—†	—
r_b	7.0074	49.104	—†	—

† Difficult to make precise measurements near the bottom of the cell.

Thus

$$\bar{v} = \frac{A - B}{A\rho_2 - B\rho_1} \quad \text{and} \quad M = \frac{A}{1 - \bar{v}\rho_1}$$

so from the values of A and B, $\bar{v}$ and M can be calculated. The density of the solvent in the second experiment is:

$$\rho_2 = 0.2\rho_{H_2O} + 0.8\rho_{D_2O} = (0.2 \times 0.997) + (0.8 \times 1.105) = 1.083 \text{ kg dm}^{-3}.$$

Noting that $\omega = (2\pi \text{ r.p.m.}/60)^2 = 2.8 \times 10^6 \text{s}^{-2}$ and that the slopes have to be multiplied by 10^4 to bring them into units of m^{-2}, we have

$$A = \frac{2 \times 8.31 \times 298 \times 2.36 \times 10^4}{2.8 \times 10^6} = 41.8 \text{ kg mol}^{-1}$$

and

$$B = \frac{2 \times 8.31 \times 298 \times 1.82 \times 10^4}{2.8 \times 10^6} = 32.2 \text{ kg mol}^{-1}$$

so that

$$A\rho_2 = 41.8 \times 1.083 = 45.3$$

and

$$B\rho_1 = 32.2 \times 0.997 = 32.1$$

$$\therefore \quad \bar{v} = \frac{41.8 - 32.2}{45.3 - 32.1} = 0.727 \text{ dm}^3 \text{ kg}^{-1} \text{ (or cm}^3 \text{ g}^{-1}\text{)}$$

(a) The partial specific volume of immunoglobulin in water at 298 K is

$$0.727 \text{ cm}^3 \text{ g}^{-1}.$$

(b) Now $(1 - \bar{v}\rho_1) = (1 - 0.727 \times 0.997) = 0.275$, hence

$$M = \frac{41.8}{0.275} = 152 \text{ kg mol}^{-1}$$

The molecular weight of this immunoglobulin is 152 000.

(c) The sample contains some high molecular weight contamination as can be seen in the upwards curvature of the ln $c(\Delta I)$ vs. r^2 plot. This contamination may be aggregated protein, which is accumulated near the cell bottom in the spinning cell.

Since the protein concentration is very low in the experiment no extrapolation to zero concentration is necessary.

The temperature dependence is taken into account in the equation used in the calculations, since the equations contain a term in T.

Approach to Equilibrium

If sedimentation equilibrium is reached, the net flow of material is zero in all parts of the cell. As has been shown previously it is very convenient to calculate molecular weight from the equilibrium concentration distribution of a solute in the spinning ultracentrifuge cell.

It takes several hours (depending on the length of the solution column in the cell) to reach equilibrium. However, there are two points in the cell, the meniscus and the bottom, where obviously there is no transport of matter in any phase of the experiment. From the concentration distribution function near to those surfaces the molecular weight can be calculated, before the total equilibrium is reached if the angular velocity ω is known. This method is named after Archibald, who first pointed out the possibility of its application.

The advantage of the method is the short time required for the experiment. The disadvantage is its inferiority in accuracy (because the precision of the optical method is poorest near the extremities) and the computation of the results is complicated and time consuming.

For polydisperse systems the method will result in low molecular weight species collecting at the meniscus and higher molecular weight species collecting at the bottom of the cell. (This follows directly from the molecular weight dependent transport of matter by sedimentation.) By extrapolation to zero time, the weight average molecular weight can be obtained. The Archibald method has been used less frequently since short column equilibrium experiments became popular. These allow *more* precise results from runs of reasonable duration.

PROBLEMS

1. The binding of a substrate (ADP) to the enzyme pyruvate kinase was studied by measuring the quenching of the protein fluorescence. The following results were obtained.

Total concentration of ADP (mmol dm^{-3})	0.29	0.36	0.42	0.53	0.84	1.18	1.89
mol ADP bound/mol protein	1.25	1.35	1.62	1.82	2.26	2.62	3.03

The enzyme concentration was 4 μmol dm^{-3} throughout the titration. Calculate the number of substrate binding sites on the enzyme and the dissociation constant, from these data.

2. Consider the binding of substrate (S) to enzyme (E). If an inhibitor, I, which can bind to the enzyme (but not to the ES complex) is now added to the system we have two simultaneous equilibria to be considered:

$$E + S \rightleftharpoons ES \qquad K_s = \frac{[E][S]}{[ES]},$$

and $$E + I \rightleftharpoons EI \quad K_i = \frac{[E][I]}{[EI]}.$$

(*a*) Evaluate the concentrations of [E], [EI], and [ES] as fractions of $[E]_{total}$ for the case where

$$[E]_{total} = 1\ \mu mol\ dm^{-3}, \quad [S]_{total} = 150\ \mu mol\ dm^{-3},$$
$$K_s = 50 \times 10^{-6}\ (mol\ dm^{-3}), \quad K_i = 75 \times 10^{-6}\ (mol\ dm^{-3}).$$

$[I]_{total} = 150\ \mu M$

(*b*) Discuss the situation (i.e. examine the algebra) when $[E]_{total}$ is comparable with $[S]_{total}$ and $[I]_{total}$ in this system.

3. If the binding of a ligand to a macromolecule is very tight, we can determine the number of ligand binding sites by a direct titration method. Such a case is provided by the binding of NADPH to isocitrate dehydrogenase from bovine heart mitochondria, which is monitored by the increase in fluorescence of NADPH when it is bound to the enzyme. The following data were obtained when NADPH was added to a 0.0135 g dm^{-3} solution of the enzyme (a small correction has been made for the fluorescence of free NADPH). The molecular weight of the enzyme is 90 000.

NADPH added (μmol dm^{-3})	0.07	0.14	0.21	0.28	0.35	0.50	0.75
Fluorescence increase	2.3	4.7	7.0	9.3	9.9	10	10

Determine the number of sites for NADPH on the enzyme.

4. The following data were obtained for the binding of Mn^{2+} to the enzyme phosphorylase a:

$[Mn^{2+}]_{total}$ (μmol dm^{-3})	110	190	290	500	750
$[Mn^{2+}]_{bound}$ (μmol dm^{-3})	75	125	175	250	300

The enzyme concentration is 100 μmol dm^{-3}. By means of a suitable plot calculate *n* and *K*.

5. The binding of NAD^+ to yeast glyceraldehyde-3-phosphate dehydrogenase was studied by equilibrium dialysis with the following results:

$[NAD^+]_{total}$ (μmol dm^{-3})	41	78	132	187	230	285	374	474
$[NAD^+]_{free}$ (μmol dm^{-3})	13	21	30	39	48	68	125	211

The enzyme concentration is 71 μmol dm^{-3}. What type of binding is being observed in this case?

6. The enzyme aspartate transcarbamoylase is inhibited by CTP. Measurements of the binding of CTP to the enzyme were made by equilibrium dialysis.

Experiment A

$[CTP]_{total}$ (μmol dm^{-3})	0.78	1.29	2.35	4.63	11.65
$[CTP]_{free}$ (μmol dm^{-3})	0.24	0.43	1.00	2.56	8.86

Experiment B

$[CTP]_{total}$ (μmol dm^{-3})	144.9	172.0	203.7	269.5	416.0
$[CTP]_{free}$ (μmol dm^{-3})	27.9	43.0	65.7	122.5	260.0

The concentrations of enzyme in experiments A and B were 0.27 and 9 g dm^{-3} respectively. Aspartate transcarbamoylase has been shown to consist of six catalytic and six regulatory subunits with a total molecular weight of 300 000. What can you conclude from these data?

•7. A solution containing 0.5 g ribonuclease in 0.1 dm^3 of 0.2 mol dm^{-3} sodium chloride exerted an osmotic pressure of 0.0097 atm at 298 K. The membrane is permeable to all species except ribonuclease. Determine the molecular weight of ribonuclease. How and why (qualitatively) would the result have been different if the experiment had been carried out in the presence of a much smaller amount of sodium chloride? (R = 0.082 dm^3 atm mol^{-1} K^{-1}.)

8. Calculate the difference in osmotic pressure between capillary blood and the tissue fluid at 310 K. Assume that this difference arises because of the protein in the blood (concentration = 7 per cent; molecular weight 66 000), which is absent in the tissue fluid. (Ignore the Donnan effect.) (R = 0.082 dm^3 atm mol^{-1} K^{-1}.)

9. The osmotic pressure of blood plasma is 7.6 atm at 310 K. What is the total concentration of dissolved species present? Use this result to calculate the freezing point depression of the plasma. What would the freezing point depression be if the plasma were dialysed to remove small solutes? The protein concentration and molecular weight can be assumed to be the same as in Problem 6.

   Isotonic saline is that concentration of NaCl which prevents osmotic disruption of the blood cells. Comment on the fact that isotonic saline is 0.95 per cent NaCl (by weight).

10. Calculate the difference in osmotic pressure between the blood plasma of a diabetic patient and that of a healthy individual if the difference is assumed to be solely due to the higher level of glucose in the plasma of the diabetic patient. (The plasma glucose levels are 1.80 and 0.85 g dm^{-3} respectively, T = 310 K.)

11. The antifreeze protein from an Antarctic fish was isolated and found to have a molecular weight of 17 000. What should the freezing point depression of the serum be if the protein solution behaves ideally? (It can

be assumed that the protein concentration is 10 g dm^{-3}, and that the serum has been dialysed to remove small molecules.) Compare your result with the observed freezing point depression of 0.6 K and with the freezing point depression of a 10 g dm^{-3} solution of lysozyme (molecular weight 14 500) which is 0.0014 K.

12. Consider a protein of molecular weight 60 000, which carries six negative charges at pH 7.4, placed in a cell enclosed by a membrane permeable to all ions except the cell protein. Assume the protein (at a concentration of 60 g dm^{-3}) is added as its sodium salt and 0.15 mol dm^{-3} NaCl is also added to the protein side. The volume of water on the other side of the membrane is the same as the volume of solution on the protein side. Calculate the equilibrium concentrations of ions on both sides of the membrane.

    This situation approximates to that in the kidney. The observed concentrations of Na^+ and Cl^- ion on the protein (plasma) side are 0.120 mol dm^{-3} and those on the filtrate (water) side are 0.6 mmol dm^{-3}. Comment.

13. The concentrations of Na^+ and K^+ ions in tissues are about 11 and 92 mmol dm^{-3} respectively (inside) and 140 and 4 mmol dm^{-3} respectively (outside). Calculate the free energy requirement for maintenance of each of these ion gradients. (T = 310 K.)

14. In a study of the sedimentation of pancreatic α-amylase at 298 K and a rotor speed of 50 000 r.p.m. the following set of data was obtained.

t(s)	0	360	720	1080	1440	1800
boundary (cm)	6.236	6.267	6.298	6.330	6.362	6.393

    (*a*) Noting that the ratio of viscosities of the solvent at 298 and 293 K is 0.8872 calculate the $s_{20,w}$ sedimentation coefficient.

    (*b*) Use this value to calculate the molecular weight of α-amylase. ($D_{293,w}$ = 7.62×10^{-11} m^2 s^{-1}; $\bar{v}$ = 0.728 cm^3 g^{-1}; R = 8.31 J K^{-1} mol^{-1}.) (Assume the solution density is 1.000 kg dm^{-3}.)

15. The equilibrium concentration distribution in a sedimentation equilibrium experiment for a given system at a given temperature is determined by the rotor speed. In a conventional sedimentation experiment the optimal ratio of the concentration at the meniscus to that at the cell bottom is 1 : 4 for precise calculation. In the case of a meniscus depletion experiment the concentration at the meniscus should be almost zero.

    (*a*) Calculate the optimal rotor speed for a conventional (low speed) sedimentation equilibrium experiment for a protein sample of molecular weight 100 000 at 293 K if the position of the meniscus r_m and the bottom of the cell r_b are 6.7 and 7.0 cm respectively from the axis of rotation. ($\bar{v}$ = 0.75 cm^3 g^{-1}; density of the solution 1.00 kg dm^{-3}.)

(*b*) Calculate the optimal rotor speed for a meniscus depletion experiment for the same system. (A concentration ratio at the meniscus to that at the bottom of the cell $c_m : c_b$ of 1 : 1000 is required.)

(*c*) Make logarithmic plots for the required r.p.m. values vs. molecular weight for both types of equilibrium.

16. A meniscus depletion (high speed) sedimentation equilibrium experiment was carried out at 277 K with a ceruloplasmin sample of 0.02 per cent initial concentration. Rayleigh interferometer optics were used and the rotor speed was 16 000 r.p.m. To reduce the time required to reach equilibrium a 0.3 cm solution column was used in a double sector cell. The following set of data was obtained by measuring the fringe displacement as a function of radial distance in the interference pattern.

No.	r_i (cm)	Δl_i (μm)	No.	r_i (cm)	Δl_i (μm)
r_m	6.7083	0	12	6.8650	65
2	6.7150	0	13	6.8800	107
3	6.7300	0	14	6.8950	194
4	6.7450	0	15	6.9100	347
5	6.7600	0	16	6.9250	631
6	6.7750	0	17	6.9400	1141
7	6.7900	−5	18	6.9550	2059
8	6.8050	5	19	6.9700	4230
9	6.8200	10	20	6.9850	12330
10	6.8350	19	21	7.0000	38170
11	6.8500	34	22	7.0150	—
			r_b	7.0270	—

(*a*) Calculate the molecular weight of ceruloplasmin.

(*b*) Comment on the homogeneity of the sample.

($\bar{v}$ of ceruloplasmin at 277 K is 0.713 $cm^3\ g^{-1}$ and the density of the solution was 1.00 kg dm^{-3} (R = 8.31 J K^{-1} mol^{-1}).)

Solutions to Problems

1. Since ADP is always present in large excess over enzyme, we can treat $[ADP]_{total}$ as $[ADP]_{free}$. Thus a plot of

$$\left(\frac{\text{no. of moles ADP bound}}{[\text{ADP}]_{\text{total}}}\right)$$

$$\text{vs. (no. of moles ADP bound)}$$

would give a line of slope $(-1/K_d)$ and intercept on the x axis of n. From this we find that $K_d = 6.4 \times 10^{-4}$ (mol dm^{-3}), $n = 4$. The four sites are equivalent and independent because the binding plot appears linear. (Pyruvate kinase contains 4 subunits, each of which can bind one ADP molecule).

2. (*a*) In the case quoted, both S and I are present in vast excess over E, and their *free* concentrations are effectively equal to their total concentrations. Thus

$$\frac{[\text{E}]}{[\text{ES}]} = \frac{1}{3}, \frac{[\text{E}]}{[\text{EI}]} = \frac{1}{2},$$

i.e. $[\text{E}] = 0.17[\text{E}]_{\text{total}}$; $[\text{ES}] = 0.5\,[\text{E}]_{\text{total}}$; $[\text{EI}] = 0.33[\text{E}]_{\text{total}}$.

(*b*) When $[\text{E}]_{\text{total}}$ becomes comparable with $[\text{S}]_{\text{total}}$ and $[\text{I}]_{\text{total}}$, we can no longer make the assumption above. In this case we have to solve two simultaneous quadratic equations to evaluate the concentrations of all the species. This involves the solution of a cubic equation (usually performed by computer).

3. From plot of fluorescence increase against NADPH added, we can deduce that the enzyme can bind 0.3 μmol dm^{-3} NADPH. As the enzyme concentration is 0.15 μmol dm^{-3}, this means there are *two* binding sites for NADPH on the enzyme. From other work the enzyme is known to consist of two subunits. The binding of NADPH is clearly very tight indeed, so that no appreciable dissociation of the enzyme–NADPH complex is observed under these conditions.

4. From either a Scatchard or Hughes–Klotz plot we deduce that there are 4 binding sites (i.e. $n = 4$) and that $K = 1.48 \times 10^{-4}$ (mol dm^{-3}). The sites appear equivalent and independent. (From other work, the enzyme is known to consist of 4 subunits (i.e. each subunit binds one metal ion)).

5. From a Scatchard or Hughes–Klotz plot we can deduce that the binding shows features typical of positive co-operativity (this should be compared with the binding of NAD^+ to the rabbit muscle enzyme which shows features of negative co-operativity). From the extrapolated value of the intercept it is likely that there are four binding sites on the enzyme (which is, in fact, a tetramer).

6. The binding plots show features typical of negative co-operativity. Extrapolation suggests there are six binding sites for CTP per enzyme molecule (i.e. one per regulatory subunit).

7. Molecular weight = 12 600.

In the presence of a smaller concentration of sodium chloride, the Donnan effect will lead to an unequal distribution of ions across the membrane (unless the protein carries no net charge, i.e. is at its isoelectric point). This would lead to an erroneous molecular weight for the protein. The Donnan effect is negligible if the concentration of salt is much larger than that of the protein.

8. Osmotic pressure difference = 0.027 atm.

This difference is of great importance in living systems. The balance of this pressure and the hydrostatic pressure in the capillary bed governs the flow of water and small ions from the capillaries to the tissues. A lowering of the protein level in the capillaries (e.g. by starvation) is one of the factors which leads to an accumulation of water in the tissues (a condition known as hypoproteinoemic oedema).

9. The total concentration of dissolved species is 0.3 mol dm^{-3}. The freezing point depression of the plasma is 0.56 K. If all small solutes were removed the freezing point depression exerted by the protein would be 0.002 K.

Now 0.95 per cent NaCl is 0.16 mol dm^{-3}, but since NaCl is dissociated to give two ions, the total concentration of dissolved species is 0.32 mol dm^{-3}, i.e. the osmotic pressure of this solution is the same as that exerted by the plasma.

10. Osmotic pressure difference = 0.134 atm.

This large osmotic pressure difference has pronounced effects in the diabetic patient. The high osmotic pressure of extracellular fluids necessitates a large urine volume, and would lead to loss of water from the tissues. A severely diabetic patient remains thirsty in spite of drinking large amounts of water and would suffer severe dehydration if he were untreated.

11. Freezing point depression should be 0.0011 K.

The freezing point depression is some 550-fold greater than this and indicates the effectiveness of the 'anti-freeze protein'. The exact mechanism of this remains unsolved (see text).

The calculated freezing point depression for lysozyme (0.0013 K) is in close agreement with the observed value and indicates that the solution is behaving almost ideally.

12. The equilibrium concentrations are: inside $[Na^+]$ = 79.5 mmol dm^{-3}; $[Cl^-]$ = 73.5 mmol dm^{-3}; outside $[Na^+] = [Cl^-]$ = 76.5 mmol dm^{-3}. This result shows that the Donnan effect leads to a small difference between the ion concentrations on the two sides of the membrane. In the kidney, however, there is an accumulation of Na^+ and Cl^- on the plasma side of the membrane. This 'active transport' requires the expenditure of energy

(ATP hydrolysis), and ensures the minimum loss of ions by excretion.

13. Using the relation $|\Delta G| = RT \ln([c]_{inside}/[c]_{outside})$ for each ion the free energy requirements are 6.6 kJ g ion^{-1} (Na^+) and 8.1 kJ g ion^{-1} (K^+). The accumulation of K^+ at the expense of Na^+ is of crucial importance in living systems, e.g. in the conduction of nerve impulses.

14. (*a*) Calculate the values of ln *r*. A plot of ln *r* vs. *t* gives a straight line with slope = 1.3374×10^{-5} s^{-1}. Noting that $\omega^2 = 2.7410 \times 10^7$ s^{-2} then from eqn (10.18)

$$s_{298,w} = 5.027 \times 10^{-13} \text{ s.}$$

The correction for viscosity has to be made

$$s_{293,w} = (\eta^\circ_{298}/\eta^\circ_{293}) \cdot s_{298,w}$$

from which

$$s_{293,w} = 4.46 \times 10^{-13} \text{ s.}$$

(*b*) From the Svedberg equation $M = 52\,400$.

15. (*a*) $d(\ln c) = \ln c_b - \ln c_m$ and $dr^2 = r_b^2 - r_m^2$ then for $c_b/c_m = 4$, r.p.m. $= 2.45 \times 10^6/\sqrt{M}$.

If $M = 100\,000$ then r.p.m. $= 7.74 \times 10^3$

(*b*) If $c_b/c_m = 1000$, r.p.m. $= 5.49 \times 10^6/\sqrt{M}$. If $M = 100\,000$ then r.p.m. $= 1.73 \times 10^4$

(*c*) Calculate the r.p.m. values for molecular weights of 10,000 and 1000 000 as typical

M	High speed (r.p.m.)	Low speed (r.p.m.)
10 000	5.49×10^4	2.45×10^4
1 000 000	5.49×10^3	2.45×10^3

Use then the results from (*a*) and (*b*) to give the required plot. Note its linearity. This plot is useful for selection of rotor speeds in general.

16. (*a*) Plot $\ln \Delta I_i$ versus r_i^2 and obtain the slope = 2.83.

Note that

$$\frac{d \ln \Delta I_i}{dr^2} \equiv \frac{d \ln c}{dr^2}.$$

$$M = 128\,000$$

(*b*) The upwards curvature in the $\ln \Delta I_i$ versus r_i^2 plot near to the bottom of the cell indicates that some high molecular weight contamination and aggregates are present in the sample.

Appendix–1
Mathematics—A Review

FRACTIONS, MULTIPLES AND POWERS

Common Fractions

You will be familiar with the expression of fractional terms in the 'common' form of $\frac{\text{numerator}}{\text{denominator}}$. In handling such fractions remember the following points:

1. Common fractions with the same denominator can be added or subtracted by adding or subtracting their numerators.

   To add or subtract common fractions possessing different denominators, the lowest common multiple of their denominators must be obtained.

2. To multiply common fractions, multiply the numerators to obtain the new numerator, and multiply the denominators to obtain the new denominator; then simplify if possible.
3. To divide common fractions, invert the divisor and multiply.
4. $\frac{0}{x}$ equals zero, provided that x does not itself equal 0; however, $\frac{x}{0}$ is meaningless, and division by o is 'banned'.
5. You will *retain* the value of a fraction if you multiply both its numerator and denominator (or divide both) by the same term,

   thus, $$\frac{a}{b} = \frac{Ka}{Kb} \left(\text{but this does } not \text{ equal } K\frac{a}{b} \right)$$

   You *alter* the value of a fraction if you:

   (i) add a constant to both numerator and denominator

$$\frac{a}{b} \neq \frac{a+K}{b+K}$$

(ii) raise both numerator and denominator to the same power or reduce each to the same root,

$$\frac{a}{b} \neq \frac{a^2}{b^2} \quad \text{and} \quad \frac{a}{b} \neq \frac{\sqrt{a}}{\sqrt{b}}$$

Decimal Fractions

Addition and subtraction of decimal fractions present no difficulty if the figures are tabulated so that their decimal points fall in a vertical line. Multiplication and division are best carried out by the use of a slide rule, or logarithms.

Expression of a number *y* as a multiple of another number *x*

If $y = ax$ where a is a constant, so long as the value of x is known, then y is defined by this equation as a function of x.

By expressing a series of very large numbers as multiples of a common denominator x which itself has a large value, the numbers are reduced to more manageable proportions. (Similarly for a series of very small numbers, when x would be made a small number.) In this way a series of numbers can be 'scaled up' or 'scaled down' by the factor x.

Expression of a number *y* as a power of another number *x*

If $y = x^i$, this is a shorthand way of stating that if x is multiplied by itself i times, the product will equal y. The term x^i (described as x to the power of i), consists of the *base* x and its *index* (or *power or exponent*) i. Since the index i need not necessarily be an integral number, the value of y can be expressed as an exponential term whatever is the value of y, and whatever number is chosen as the base x; that is, whatever is the defined value of x, it is possible by raising it to a certain power i (possibly negative or fractional) to equal the value of y.

Multiplication and division of numbers are much simplified by first converting them into exponential terms having the same base, for:

(i) to multiply exponential terms having a common base, one adds their indices,

$$x^a \times x^b = x^{(a+b)}$$

(ii) to divide exponential terms with a common base, one subtracts their indices,

$$x^a \div x^b = x^{(a-b)}$$

It makes no difference if some of the indices have a negative value, for they are added or subtracted algebraically,

$$x^a \times x^{-b} = x^{(a-b)}$$

Note the following relationships:

$$x^{-a} = \frac{1}{x^a} \qquad x^0 = 1$$

$$x^{1/a} = \sqrt[a]{x} \qquad x^{a/b} = \sqrt[b]{x^a}$$

Expression of any number as the product of two numbers, one of which is an integral power of 10

This method of expressing a number is usually employed when the number is very large or very small (having a large number of noughts immediately preceding or following its decimal point); e.g.

$$3\,670\,000\,000 = 3.67 \times 10^9$$

$$0.0000\,443 = 4.43 \times 10^{-5}$$

The proper index to the base 10 is given by the number of places the decimal point must be shifted to the left (positive index), or to the right (negative index), to arrive at the number by which the power of 10 is multiplied.

Multiplication and division of large or small numbers are facilitated by handling them in this form; for example,

$$(3.67 \times 10^9) \times (4.43 \times 10^{-5}) = (3.67 \times 4.43) \times 10^{(9-5)} = 16.26 \times 10^4$$

$$= \underline{1.626 \times 10^5}$$

$$\frac{3.67 \times 10^9}{4.43 \times 10^{-5}} = \frac{3.76}{4.43} \times 10^{(9+5)} = 0.848 \times 10^{14} = 8.48 \times 10^{13}$$

The conventional way of tabulating values which are all multiples of the same product of 10 should be explained, for it will be used in later chapters both in Tables and for defining the scales employed in plotting graphs. According to this convention, the heading (or scale) reports the power of 10 by which the measured quantities *have been* multiplied to obtain the listed values. For example, consider the following:

TABLE 1.1

$10^4\ K_{eq}$	10^{-4} mass/g
1.67	2.2
3.32	3.6

This must be translated as meaning that the reported equilibrium constants are 1.67×10^{-4} and 3.32×10^{-4}, and that the masses are 22 000 and 36 000 g.

LOGARITHMS

If $y = x^4$, then so long as the magnitude of x is defined, the value of y can be represented by the index i. This indeed is what is done when a number is represented by its logarithm, for a logarithm is another name for the index of an exponential term. Thus the relationship between y and x, defined by the equation $y = x^4$, may also be stated as follows, 'i is the logarithm to the base x of the number y', and written in 'shorthand' as

$$i = \log_x y$$

The identity of the logarithm with the exponential index is sometimes obscured by the conventional way in which logarithms are written. For example, 0.02 equals 10^{-17} but $\log_{10} 0.02$ equals $\bar{2}.3$.

The logarithm is written as a decimal number composed of two parts:

(i) the ***characteristic***, which precedes the decimal point and is therefore either zero or an integral number. It may have a positive or a negative value. When it has a negative value, this is indicated by a superscript bar, e.g. $\bar{2}$;

(ii) the ***mantissa***, which is that part of the logarithm that follows the decimal point. It is *always* positive.

Thus the logarithm to the base 10 of 2, is

$$\underbrace{0}_{\text{characteristic}}.\underbrace{3010}_{\text{mantissa}}$$

Because the mantissa of a logarithm is always positive, to represent $10^{-1.7}$ as a logarithm to the base 10, you cannot simply demote the index; i.e log 0.02 cannot be written as – 1.7, for this would mean that the mantissa would be negative (– .7) and this is forbidden. But –1.7 is 0.3 more positive than –2.0 (i.e. $-1.7 = -2 + 0.3 = \bar{2}.3$); thus, $\log_{10} 0.02 = \bar{2}.3$.

Bases for Logarithms

Although *any* number could be made the base for a set of logarithms, three numbers are most often employed in practice:

(a) *10*—logarithms to the base 10 are the most frequently used of any and are called *common* logarithms (sometimes ***Briggsian*** logarithms after their inventor.) In this book we shall symbolize these logarithms as log, e.g. log 2 = 0.3010;

(b) 2—logarithms to the base 2 ($\log_2$) have a very evident application to processes wherein some property increases by doubling (e.g. binary fission effecting the multiplication of bacteria);

(c) *'e' (the 'natural base')*—logarithms to the base e ($\log_e$ or ln) are also known as ***Napierian*** logarithms. Invented to describe processes that increases or decrease in a naturally exponential manner, these logarithms have as their base the number symbolized by e, which is the outcome of the infinite series,

$$1 + \frac{1}{1} + \frac{1}{1 \times 2} + \frac{1}{2 \times 3} + \frac{1}{2 \times 3 \times 4} + \text{etc.}$$

To five places of decimals e equals 2.71828 and,

$$\ln x = 2.303 \log x$$

Use of Tables of Common Logarithms

A Table of Logarithms to the base 10 lists mantissas only; the characteristic of the common logarithm of a number must be obtained by inspection of that number.

To determine the common logarithm of a number, ignore any decimal point and from the Table of Logarithms read off that mantissa which corresponds to the first four figures of the number (reading from left to right). Insert a decimal point to the left of this mantissa and precede this with a characteristic calculated according to the following simple rules:

(a) for any number greater than 1, the characteristic is positive and is one *less* than the number of figures preceding the decimal point;

(b) for any number less than 1, the characteristic is negative and is one *plus* the number of noughts immediately following the decimal point.

Alternatively, if the small number is written as a number between 1 to 10 multiplied by a negative power of 10, then the characteristic is identical with the negative index of 10.

For example, 2000 'looked up' in the Table of Logarithms gives the mantissa 3010. Therefore,

$$\log 200 = 2.3010 \qquad \log 0.2 = \bar{1}.3010$$
$$\log 2 = 0.3010 \qquad \log 2 \times 10^{-5} = \bar{5}.3010$$

The number whose value is given as the logarithm to a stated base is itself called the *antilogarithm*, i.e. antilogarithm = $(\text{base})^{\text{logarithm}}$. Thus, 'the antilog of 0.3010' means 'that number whose log is 0.3010', i.e. 2.0.

To determine the antilogarithm of a given common logarithm, the number corresponding to its mantissa is read off from the Table of Antilogarithms (Appendix). The actual antilogarithm is then obtained by inserting a decimal point into this number in a position defined by the characteristic of the logarithm. Obviously, the rules that governed the assignment of the characteristic to the logarithm in the first place operate in reverse when the characteristic is used to locate the decimal point in the antilogarithm, so that:

(a) if the characteristic is *positive*, the number of figures that must precede the decimal point in the antilogarithm is one *greater* than the characteristic;

(b) if the characteristic is *negative*, then between the decimal point and the number 'read off' from the Table of Antilogarithms there must be inserted a number of noughts which is one *less* than the characteristic.

EXAMPLE

Determine the antilogarithms of (i) 2.5529, and (ii) $\bar{3}$.5529.

First read the number corresponding to the mantissa 5529 in the Table of Antilogarithms. This turns out to be 3572.

(i) If the logarithm is 2.5529, the characteristic is 2 (and positive), so the antilogarithm will have (2 + 1) = 3 figures preceding its decimal point.

$$\therefore \qquad \text{antilog } 2.5529 = 357.2$$

(ii) If the logarithm is $\bar{3}$.5529, the characteristic is – 3, and the antilog will have (3 – 1) = 2 noughts immediately following its decimal point.

$$\therefore \qquad \text{antilog } \bar{3}.5529 = 0.003\,572 = 3.572 \times 10^{-3}$$

If you should ever be in doubt about the application of these 'rules', translate the logarithm into its exponential form. For example, if log $x = \bar{3}.5529$

$$x = 10^{0.5529} \times 10^{-3}$$

whence $x = (\text{antilog } 0.5529) \times 10^{-3} = 3.572 \times 10^{-3}$

Multiplication and Division Using Logarithms

Because logarithms are exponents, logarithms to the same base conform to the same rules as indices to the same base. Therefore to multiply two numbers, one simply adds their common logarithms and determines the antilogarithm of the sum; i.e.

$$a \times b = \text{antilog } (\log a + \log b)$$

Similarly, to divide two numbers, one substracts their common logarithms and determines the antilogarithm of their difference; i.e.

$$a \div b = \text{antilog } (\log a - \log b)$$

(Remember to subtract the log denominator from the log numerator.)

Powers and roots are similarly easily determined by multiplying and dividing their logarithms.

Remember therefore,

$$\log ab = \log a + \log b$$

$$\log \frac{a}{b} = \log a - \log b$$

$$\log \frac{1}{a} = -\log a$$

$$\log a^n = n \log a$$

$$\log \sqrt[n]{a} = \frac{\log a}{n}$$

EXAMPLE

(i) What is the logarithm of 6.35 × 10^{-4}?

$$\log ab = \log a + \log b$$

$$\therefore \log 6.35 \times 10^{-4} = (\log 6.35 + \log 10^{-4}) = 0.8028 + (-4)$$
$$= \bar{4}.8028$$

(ii) Calculate $\dfrac{17.53 \times 13.76 \times 0.356}{5.41 \times 0.022}$ ***using logarithms***

(a) *Add* the logs of the numerators

log 17.53	1.2437
log 13.76	1.1386
log 0.356	$\bar{1}$.5514
Sum	1.9337

(b) *Add* the logs of the denominators

log 5.41	0.7332
log 0.022	$\bar{2}$.3424
Sum	$\bar{1}$.0756

(c) *Subtract*: (summed logs of numerators) – (summed logs of denominators)

	1.9337
	$\bar{1}$.0756
Difference	2.8581

From the Table of Antilogarithms the mantissa 8581 = 7213

Since the characteristic is +2

Antilog 2.8581 = 721.3 = Answer

Arithmetic, Geometric and Logarithmic Series

Consider the following series of numbers,

0, 10, 20, 30, 40, 50, 60, 70, etc.

The numbers form an ***arithmetic*** progression or series, in which the *difference* between two successive numbers is constant. In the present example, the increment is 10 and, irrespective of their location in the series, the difference between two successive numbers is always 10. (An arithmetic series will be obtained whatever is the size of the increment so long as this is constant.)

Now consider the following series,

1, 10, 100, 1000, 10 000, 100 000, etc.

These numbers form a ***geometric*** progression, or series, in which it is the *ratio* between two successive numbers that is constant. Thus in the example given, whatever is their place in the geometric progression one number is always 10 times as great as the preceding number. Note, however, that the arithmetic increment (i.e. the difference) between two successive numbers depends on their situation in the series, i.e. from 10 to 100 is an increase of 90, but from 10 000 to 100 000 is an increase of 90 000.

The geometric progression consists of numbers that increase 'exponentially' in magnitude. This is made plain by rewriting the numbers of a geometric series in their exponential form, i.e.

1, 10, 100, 1000, 10 000, 100 000, etc.

may be rewritten as,

$$10^0,\ 10^1,\ 10^2,\ 10^3,\ 10^4,\ 10^5,\ \text{etc.}$$

Employing the common logarithms of these numbers, we transmute the geometric series into the following arithmetic series:

logs are: 0, 1, 2, 3, 4, 5, etc.

Even if we selected a different base for the logarithms, the geometric series would in its logarithmic form be an arithmetic progression. For example, taking logarithms to the base 2,

the geometric series:	1,	10,	100,	1000,	10 000,	100 000, etc.
becomes:	0,	3.322,	6.644,	9.966,	13.288,	16.610, etc.

This $\log_2$ series is arithmetic, an increment of 3.322 in the value of $\log_2$ representing a tenfold increase in the value of the antilogarithm; i.e.

$$\log_2 10 = 3.322$$

This makes sense of the 'classical' but cryptic definition of logarithms as 'a series of numbers in arithmetic progression corresponding to another series of numbers (their antilogarithms) in geometric progression'.

Scales whose divisions are constructed so that they form an arithmetic progression are used when plotting parameters that increase arithmetically in value, e.g. time. Scales whose equal divisions mark steps in a geometric progression are useful when plotting parameters whose values increase exponentially.

In general, when any quantity y varies with another quantity x in such a way that the rate of change in y is always proportional to the value of y, it is said to vary in an *exponential* manner. The equation $y = Ze^{kx}$, where Z and k are constants and e is the 'natural base', defines the behaviour of systems that conform to what is known as the *law of continuous growth*, e.g. exponentially growing bacterial cultures. On the other hand, the similar exponential function which has a negative index $y = Ze^{-kx}$, describes systems that conform to the '*law of decay*', e.g. decrease of radioactivity of DNA labelled with ^{32}P.

GRAPHIC REPRESENTATION OF THE RELATIONSHIP BETWEEN TWO QUANTITIES *x* AND *y*

The graphs employed in this book possess two axes which are mutually perpendicular. It is conventional to refer to the horizontal axis as the *abscissa* (or x axis) and to the vertical axis as the *ordinate* (or y axis). The point of intersection of these axes is the *origin*.

The position of any point in the plane of the graph is defined by its perpendicular distances from the ordinate and abscissa. These distances, measured in the scale units into which the axes are divided, are the ***coordinates*** of the point. Any relationship between values of y and corresponding values of x can be represented by a line on this graph. The shape of the curve followed by this line defines the relationship between x and y, and can be expressed as an equation. With a little experience, it is possible from the form of the equation that relates x and y to predict the appearance of the curve that will be obtained when values of y are plotted against corresponding values of x.

Equation of a Straight Line

Whenever the relationship between x and y can be expressed in the following form

$$y = mx + c$$

where m and c are constants, one can be certain that if y is plotted against x, a straight line will be obtained whose slope is m, and whose intercept on the y axis equals c (for c is the value of y when x is 0).

For example, the Lineweaver–Burk equation relates the initial velocity of an enzyme-catalysed reaction (v_0) to the concentration of its substrate [S] as follows,

$$\frac{1}{v_0} = \frac{K_m}{V_{max}} \cdot \frac{1}{[S]} + \frac{1}{V_{max}} \quad \text{(where } K_m \text{ and } V_{max} \text{ are constants)}$$

This equation has the form $y = mx + c$ (where $y = 1/v_0$, $x = 1/[S]$, $m = K_m/V_{max}$ and $c = 1/V_{max}$). This means that when values of $1/v_0$ are plotted against values of $1/[S]$, a straight line is obtained of slope K_m/V_{max}, which intersects the $1/v_0$ ordinate to make an intercept which is equal to $1/V_{max}$

Equation of a Standard Exponential Curve

If we plot values of y against corresponding values of x when $y = e^x$, for positive values of x we obtain the standard exponential curve.

This natural exponential curve for the expression $y = e^x$ crosses the ordinate at $y = 1$ (when $x = 0$), and rises progressively more steeply as x increases.

If we plot these values of y as their logarithms, we obtain a straight line. This comes about, since given that $y = e^x$, then $\ln y = x$, which is the equation of a straight line of slope = 1 stemming from the origin (when $x = 0$, $\ln y = 0$). If any other base (n) is used for the logarithmic expression of values of y,

$$\log_n y = \frac{x}{\text{constant}}$$

and a straight line is obtained which again passes through the origin but has a slope of 1/constant.

With a little thought, you will recognize that the use of a 'semi-log' plot to transform an exponential curve into a straight line is made possible by the fact that the logarithms of the numbers in a geometric (exponential) series are themselves members of an arithmetic series.

Frequently, instead of calculating the logarithms of the 'y' term, and plotting these values of log y on' ordinary' graph paper whose x and

y axes are both subdivided arithmetically, values of y itself are directly plotted on 'semi-logarithmic' graph paper whose x axis is scaled arithmetically but whose y axis is scaled logarithmically. Inspection of this type of graph paper is sufficient to convince one of an important feature of any logarithmic scale, namely that it is in a sense distorting, in that it gives prominence to differences between lesser numbers but becomes progressively more 'squashed' as the scale is ascended.

The equations of many regular shaped curves could be listed here, e.g. hyperbolae, parabolae, etc., but this information is to be found in every elementary textbook of coordinate geometry.

Appendix–2
SI Units and their Usage

In calculations involving several measured quantities, it is essential that all are expressed in mutually consistent units; i.e. units based on the same fundamental system of measurement. The system that formerly was

TABLE 2.1

Physical quantity	Name of SI unit	Symbol
length	metre	m
mass	kilogramme	kg
time	second	s
electric current	ampere	A
thermodynamic temperature	kelvin	K
luminous intensity	candela	cd
amount of substance	mole	mol

most generally employed in chemical and biological work was the centimeter–gram–second (c.g.s.) system, but we are now adopting for all scientific work the Système International d'Unités, known as SI. This is an internationally agreed form of the metre–kilogramme–second (m.k.s.) system of measurement which allots basic (primary) units to the seven physical quantities listed in Table 2.1.

The size of each of these primary (base) units is defined as an invariant, internationally agreed standard, and they are in turn employed to construct a variety of derived units. The derived SI units relevant to biological studies are listed in Table 2.2.

Choice of Units of Appropriate Size

For certain purposes, the primary and derived SI units could prove inconveniently large or inordinately small. In these circumstances, sec-

TABLE 2.2 Special Names and Symbols for SI Derived Units

Physical quantity	Name of SI unit	Symbol for SI unit	Definition of SI unit	Equivalent in SI units
energy	joule	J	$m^2\ kg\ s^{-2}$	N m
force	newton	N	$m\ kg\ s^{-2}$	$J\ m^{-1}$
pressure	pascal	Pa	$m^{-1}\ kg\ s^{-2}$	$N\ m^{-2}$
power	watt	W	$m^2\ kg\ s^{-3}$	$J\ s^{-1}$
electric charge	coulomb	C	A s	$J\ V^{-1}$
electric potential difference	volt	V	$m^2\ kg\ s^{-3}\ A^{-1}$	$J\ C^{-1}$
electric resistance	ohm	Ω	$m^2\ kg\ s^{-3}\ A^{-2}$	$V\ A^{-1}$
electric conductance	siemens	S	$m^{-2}\ kg^{-1}\ s^3\ A^2$	Ω^{-1}
electric capacitance	farad	F	$m^{-2}\ kg^{-1}\ s^4\ A^2$	$C\ V^{-1}$
luminous flux	lumen	lm	cd sr	
illumination	lux	lx	m^{-2} cd sr	$lm\ m^{-2}$
frequency	hertz	Hz	s^{-1}	

ondarily modified units might be used. The size of any secondary unit will be defined as a multiple of the primary or derived unit. To obtain a convenient range of secondary units from a primary or derived unit this is multiplied by different powers of 10. The size of the secondary unit is then indicated by attaching a modifying prefix to the name of the primary or derived unit. These prefixes (listed in Table 2.3) indicate the power of 10 by which the primary or derived unit has been multiplied.

TABLE 2.3 SI Prefixes

Multiple	Prefix	Symbol	Multiple	Prefix	Symbol
10	deca	da	10^{-1}	deci	d
10^2	hecto	h	10^{-2}	centi	c
10^3	kilo	k	10^{-3}	milli	m
10^6	mega	M	10^{-6}	micro	μ
10^9	giga	G	10^{-9}	nano	n
10^{12}	tera	T	10^{-12}	pico	p
			10^{-15}	femto	f
			10^{-18}	atto	a

General Observations on the Usage of SI Units and Their Symbols

Symbols will consist of capital letters when the units that they represent are named after persons, though when the names of these units are written in full they are not given initial capital letters.

Example: a force 50 newton(s) = 50 N

Symbols must only be written in their singular form:

Example: 2.4 mol, *NOT* 2.4 mols

No point (i.e. full stop) should be inserted after the symbol (unless this happens to conclude a sentence).

Example: six milligrammes of NaCl = 6 mg NaCl, *NOT* 6 mg. NaCl

When two or more unit symbols are combined to create a derived unit symbol a space is left between them.

Example: 1 C = 1 A s

No space is left between a prefix (indicating a power of 10) and the symbol to which it applies. This rule is particularly important in guarding against confusion between m (meaning milli) and m (meaning metre).

Example: One millisecond = 1 ms
One metre per second = 1 m s^{-1}

Compound prefixes should not be used; use only one multiplying prefix

Example: 10^{-9} m = 1 nm, *NOT* 1 mμm

A combination of prefix and symbol for a unit is regarded as a single symbol. Thus when a modified unit is raised to a power of 10 the power applies to the whole unit including the prefix.

Examples: 1 litre = 1 dm^3 = 1 $(dm)^3$ = 10^{-3} m^3, *NOT* 10^{-1} m^3
1 cm^3 = 1 $(cm)^3$ = 10^{-6} m^3, *NOT* 10^{-2} m^3
∴ 1 cm^3 = 10^{-3} dm^3

The decimal sign may either be a full point *on the line.*

Example: write 1 352 670.47

Use of the solidus (i.e. stroke) is discouraged in favour of the negative index when writing symbols of reciprocal units.

Example: preferable to write N m^{-2} rather than N/m^2

The solidus must in any event never be used more than once in any unit.

Example: the molar gas constant R = 8.314 J K^{-1} mol^{-1}, *NOT* 8.314 J/K/mol.

It is in fact convenient to keep the solidus (a) for division of a physical quantity by another physical quantity, e.g. *PV*/*RT*, or (b) for the division of a physical quantity by its units, e.g. R/J K^{-1} mol^{-1} = 8.314.

The degree sign is omitted when recording temperature in the SI unit (kelvin).

Example: the f.p. of water is 273.15 K, *NOT* 273.15 °K

Some common non-SI units are likely to remain in use but should no longer be employed in a precise scientific context; e.g. min., hr., cal., Curie, atm., *etc.*

Some Implications of the Adoption of SI Units

Volume: The SI unit of volume is the cubic metre, m^3. The old definition of the litre (leading to the value 1.000 028 dm^3) was abolished in 1964, and the litre was redefined as being exactly equal to the cubic decimetre. But, although it is recognized that the litre will remain in common usage, it is still recommended that both the litre and millilitre are abandoned in exact scientific work, their respective volumes being represented as the corresponding fractions of a cubic metre;

i.e. 1 litre (symbol, L) = 1 dm^3 = 10^{-3} m^3
1 millilitre (symbol, mL) = 1 cm^3 = 10^{-6} m^3

Mass: The primary SI unit of mass is the kilogramme, kg. This means that the SI rules regarding usage of prefixes confer a special status on the gramme.

Example: 10^{-6} kg cannot be written as 1 μkg but can quite properly be written as 1 mg.

[Note the agreed spelling: gramme, *NOT* gram.]

Amount of substance: The primary SI unit is the mole, given the symbol mol.

The mole is defined as 'the amount of substance of a system which contains as many elementary entities as there are atoms in 0.012 kg of carbon 12'.

The elementary unit must be specified and may be an atom, a molecule, an ion, an electron, a photon, etc., or a specified group of such entities.

Since 0.012 kg of carbon 12 contains the Avogadro number (6.022 $\times 10^{23}$) of atoms of carbon, this is the number of elementary units (of a specified kind) contained in 1 mol. This means that terms such as gram-equivalents, gram-molecules, gram ions, etc., are all obsolete and are abandoned in favour of the mol.

Example: 1 mole of electrons has a mass of 5.4860×10^{-7} kg and carries a negative electric charge equal to 96487 C.

Concentration: Previously, molality (symbol, m) was used to indicate the number of moles of solute in 1000 g of solvent. Fortunately, this is easily transmuted into primary SI units; what was termed a 1 molal solution has a concentration of 1 mol kg^{-1}. Because the symbol m (for molality) could be confused with m (for metre), both the term molal, and use of the symbol m to represent molality, should be abandoned in favour of mol kg^{-1}.

We face greater difficulties in determining the appropriate SI unit in which to express concentrations in mole/unit volume terms. Previously, molarity (symbol, M) was used to indicate the amount of solute (in moles) dissolved in 1 litre of solution. Therefore what was termed a 1 M solution has a concentration of 10^3 mol m^{-3} = 1 kmol m^{-3} = 1 mol dm^{-3}.

Because the symbol M is already employed in SI to represent the prefix mega, use of the term molarity and its representation by the symbol M should be abandoned in favour of kmol m^{-3} or mol dm^{-3}. Furthermore, the word 'molar' before the name of an extensive quantity is to be restricted to the meaning 'divided by amount of substance', where the amount of substance need not be expressed as a function of 1 mol. It would therefore be wrong to continue to speak of molar concentrations.

For ease of 'transliteration' of old-style molarities into their SI equivalents, it is certainly easiest to employ the mol dm^{-3} nomenclature.

i.e. 1 μmole/ml = 1 μmol cm^{-3}

1 M = 1 mol dm^{-3} = 1 mol l^{-1}

Force: The derived SI unit of force is the newton, N (Table 2.2).

Pressure: The derived SI unit of pressure is the pascal, Pa (Table 2.2). Pressures should therefore be expressed either in Pa or in N m^{-2} (since 1 Pa = 1 N m^{-2}). Equivalents of other units are:

1 lbf/in^2 = 6894.76 Pa

1 mmHg = 133.322 Pa

1 millibar = 100 Pa

1 atm. = 101 325 Pa or 101.325 kPa (k = kilo)

Energy: The derived SI unit of energy is the joule, J (Table 2.2). Equivalent of other units are:

1 erg = 10^{-7} J

1 litre-atm. = 101.328 J

1 $calorie_{(thermochemical)}$ = 4.184 J

Radioactivity: The curie (symbol, Ci) is redundant,

$$1 \text{ Ci} = 3.7 \times 10^{10} \text{ s}^{-1}$$

Frequency: The derived SI units of frequency is the hertz, Hz (Table 2.2).

Example: an n.m.r. machine using 60 megacycles/sec. should not be rated as 60 MHz; i.e., 60 megahertz.

Temperature: The SI unit of temperature is the kelvin, K (1 K = 1/273.16 of the thermodynamic temperature of the triple point of water). It is envisaged that although the Celsius (i.e. centigrade) scale will be retained for everyday, domestic purposes it will be discontinued for exact scientific work.

Example: 0°C = 273.15 K, so that *T*/K = (t/°C + 273.15)

Luminous intensity: The primary SI unit is the candela, cd (Table 2.1).

Luminous flux: The derived SI unit is the lumen, lm (Table 2.2).

Illumination: The derived SI unit is the lux, lx (Table 2.2). Other units should be abandoned;

e.g. 1 foot candle = 10.7639 lx

Table 2.4 lists the SI equivalents of most of the units encountered in biochemical texts, while useful physical constants (in SI units) are listed in Table 2.5.

TABLE 2.4 Conversion Table for Translation of Common Units into Their SI Equivalents

Unit	SI equivalent
ampere, A	1 A
angstrom, A	100 pm = 10^{-10} m
atmosphere, standard; atm.	101 325 Pa
bar, b	10^5 Pa
calorie (international table); cal	4.1868 J
calorie 15 °C; cal_{15}	4.1855 J
calorie, thermochemical	4.184 J
candela, cd	1 cd
centigrade (Celsius) degree, °C	(t/°C + 273.15) K
centimetre, cm	10^{-2} m
coulomb, C	1 C
cubic centimetre, cm^3	1 cm^3 = 10^{-6} m^3
cubic decimetre, dm^3	1 dm^3 = 10^{-3} m^3 = 1 litre
cubic foot, ft^3	0.028 316 8 m^3
cubic inch, in^3	16.3871 cm^3
cubic metre, m^3	1 m^3
curie, Ci	3.7×10^{10} s^{-1}
cycle/second, c/s	1 Hz
degree (angle),	$\pi/180$ rad
degree centigrade (degree Celcius), °C	(t/°C + 273.15) K
degree Fahrenheit, °F	(t/°F + 459.67) K
drachm (apothecaries)	3.887 93 g
drachm, fluid	3551.63 mm^3
dram (avoirdupois)	1.771 85 g
dyne, dyn	10^{-5} N
electron volt, eV	1.6021×10^{-19} J
erg	10^{-7} J
farad	1 F
fluid ounce, fl oz	28.4131 cm^3
foot, ft	0.3048 m
foot-candle, lm/ft^2	10.7639 lx
foot-lambert	3.426 26 cd m^{-2}

(Contd.)

TABLE 2.4 *(Continued)*

Unit	SI equivalent
foot of water (pressure)	2989.07 Pa
foot pound-force, ft lbf	135 582 J
gallon, gal	4.546 09 dm^3
gramme, g	10^{-3} kg
henry, H	1 H
hertz, Hz	1 Hz
hour, h	3600 s
inch, in	25.4 mm
inch of water (pressure)	249.089 Pa
joule, J	1 J
kilowatt, kW	1 kW
kilowatt hour, kW h	3.6 MJ
litre, l	1 dm^3 = 10^{-3} m^3 = 1 l
litre atmosphere	101.328 J
lumen, lm	1 lm
lumen/sq. ft, lm/ft^2	10.7639 lx
lumen/sq. metre, lm/m^2	1 lx
lux, lx	1 lx
micron, μ	1 μm
millibar	100 Pa
millilitre	1 cm^3 = 10^{-6} m^3 = 1 ml
millimetre of mercury, mmHg	133.322 Pa
millimetre of water	9.806 65 Pa
minim	59.1939 mm^3
molal, m	1 mol kg^{-1}
molar, M	1 mol dm^{-3} = 1 mol l^{-1}
mole	1 mol
newton, N	1 N
ohm	1 Ω
ounce, oz	28.3495 g
ounce, apothecaries	31.1035 g
ounce fluid	28.4131 cm^3

TABLE 2.4—*Continued*

Unit	SI equivalent
pascal, Pa	1 Pa = 1 N m^{-2}
pint, pt	0.568 261 dm^3
poise, P	0.1 kg m^{-1} s^{-1}
poiseuille, Pl	1 N s m^{-2} = 1 Pl
pound, lb	0.453 592 37 kg
pound-force, lbf	4.448 22 N
pound-force/sq. in, lbf/in^2	6894.76 Pa
pound/sq. in, lb/in^2	703.070 kg m^{-2}
rad (100 erg/g)	0.01 J kg^{-1}
radian	1 rad
siemens, S	1 S
square foot, ft^2	0.092 903 m^2
square inch, in^2	645.16 mm^2
stokes, St	10^{-4} m^2 s^{-1}
therm	105.506 MJ
ton of refrigeration	3516.85 W
torr	133.322 Pa
volt, V	1 V
watt, W	1 W

Conventions for Labelling Axes on Graphs and Heading Columns in Tables of Quantities

To avoid repetitious use of defining units, all physical quantities, constants, etc., are tabulated as columns of pure numbers. This in turn means that the heading of each column must itself be a pure number, e.g. the quotient of the measured physical quantity and the symbol for the unit of measurement.

Examples: temperatures should be listed under the column heading T/K, pressures might be recorded under the heading P/N m^{-2} or P/Pa, and changes in Gibbs free energy would be tabulated under the heading ΔG/J mol^{-1}.

The same principle applies to the labelling of axes on graphs.

TABLE 2.5 **Values in SI Units of Some Useful Physical Constants**

Physical constant	Symbol	Value
Avogadro constant	L (or, N_A)	$6.022\ 52 \times 10^{23}$ mol^{-1}
Boltzmann constant	k	$1.380\ 54 \times 10^{-23}$ J K^{-1}
Gas constant	$R = Lk$	8.3143 J K^{-1} mol^{-1}
charge of electron	e	$1.602\ 10 \times 10^{-19}$ C
Faraday constant	$F = Le$	9.6487×10^{4} C mol^{-1}
Planck constant	h	6.6256×10^{-34} J s
Molar volume of ideal gas at 273.15 K and 101 325 Pa		22.4136 dm^3 mol^{-1}

Until one gets used to this convention it could prove confusing; just remember that the unit of measurement is always indicated by putting its symbol after a solidus (i.e. stroke) inserted immediately following the symbol for the physical quantity being tabulated or plotted.

To avoid terms containing an incovenient number of digits, the actually tabulated or plotted numbers might be the measured values multiplied by a convenient (constant) power of 10. The power of 10 that is in this case written into the heading or legend (immediately preceding the symbol for the physical quantity) is that number by which the measured quantities have been multiplied to yield those more convenient numbers that are listed or plotted. This of course does not preclude the concurrent use of the most suitable size of unit obtained by modifying the SI unit with one or other of the prefixes listed in Table 2.3.

Examples: (i) An entry 3.6 under the heading $10^2 k$ means that the value of k is 0.036; the same entry (3.6) under the heading $10^{-2} k$ means that the value of k is 360.

(ii) A substrate concentration of 0.000 12 mol dm^{-3} can be listed as 1.2 under a heading 10^4[S]/mol dm^{-3} or as 0.12 under a heading [S]/mmol dm^{-3}.

t/°C	T/K	$10^4 k$/s^{-1}	$10^3/T$ in K^{-1}	log k
15	288	2.51	3.472	$\bar{4}.3997$
20	293	4.57	3.412	$\bar{4}.6599$
25	298	8.22	3.356	$\bar{4}.9149$

Handling Physical Quantities in Calculations

The following rules must be obeyed:

(1) Every measured quantity must be represented as a *number* times a *unit*; although one could write in the multiplication sign, this is usually omitted.

(2) Only quantities that have the same dimensions can be added or subtracted from each other. For example, one can add two energy terms but one cannot sensibly add an energy term and a pressure term.

(3) Be consistent in the choice of units; wherever it appears, the one physical quantity must be measured in identical units. To a large extent this is ensured by strict adherence to the rules of SI, so that, for example, with all energy terms expressed in joules there can be no confusion as might occur when non-coherent energy terms were employed in the one calculation, e.g. ergs and calories.

In multiplying and dividing quantities, their units are treated in precisely the same way as the accompanying numbers, and the product or quotient of the unit must be appended to the final numerical answer.

The way in which units can be multiplied and divided can be illustrated by deriving the units of the molar gas constant R whose value is defined by the following idea gas equation:

$$R = \frac{PV}{nT} \quad \text{where} \begin{cases} R = \text{molar gas constant} \\ P = \text{pressure of gas in N m}^{-2} \\ V = \text{volume of gas in m}^3 \\ n = \text{quantity of gas in mol} \\ T = \text{temperature in K} \end{cases}$$

i.e. $R = \left(\frac{PV}{T}\right) \text{ mol}^{-1}$ and the units of $R = \frac{\text{N m}^{-2} \times \text{m}^3}{\text{K}} \text{ mol}^{-1}$

$$= \text{N m K}^{-1} \text{ mol}^{-1}$$

But $1 \text{ N} = 1 \text{ J m}^{-1}$, whence $1 \text{ Nm} = 1 \text{ J}$

$\therefore$ R is measured in the units $\text{J K}^{-1} \text{ mol}^{-1}$

Finally it should be emphasized that in multiplying and dividing units during the course of a calculation, it is extremely hazardous to ignore the units you have arrived at a numerical result and only then 'conjure up' an appropriate unit to define this value.

Appendix–3

Some Useful Constants and Conversion Factors

Planck's constant	k	6.63×10^{-34} J s
Speed of light in a vacuum	c	3×10^{8} m s^{-1}
Avogadro number	N_A	6.02×10^{23} mol^{-1}
Gas constant	R	8.31 J K^{-1} mol^{-1}
Faraday constant	F	96 500 C mol^{-1}
Curie	Ci	3.7×10^{10} s^{-1}
Svedberg	S	10^{-13} s
Molar volume of ideal gas at 273.15 K and 1 atm	=	22.41 dm^3 mol^{-1}
1 electron volt	=	96.5 kJ mol^{-1}
1 calorie	=	4.18 J
0°C (centigrade)	=	273.15 K
1 atm (760 mm Hg)	=	101.325 kPa
ln x	=	2.303 $\log_{10} x$